Interactive Dynamics of Convection and Solidification

NATO ASI Series

Advanced Science Institutes Series

A Series presenting the results of activities sponsored by the NATO Science Committee, which aims at the dissemination of advanced scientific and technological knowledge, with a view to strengthening links between scientific communities.

The Series is published by an international board of publishers in conjunction with the NATO Scientific Affairs Division

A Life Sciences
B Physics

Plenum Publishing Corporation
London and New York

C Mathematical and Physical Sciences
D Behavioural and Social Sciences
E Applied Sciences

Kluwer Academic Publishers
Dordrecht, Boston and London

F Computer and Systems Sciences
G Ecological Sciences
H Cell Biology
I Global Environmental Change

Springer-Verlag
Berlin, Heidelberg, New York, London, Paris and Tokyo

NATO-PCO-DATA BASE

The electronic index to the NATO ASI Series provides full bibliographical references (with keywords and/or abstracts) to more than 30000 contributions from international scientists published in all sections of the NATO ASI Series.
Access to the NATO-PCO-DATA BASE is possible in two ways:

– via online FILE 128 (NATO-PCO-DATA BASE) hosted by ESRIN, Via Galileo Galilei, I-00044 Frascati, Italy.

– via CD-ROM "NATO-PCO-DATA BASE" with user-friendly retrieval software in English, French and German (© WTV GmbH and DATAWARE Technologies Inc. 1989).

The CD-ROM can be ordered through any member of the Board of Publishers or through NATO-PCO, Overijse, Belgium.

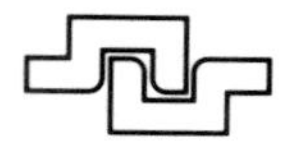

Series E: Applied Sciences - Vol. 219

Interactive Dynamics of Convection and Solidification

edited by

S. H. Davis

Department of Engineering Sciences and Applied Mathematics,
Northwestern University, Evanston, Illinois, U.S.A.

H. E. Huppert

Institute of Theoretical Geophysics,
Department of Applied Mathematics and Theoretical Physics,
University of Cambridge, Cambridge, U.K.

U. Müller

Kernforschungszentrum Karlsruhe GmbH,
Karlsruhe, Germany

and

M. G. Worster

Department of Engineering Sciences and Applied Mathematics
and Department of Chemical Engineering,
Northwestern University, Evanston, Illinois, U.S.A.

Kluwer Academic Publishers

Dordrecht / Boston / London

Published in cooperation with NATO Scientific Affairs Division

Proceedings of the NATO Advanced Study Institute on
Interactive Dynamics of Convection and Solidification
Chamonix, France
8–13 March, 1992

Library of Congress Cataloging-in-Publication Data

Interactive dynamics of convection and solidification / edited by
Davis, S.H. ...[et al.].
p. cm. -- (NATO ASI series. Series E, Applied sciences ; vol.
219)
"Published in cooperation with NATO Scientific Affairs Division."
ISBN 0-7923-1910-9 (acid free paper)
1. Heat--Convection--Congresses. 2. Solidification--Congresses.
3. Phase transformations (Statistical physics)--Congresses.
I. Davis, Stephen H. II. North Atlantic Treaty
Organization. Scientific Affairs Division. III. Series: NATO ASI
series. Series E, Applied sciences ; no. 219.
QC326.I58 1992
536'.42--dc20 92-24978

ISBN 0-7923-1910-9

Published by Kluwer Academic Publishers,
P.O. Box 17, 3300 AA Dordrecht, The Netherlands.

Kluwer Academic Publishers incorporates the publishing programmes of
D. Reidel, Martinus Nijhoff, Dr W. Junk and MTP Press.

Sold and distributed in the U.S.A. and Canada
by Kluwer Academic Publishers,
101 Philip Drive, Norwell, MA 02061, U.S.A.

In all other countries, sold and distributed
by Kluwer Academic Publishers Group,
P.O. Box 322, 3300 AH Dordrecht, The Netherlands.

Printed on acid-free paper

Printed in the Netherlands

This book contains the proceedings of a NATO Advanced Research Workshop held within the programme of activities of the NATO Special Programme on Chaos, Order and Patterns as part of the activities of the NATO Science Committee.

Other books previously published as a result of the activities of the Special Programme are:

ABRAHAM, N.B., ALBANO, A.M., PASSAMANTE, A. and RAPP, P.E. (Eds.) - *Measures of Complexity and Chaos* (B208) 1990 Plenum Publishing Corporation. ISBN 0-306-43387-7.

BUSSE, F.H. and KRAMER, L. (Eds.) - *Nonlinear Evolution of Spatio-Temporal Structures in Dissipative Continuous Systems* (B225) 1990 Plenum Publishing Corporation. ISBN 0-306-43603-5.

CHARMET, J.C., ROUX, S. and GUYON, E. (Eds.) - *Disorder and Fracture* (B235) 1991 Plenum Publishing Corporation. ISBN 0-306-43688-4.

MARESCHAL, M. (Ed.) - *Microscopic Simulations of Complex Flows* (B236) 1991 Plenum Publishing Corporation. ISBN 0-306-43687-6.

MacDONALD, G.J. and SERTORIO, L. (Eds.) - *Global Climate and Ecosystems Change* (B240) 1991 Plenum Publishing Corporation. ISBN 0-306-43715-5.

CHRISTIANSEN, P.L. and SCOTT, A.C. (Eds.) - *Davydov's Soliton Revisted: Self-Trapping of Vibrational Energy in Protein (B243) 1991 Plenum Publishing Corporation. ISBN 0-306-43734-1*

HOLDEN, A.V., MARKUS, M. and OTHMER, H.G. (Eds.) - *Nonlinear Wave Processes in Excitable Media* (B244) 1991 Plenum Publishing Corporation. ISBN 0-306-43800-3.

LING-LIE CHAU, NAHM, W. (Eds.) - *Differential Geometric Methods in Theoretical Physics: Physics and Geometry* (B245) 1991 Plenum Publishing Corporation. ISBN 0-306-43807-0.

ATMANSPACHER, H. and SCHEINGRABER, H. (Eds.) - *Information Dynamics* (B256) 1991 Plenum Publishing Corporation. ISBN 0-306-43912-3.

BABLOYANTZ, A. (Ed.) - *Self-Organisation, Emerging Properties, and Learning* (B260) 1991 Plenum Publishing Corporation. ISBN 0-306-43930-1.

PELITI, L. (Ed.) - *Biologically Inspired Physics* (B263) 1991 Plenum Publishing Corporation. ISBN 0-306-44000-8.

BISHOP, A.R., POKROVSKY, V.L. and TOGNETTI, V. (Eds.) - *Microscopic Aspect of Nonlinearity in Condensed Matter* (B264) 1992 Plenum Publishing Corporation. ISBN 0-306-44001-6.

JIMENEZ, J. (Ed.) - *The Global Geometry of Turbulence: Impact of Nonlinear Dynamics* (B268) 1992 Plenum Publishing Corporation. ISBN 0-306-44014-8.

MOSEKILDE, E. and MOSEKILDE, L. (Eds.) - *Complexity, Chaos and Biological Evolution* (B270) 1992 Plenum Publishing Corporation. ISBN 0-306-44026-1.

ROY, A.E. (Ed.) - *Predictability, Stability and Chaos in N-Body Dynamical Systems* (B272) 1992 Plenum Publishing Corporation. ISBN 0-306-44034-2.

BEN AMAR, M., PELCE, P. and TABELING, P. (Eds.) - *Growth and Form: Nonlinear Aspects* (B276) 1992 Plenum Publishing Corporation. ISBN 0-306-44046-6.

JENA, P., KHANNA, S.N. and RAO, B.K. (Eds.) - *Physics and Chemistry of Finite Systems: From Clusters to Crystals* (C374) 1992 Kluwer Academic Publishers. ISBN 0-7923-1818-8

SEGUR, H. (Ed.) - *Asymptotics Beyond All Orders* (B284) 1992 Plenum Publishing Corporation. ISBN 0-306-44112-8

PREFACE

The phase transformation from liquid to solid is a phenomenon central to a wide range of manufacturing and natural processes. The presence of phase transformation can drive convection in the melt through the liberation of latent heat, the rejection of solute, and the change of density upon freezing. The fluid mechanics itself can play a central role; the phase transformation can be strongly altered by convective transport in the liquid through the modification of the thermal and solutal environment of the solid–liquid interface; these local fields control the freezing characteristics at the interface. The convection can be generated naturally by buoyancy forces arising from gradients of temperature and concentration in the liquid, by density changes upon freezing, and by thermocapillary and solutocapillary forces on liquid–solid interfaces. The interactive coupling between solidification and convection forms the subject of this volume.

Such coupled processes are significant on a large range of scales. Among the applications of interest are the manufacture of single crystals, the processing of surfaces using laser or molecular beams, and the processes of soldering and welding. One wants to understand and predict macrosegregation in castings, transport and fractionation in geological and geophysical systems, and heat accumulation in energy redistribution and storage systems.

This volume contains papers presented at the NATO Advanced Research Workshop on "Interactive Dynamics of Convection and Solidification" held in Chamonix, France, March 8–13, 1992. The workshop was aimed at encompassing the current activities in the field, both theoretical and experimental, and focused upon the fundamental understandings of the physical mechanisms involved. The topics addressed included basic topics of the role of convection in determining interface morphology, eutectic, cellular and dendritic microstructures, the formation of mushy zones, and the non-equilibrium effects during rapid solidification.

This meeting brought together fundamental and applied scientists who broadly study a question of importance, viz. "what determines the internal chemical and mechanical structure of the solidified material?" The answer involves all types of issues of pattern and scale selection, of interactive instabilities, frontogenesis, localized structures, etc. In short, all of the major topics of modern dynamics fit into the context of phase transformation coupled with mass, momentum and heat transport. Control of microstructure and macrostructure is tied to the understanding of its formation and evolution.

The workshop was divided according to the scale of the phenomena involved. On the MICROSCALE, there were discussions of morphological instabilities, their finite-amplitude evolutions, and the development of cells and dendrites. On the MESOSCALE, there were discussions of "developed" cells and dendrites, and of mushy zones. On the MACROSCALE, there were discussions of full "systems" of solidifying/melting and convecting material. Here the influence of flow on crystal-growing processes, processing by beams, spincasting, macrosegregation in casting and geological systems and energy conversion in engineering systems were all addressed. The final half day of the meeting was devoted to a GRAND ROUNDTABLE; an organized discussion led by a group of scientists from different fields. The discussion aimed at highlighting the ideas and techniques developed in regard to disparate applications that might be useful elsewhere. There was an explicit attempt to create cross-fertilization of fields that might lead to a common basis for the development of new insights into this exciting and important area of fundamental research.

The workshop was sponsored by the North Atlantic Treaty Organization, the National Aeronautics and Space Administration and by the Kernforschungszentrum Karlsruhe. The organizing committee was S.H. Davis (General Chairman), H.E. Huppert, U. Müller and M.G. Worster. P. Ehrhard was in charge of local organization, V. Lallemand was the conference secretary on site and J. Piehl the secretary at Northwestern University. These Proceedings were supported financially by NATO.

S.H. Davis – Northwestern University
H.E. Huppert – University of Cambridge
U. Müller – Kernforschungszentrum Karlsruhe
M.G. Worster – Northwestern University

May 11, 1992

TABLE OF CONTENTS

MORPHOLOGICAL STABILITY AND CONVECTIVE FLOW: SOME OLD AND NEW PROBLEMS

D T J HURLE
H H Wills Physics Laboratory
University of Bristol, Tyndall Avenue
Bristol BS8 1TL, UK

ABSTRACT. The motivation for the study of the effects of convective flow on the morphological stability of a crystal-melt interface is the desire to be able to grow alloy single crystals which are free from cellular sub-structure. There is a particular and pressing need to be able to grow homogeneous high-quality single crystals of ternary III-V compounds such as $Ga_{1-x}In_xAs$ for use as substrates in the fabrication of a range of opto-electronic devices. In this paper, the theory of coupling between natural convection and morphological instability is briefly reviewed together with some relevant experimental evidence. Ways in which the theory might be usefully extended to include the effects of additional control parameters which could advantageously affect morphological stability are indicated.

1. INTRODUCTION

Detailed observation of a pattern of cellular microsegration during directional crystallisation from a melt and its explanation as due to an instability in the shape of the crystal-melt interface caused by conditions of constitutional supercooling in the melt was first made almost 40 years ago by Rutter and Chalmers. [1]. A rigorous description using linear perturbation theory, of this morphological instability was later provided by Mullins and Sekerka [2].

Requirements for high-quality, homogeneous single crystals of a range of materials to meet the needs of the emerging solid state electronics industry led to detailed studies of the phenomenon in semiconductors [3-5] and laser host-lattice materials [6] as well as in the low melting point metals studied originally [1,7]. It soon became evident that avoiding the formation of a cellular structure was a vital pre-requisite to obtaining single crystals of adequate quality. To this end lower growth rates and improved melt stirring were employed. The original theory predicting the onset of constitutional supercooling in a quiescent melt was extended by Hurle [8] to include the effects of crystal rotation during Czochralski growth.

S. H. Davis et al. (eds.), Interactive Dynamics of Convection and Solidification, 1–14.

The problem first became acute in about 1960 when, for the first time, semiconductor crystals were heavily doped in order to make tunnel ('Esaki') diodes. At this time more radical solutions to the problem were being sought. J B Mullin and the author in unpublished work, performed some experiments with Ge and with InSb Czochralski growth in which they passed a large DC current through the crystal-melt interface in an attempt to stabilise a planar interface. They argued that if a cellular structure formed then the current would be focused strongly in the cell boundary grooves because the electrical conductivity of the molten semiconductor was many times greater than that of the crystal at the melting point (of the order of 20:1 in the case of germanium). Accordingly, they argued, if the sense of the current flow was such as to produce Peltier cooling at the crystal-melt interface then this effect would abstract latent heat preferentially in the cell boundaries causing this part of the interface to advance more rapidly thereby, at least partially, eliminating the cellular structure. The large Peltier coefficient (P)of these two semiconductors made this seem feasible; certainly it looked possible to produce a Peltier heat sink per unit area of interface of magnitude PI (where I is the current density) which exceeded the latent heat source Lv (where L is the latent heat per unit volume and v is the normal growth rate).

Unfortunately of course, they were not smart enough to know anything about scaling analysis, and so did not appreciate that the length scale associated with the cellular structure (D/v, where D is the solute diffusion coefficient) was much too small for significant temperature differential to be obtained, the latter having a scaling length of K/v (where K is the thermal diffusivity).

However, in the course of these experiments, it was discovered that significant ion migration occurred such that the concentration of Sb at the interface during the growth of InSb could be shifted from the stoichiometric 50:50 composition to one approaching the InSb-Sb eutectic (approx.30:70) with the resultant growth of a eutectic structure. This effect is potentially very much more effective for controlling the microstructure. Unfortunately the experimental programme had to be terminated before interface stabilisation could be demonstrated unambiguously.

The influence of natural convective flow on microsegregation was not really considered by crystal growers until it became apparent in the mid 1960's that time-periodic or turbulent flow was the cause of solute striations in metal and semiconductor crystals [9,10]. At this time a number of us made tentative approaches to fluid dynamicists only to find that they shook their heads sadly when, in reply to their query as to the Prandtl number of our fluid, we gave the answer 10^{-2}! We tended to be get referred to their astrophysicist friend. However, things have changed dramatically since that time.

By contact with these hydrodynamicists, crystal growers acquired some familiarity with that classic problem of hydrodynamic stability -

the Bénard cell, from which it occurred to Coriell et al.[11] and independently to Hurle, Jakeman and Wheeler [12] that some coupling of morphological and convective instability might occur; again with the hope that conditions might be found which would produce a stabilisation of a planar interface. It is in this concept that the present conference has its roots.

The benefits which would accrue to solid state electronics if interface stabilisation could be achieved such that it became possible to grow homogeneous single crystals of IV-IV and III-V solid solutions such as Ge-Si and $Ga_xIn_{1-x}As$ remain considerable.
This is particularly the case for opto-electronic devices based on III-V compounds where ternary, and even quaternary, III-V epitaxial layers of compositions tuned to the particular wavelength of the device are grown onto substrates, usually required to have a larger band gap than the layer in order that it be transparent to incident radiation of the selected wavelength. To obtain a high quality layer, it is necessary to match its lattice parameter (and ideally its expansion coefficient) to that of the substrate. At present only binary substrates such as GaAs, InP and GaSb are available so that simultaneous matching of lattice parameter and selection of band gap is possible only by use of quaternary layers and then only within certain limits. The ability to provide ternary substrates covering the whole range of compositions between the end members (eg $Ga_{1-x}In_xAs$ with $0<x<1$) would afford much greater flexibility in the design of the device by allowing lattice matching to be achieved by adjusting the substrate leaving only the opto-electronic properties to be tailored in the epitaxial layer.

Some progress to this end has been achieved recently using conventional liquid encapsulation Czochralski techniques with careful matching of the seed crystal composition and very low growth speeds [13] but this is unlikely to yield material covering the whole composition range. Maximum allowable growth rates are very low. Fluctuations due to melt turbulence, limitations of temperature servo-control and mechanical disturbances make such very low steady growth speeds unattainable.

In section 2 of this paper, the theory of coupling between natural convective instability and morphological instabiliy of melting, stationary and freezing interfaces is briefly reviewed but with the case of the freezing interface left largely to the following paper by S H Davis[14]. In section 3 some relevant experimental evidence is presented and discussed and in section 4 ways in which the theory could usefully be extended to embrace additional control parameters is indicated.

2. REVIEW OF THEORY

2.1 Bénard convection in the presence of a thermodynamic phase boundary at the upper surface.

This configuration with cyclohexane has been studied by Davis, Müller and Dietsche[15,16]. The problem is one of Bénard convection with the upper (cooler) boundary deformable under conditons of thermodynamic equilibrium. This introduces an additional parameter into the linear theory: A =the ratio of the depth of the ice to that of the water. The case A=0 recovers the conventional Bénard problem. Davis et al [15] have shown that the onset of convection can be subcritical- and therefore hysteretic- with a sudden jump in the heat flux and in the mean position of the ice-water interface. The deformation of the, initally planar, interface can introduce sufficient asymmetry to permit the formation of an hexagonal planform. The critical Rayleigh number decreases by about 12.5% as A is increased from zero to infinity.

In summary, the deformable interface both decreases the critical Rayleigh number for onset of convection and also influences the topology of the flow.

2.2 Hydrodynamic instability with an undeformed freezing interface.

If interfacial deformation is avoided by crystallising a binary alloy under conditions for which there is no morphological instability, the segregation of solute at the crystallising interface will initially generate an exponential distribution of solute in a quiescent melt ahead of the freezing interface which can give rise to solutal convection. Indeed, since a temperature gradient is also imposed on the melt (to avoid the morphological instability), the possiblility of thermo-solutal convective effects exists. In practise, as shown by Hurle et al [17], solutal convective effects dominate (the ratio of solutal to thermal Rayleigh number is typically of the order of 10^4 or greater). It has been shown that the critical Rayleigh number for this solutal convective instability is dependent on the distribution coefficient k and on the Schmidt number $Sc=\mu/D$ where μ is the kinematic viscosity) falling to a limiting value of

$$R_c = 2(1 + Sc^{-1}) \qquad (1)$$

in the limit $k \to 0$.

This limiting value, also obtained by Young and Davis [18], is dramatically smaller than the value intuitively expected by introducing an effective fluid depth of D/v into the conventional expression for R_c which, for the equivalent boundary conditions, takes the value 320. Thus we see that a crystallising interface can render the adjacent melt exceedingly unstable with respect to rejected solute where that solute produces an unstably stratified density field.

A subtle point about this theory deserves mention. The concentration in the crystal at the interface C(0) is obtained from the flux boundary condition at the interface:

$$D\, dC/dz + v(1-k)C = 0 \qquad (2)$$

where D is the solute diffusion coefficient and z the co-ordinate normal to the interface plane.

Two distinct mathematical problems can be posed:

1) The growth velocity v is constrained to be constant even though the interface concentration, and hence the liquidus temperature, is perturbed.This requires that the thermal field be slightly and artificially adjusted. For sufficiently dilute solution, such that the liquidus temperature is changed but negligibly whilst the fluid buoyancy remains significantly altered, the change required to the thermal field is negligible.

2) The situation in which the interface remains planar but its velocity is perturbed so as to fully satisfy the coupled thermal and solute diffusion field equations. This corresponds to the coupling of the convective instability to a <u>zero wavenumber</u> morphological instability. This particular morphological instability should occur quite independently of any condition related to constitutional supercooling of the melt. It is analogous to the sub-critical instability seen by Davis et al.[15] for the undoped cyclohexane system which is driven by <u>thermal</u> convection.

The difference between these two formulations becomes serious when the non-linear finite amplitude response of the system is explored. In the limit of small k, problem 1) has been addressed by Riahi[19] and by Riley and Davis[38]. The problem has been extended to arbitrary k by Impey et al.[39]. So far as the author is aware, the most physical problem- problem 2) has not been studied to date.

The full coupling of natural convection to the classic morphological instability first studied by Mullins and Sekerka is the subject of the paper by Professor Davis [14].

2.3 Morphological instability with melting.

The analogous process of constitutional superheating can occur during melting of a binary alloy solid but, whereas the existence of constitutional supercooling is a fairly accurate guide to interface stability under most conditions of freezing, a melting interface is, in general, stable up to large degrees of constitutional superheating. Woodruff[20] and Chen and Jackson[21] have carried out linear perturbation analysis of this case and the latter found that, for given velocity and concentration and for small distribution coefficient k, the melting interface was 1/k times more stable than a freezing one. Chen and Jackson obtained experimental validation of the theory using carbon tetrabromide-hexachloroethane mixtures. However whilst the interface may remain planar under most experimental conditions, droplets of melt can nucleate in the solid ahead of the interface because of the very limited degree to which a solid can be superheated. This phenomenon was demonstrated in NaK alloys by Woodruff and Forty [22].

The analysis has been carried out only for a quiescent melt but, because the spatial wavelength of the interfacial instability, when it does occur, is very small ($\approx D_s/v$) [21] coupling to a convective mode can be expected to be even weaker than in the freezing case.

One fascinating non-steady melting situation is that studied by Huppert and Turner [23]: the melting of a vertical ice surface in a stably stratified saline solution. This modelled the melting of an iceberg. Melting produced an array of horizontal convecting fingers which spread laterally and gave rise to a scalloped morphology to the dissolving ice surface. This is essentially a double-diffusive convective instability; ie the coupling is principally between convective modes rather than to a morphological one.

3.EXPERIMENTAL EVIDENCE FOR COUPLED CONVECTIVE-MORPHOLOGICAL INSTABILITY.

3.1 Dendritic growth and mushy zones.

Vertically-upward solidification of an alloy under conditions of sufficient degree of constitutional supercooling to produce an array of extended dendrites (a so-called 'mushy' zone) can, if the rejected solute is less dense than the solute, give rise to solutal convection in the overlying melt which is subtly different from that which occurs if the crystal-melt interface is planar. A pattern of stationary toroidal convection is modified by the formation of a number of turbulent plumes which appear to originate a little below the mushy zone/melt interface and produce local erosion in the mushy zone. This has been demonstrated recently by McCay et al [24] using a confocal optical signal processing technique to observe directional solidification in a 28 wt% NH_4Cl in H_2O solution.

This phenomenon occurs in metal castings producing what are known as 'freckles' in vertically solidified ingots where eroded dendrite arms in the mushy zone get entrained in the rising plumes producing a column of fine equiaxed grains of a composition differing from that of the matrix. As can be imagined, this seriously compromises the integrity of the casting. A similar phenomena occurs when the solidifying surface is vertical. In this case the erosion channels are inclined, and, in a cylindrical ingot solidifying radially inward, produce a pattern of microsegregation resembling an 'A' in a vertical section taken through the ingot axis. Such 'A' segregates are also detrimental. The phenomenon has been studied in detail in model metal and transparent analogue systems by Copley et al.[25] amongst others.

Studies of the vertical dendritic solidifcation of ice from $NaNO_3$ solution by Huppert and Worster [26] shows a surface humpiness (their word) on a centimetric scale, much coarser than that of individual dendrites which suggests a coupled convective instability. However the system is stably stratified!

3.2. Coupling of morphological instability and Marangoni flow.

Hämäläinen [27], in some interesting experiments performed many years ago, demonstrated microsegregation patterns in solidified films of alkali halide solid solutions which he interpreted as due to the combined effects of Rayleigh-Bénard convection and a cellular freezing interface produced by constitutional supercooling. In the particular example described, the wavelength of the hexagonal convective pattern was exactly twice than of the cellular structure. The former will of course be strongly influenced by the thickness of the not-yet-frozen layer. Presumably the convective motion was driven by thermal Marangoni effects which advected solute rejected at the freezing interface. This interesting effect merits some theoretical study.

3.3 Hailstones.

Giant hailstones are encountered in certain parts of the world. Browning [28,29] has studied stones up to 10 cm diameter collected during thunderstorms in the mid-West of the USA. These stones have a marked lobe structure with lobes typically of the order of 1 cm across and extremely similar in appearance to the structure of the Huppert and Worster [26] directionally solidified ice. These stones are wafted upward by the strong updraughts produced in the thunderstorm and, no doubt, tumble as they make their erratic and delayed descent to earth. Browning has provided some explanation for the forms in terms of aerodynamic instability produced by a chance protruberance on the stone but a more complete analysis taking account of the growth dynamics is needed.

4.SOME 'NEW' PROBLEMS REQUIRING STUDY.

4.1 Introduction

The coupling of the morphological instability of a planar interface to natural buoyancy-driven convective flow has now been extensively investigated by several eminent groups but the location of regimes of strong coupling where convective flow might effectively suppress morphological instability has proved elusive. Before finally throwing in the towel on this highly desirable prize it seems worthwhile to study the possible beneficial influence of additional control variables.

4.2 Oscillation of interface position

One such additional control parameter, that has received some study, is the use of mechanical or thermal oscillation. The stabilisation of a Bénard cell brought about by periodically modulating the temperature gradient or the gravitational acceleration across it has been demonstrated and Wheeler[30] has investigated theoretically whether a similar modulation of the velocity of the interface in a solidifying binary alloy can effect stabilisation against a cellular breakdown. This was indeed shown to be possible, especially for large wavenumber instability, when the modulation amplitude was sufficient to produce melt-back during the perturbation cycle. However the effect is much weaker at small wavenumber, so that the stabilisation at the critical wavenumber did not exceed 10% for the examples considered.

All convective effects were ignored in the above. Could it be that some frequency of modulation would pull the spatial frequency of the morphological instability toward that of the convective one, thereby increasing the coupling between morphological and convective modes and giving rise to conditions under which the planar interface could be stabilised? This would surely be a difficult situation to analyse fully but would repay some study to at least test the above line of argument.

Banan[40] has recently studied the effects of axial vibration on the growth of $Ga_{1-x}In_xSb$ alloys using a vertical Bridgman technique. For a composition of $x=0.2$, a 20Hz vibration was shown to reduce the number of grain boundaries but at the price of some increase in twinning. Vibration was shown to be capable of altering the macroscopic shape of the interface and the observed structural changes could have been consequent on this. Much lower frequencies need to be explored.

4.3 Soret effect

One is looking for additional physical effects which act selectively on different parts of the crystal-melt interface as it starts to deform. One such effect is Soret diffusion which is the diffusion of solute in a binary alloy which results from an imposed temperature gradient. The mass flux is proportional to the temperature gradient and, depending on the sign of the Soret coefficient, it can transport solute to or from the interface even in the absence of a concentration gradient. Thus, for a solute with $k<1$ and a Soret solute flux away from the interface, rejected solute can be removed at the rate required for steady state crystallisation with a concentration gradient which is lower than in the absence of Soret diffusion thereby reducing the likelihood of the occurence of constitutional supercooling in the melt. This has been quantified by Verhoeven[31]. Typically it can change the flux, and hence the critical value of the Sekerka number, by up to around 10%.

There is a further effect namely that, once the interface starts to deform, a difference in thermal conductivities of crystal and melt causes a distortion of the isotherms and hence of the Soret flux. However, because the thermal scale length K/v is in general very much greater than the wavelength of the cellular structure, the distortion of the isotherms is small. The discrepancy is least for solutes with $k \ll 1$ and, for this case, large changes in the critical wavelength which depend on the sign of the Soret effect are predicted. A physical explanation of this effect has been given by Hurle[32].

4.4 Electromigration

A potentially more effective additional force is electrotransport of solute produced by application of a DC electric field normal to the growth interface as described in section 1. Linear and weakly non-linear stability analyses have been carried out recently by Wheeler et al.[33,34] These authors have shown that the marginal stability of the system is altered by applying the field in two ways: a change in the criterion for existence of constitutional supercooling resulting from the electrotransport and an effect due to the redistribution of the current flow upon interface deformation as a result of differiing electrical conductivities in crystal and melt.

Wheeler et al. [33] show that, for stationary modes, the parameter

$$G^* = G_L Dk/mC_\infty (k-1)v \tag{3}$$

(which is the critical temperature gradient for marginal stability for fixed C_∞ and v) can be expressed approximately as:

$$G^* = (1+\alpha)[k-(1+\alpha) + \alpha(\Sigma-1)/(\Sigma+1)]/(k-1) \tag{4}$$

where G_L is the melt temperature gradient and m the slope of the liquidus line. $\alpha=\mu_i E/v$ is a measure of the strength of the electromigration, E being the applied electric field and μ_i the ionic mobility. $\Sigma= \sigma_s/\sigma_L$ is the ratio of the electrical conductivities of crystal and melt respectively. Setting $\alpha=0$ recovers the Mullins and Sekerka result in the limit of zero Gibbs-Thomson effect. Increased G^* implies decreased stability. The third term in the square bracket on the RHS of equation (4) describes the effect of the redistribution of the current due to the unequal electrical conductivities of liquid and solid phases. Whether this term is stabilising or destabilising depends on the sign of k-1, α and of Σ-1. The several different regimes are described by Wheeler et al.[33].

The predictions of the analysis well describe the experimental results of Warner and Verhoeven[36] on Sn-Bi alloys but do not appear to describe the liquid phase electroepitaxy (LPEE)of GaAs [37]. This may be because the modelled conditions are significantly different from those of LPEE or it may be that convective effects need to be taken into account. Recent studies of the LPEE growth of InP by Takenaka et al.[35] suggest that fluid motion generated by the Lorentz effect is the dominant mass transport mechanism in the system.

In as much as the electrotransport modifies the solute field it will therefore also change the solutal convection and with it the coupling of that convection to the morphological instability. It would be useful therefore if the coupled convective/morphological instability analysis could be extended to cover the electric field case. Again one is looking for increased coupling between convective and morphological modes induced by reducing the difference in the wavelength of the respective uncoupled modes. At high current densities MHD effects would need to be taken into account.

4.5 Eutectic growth

Finally we draw attention to the fact that crystal growth is not confined to single phase growth but that coupled duplex phase structures such as directionally-solidified eutectics or monotectics are of potential interest and importance.

Non-faceted/non-faceted (NF-NF) eutectics growing from melts of eutecic composition do not have an O(D/v) boundary layer ahead of the eutectic interface; lateral solute redistribution is confined to a very thin boundary layer of the order of the lamellar (or rod) spacing which is typically of the order of 1 micron (compared to say 100 microns for the D/v layer). For even the most conceiveably-strong melt flow the momentum boundary layer is much, much thicker than this and flow has only a very limited effect on the growth of such structures.

However, when the melt composition differs from the eutectic composition, coupled eutectic structures can still be grown but now there is an O(D/v) boundary layer rich in the component in excess of the eutectic composition. The same is true of faceted/non-faceted (NF-F) systems even when they are grown from a melt of eutectic composition. The reason for this is that the structure is not free to smoothly and continuously change its volume fraction of the two phases comprising the eutectic solid and so the mean composition of the solid will, in general, differ from that of the melt resulting in the presence of a boundary layer of such a thickness that it is subject to major modification by melt flow. This boundary layer, present in NF-F systems, can give rise to constitutional supercooling of the melt, to cellular breakdown of the major phase and results in a so-called complex regular structure exhibiting two length scales: the finer one being that of the lamellar spacing and the coarser of O(D/v) which has the topology of a faceted cellular structure similar to that of a faceting single-phase solid such as germanium.

The impact of natural convective flow on these morphologies has been studied through a number of space experiments where the gravity level is sufficiently low that buoyancy-driven convection is largely eliminated. These experiments [41,42] have shown that the rod spacing is different in the space-grown samples, being greater in some systems and less in others. Drevet et al.[43] have shown theoretically that the effect of forced convective flow can account for the differences in rod spacing in NF-F systems. To the author's knowledge there have been no studies of the coupling of solutal convection to eutectic interfacial morphology.

5. CONCLUSIONS

The prize of interface stabilisation, sufficient to enable homogeneous solid solution single crystals to be grown, remains elusive. However the main reasons for this have become clear and the possibility that, by applying some additional external forcing, they can be overcome, remains a real one. A detailed study of some of these possibilities aimed at elucidating the physical mechanisms which might generate strong coupling appears well worthwhile.

Finally, when the single-phase crystallisation problem is fully worked out there remain the considerably more complicated problems of duplex phase growth.

REFERENCES

1.Rutter, J. W. and Chalmers, B. (1953) 'A prismatic substructure formed during solidification of metals', Canad. J. Phys. 31, 15-49.

2. Mullins, W. W. and Sekerka, R.F. (1964) 'Stabiliy of a planar interface during solidification of a binary alloy', J. Appl. Phys. 35, 444-451.

3. Bardsley, W., Callan, J.M, Chedzey, H.A. and Hurle, D.T.J.(1961) 'Constitutional supercooling during crystal growth from stirred melts. II Experimental: gallium doped germanium', Solid State Electron. 3, 142-158.

4. Bardsley, W., Boulton, J.S. and Hurle, D.T.J. (1962).'Constitutional supercooling during crystal growth from stirred melts III. Morphology of the germanium cellular structure', Solid State Electron. 5, 396-403.

5.Dikhoff, J.A.M.,(1963/4) 'Inhomogeneities in doped germanium and silicon crystals' Philips Technical Rev. 8, 195-206.

6.Bardsley, W., Cockayne, B., Green, G.W. and Hurle, D.T.J. (1963) 'Cellular structure in calcium tungstate', Solid State Electron. 6, 389-390.

7.Chadwick, G.A. (1962) 'Decanted interfaces and growth forms'. Acta Met. 10, 1-12.

8.Hurle, D.T.J.,(1961)'Constitutional supercooling during crystal growth from stirred melts. I Theoretical', Solid State Electron. 3,37-44.

9.Utech, H.P. and Flemings, M.C. (1966) 'Elimination of solute banding in indium antimonide crystals by growth in a magnetic field' J. Appl. Phys. 37, 2021-2024.

10.Chedzey, H.A. and Hurle, D.T.J. (1966)'Avoidance of growth striae in semiconductor and metal crystals grown by zone melting techniques', Nature 210, 933-934.

11.Coriell, S.R., Cordes, M.R., Boettinger, W.J. and Sekerka, R.F., (1980) 'Convective and interfacial instabilities during unidirectional solidification of a binary alloy' J. Crystal Growth 49, 13-28.

12.Hurle, D.T.J., Jakeman, E. and Wheeler, A.A. (1982). 'Effects of solutal convection on the morphological stability of a binary alloy', J. Crystal Growth 58, 163-179.

13.Bonner, W.A., Nahory, R.E., Gilchrist, H.L. and Berry, E. (1990) 'Semi-insulating single cystal $Ga_{1-x}In_xAs$: LEC growth and characterisation', Proc. Semi-insulating III-V materials, Toronto, Canada 1990,Ed. A G Milnes and C J Miner. Adam Hilger, Bristol 1990 pp 199-204

14.Davis, S. H. This conference

15.Davis, S.H., Müller, U. and Dietsche, C. (1984) 'Pattern selection in single component systems coupling Bénard convection and solidification', J. Fluid Mech. 144, 133-151

16.Dietsche, C. and Müller, U. (1985) 'Influence of Bénard convection on solid-liquid interfaces. J. Fluid Mech., 161, 249-268.

17.Hurle, D.T.J., Jakeman, E. and Wheeler, A.A. (1983)' Hydrodynamic stability of the melt during solidification of a binary alloy' Phys. Fluids, 26, 624-626.

18. Young, G.W. and Davis, S.H. (1986) 'Directional solidification with buoyancy in systems with small segregation coefficient'. Phys. Rev. B34, 3388-3396.

19.Riahi, D.N. (1988) 'Solutal convection in the melt during solidifiication of a binary alloy' Phys. Fluids 31, 27-32.

20. Woodruff, D.P. (1968) 'The stability of a planar interface during melting of a binary alloy' Phil. Mag. 17, 283-294.

21.Chen, H.S. and Jackson, K.A. (1971) 'Stability of a melting interface'. J. Crystal Growth 8, 184-190.

22.Woodruff, D.P. and Forty, A.J. (1967)'A pre-melting phenomenon in sodium-potassium alloys'. Phil. Mag. 15, 985-993.

23.Huppert, H.E. and Turner, J.S. (1980) 'Ice blocks melting into a salinity gradient'. J. Fluid Mech. 100, 367-384.

24.McCay, T.D., McCay, M.H. and Gray, P.A. (1989) 'Experimental observation of convective breakdown during directional solidification' Phys. Rev. Lettrs. 62, 2060-2063.

25.Copley, S.M., Giamei, A.F., Johnson, S. M. and Hornbecker, M.F. (1970) 'Origin of freckles in unidirectionally solidified castings'. Met. Trans. 1, 2193-2204.

26.Huppert, H.E. and Worster, M.G., (1985) 'Dynamic solidification of a binary melt', Nature 314, 703-707

27.Hämäläinen, M (1967) Constitutional supercooling in the presence of convection cells', J. Crystal Growth 1, 125-130

28.Browning, K.A. (1966) 'The lobe structure of giant hailstones', Quarterly J. Royal Meteorolog. Soc. 92, 1-14

29.Browning, K.A., (1967) 'The growth environment of hailstones', Meteorolog. Mag. 96, 202-211

30.Wheeler, A.A., (1984) 'The effect of a periodic growth rate on the morphological stability of a freezing binary alloy', J. Crystal Growth 67, 8-26.

31.Verhoeven, J.D. (1967), 'The effect of thermal diffusion on constitutional supercooling during solidification' Materials Res. Bull.1, 93-96

32.Hurle, D.T.J., (1983) 'The effect of Soret diffusion on the morphological stability of a binary alloy crystal', J. Crystal Growth 61, 463-472

33. Wheeler, A.A., Coriell,S.R., McFadden, G.B. and Hurle, D T.J. (1988) 'The effect of an electric field on the morphological stability of the crystal-melt interface of a binary alloy', J. Crystal Growth 88, 1-15

34.Wheeler, A.A., McFadden, G.B., Coriell, S.R. and Hurle, D.T.J. (1990) 'The effect of an electric field on the morphological stability of the crystal-melt interface of a binary alloy. III Weakly non-linear theory', J. Crystal Growth, 100, 78-88.

35.Takenaka,C., Kusunoki, T. and Nakajima, K. (1991) 'Solute transport mechanism during liquid phase epitaxial (LPE) growth with an applied current'. J. Crystal Growth 114, 293-298.

36. Warner, J. C. and Verhoeven, J.D. (1973) 'Effect of electrotransport on thr solidification of Sn-Bi alloys', Met. Trans. AIME 4, 1255-1261.

37. Plamondon, R. (1991) Private communication.

38. Riley, D.S. and Davis, S. H. (1989) 'Hydrodynamic stability of the melt during the solidification of a binary alloy with small segregation coefficient'. Physica D39, 231-238.

39.Impey, M.D., Riley, D.S., Wheeler, A.A. and Winters, K.H. (1991) 'Bifurcation analysis of solutal convection during directional solidification'. Phys. Fluids A3, 535-550.

40.Banan, M. (1991) 'Influence of imposed perturbations on directional solidification of $In_xGa_{1-x}Sb$ alloy semiconductor'. Ph.D thesis, Clarkson University, Potsdam, NY.

41.Müller, G. and Kyr, P. (1985) 'Directional solidification of InSb-NiSb eutectic'. Proc. 5th European Symposium on Materials Science in Microgravity-Schloss Elmau Nov. 1984. ESA SP-222 pp 141-146.

42.Larson, D.J. (1977) 'Low-gravity processing of magnetic materials'. Proc. AIAA 15th Aerospace Sciences Meeting, LA 1977.
pp123-130

43.Drevet, B., Camel, D. and Favier, J-J. (1990)'Solute boundary layer and convection in solidification of eutectic alloy'. Proc. 7th European Symposium on Materials and Fluid Sciences in Microgravity, Oxford 1989. ESA SP-295. pp101-108.

RAYLEIGH-MARANGONI BILAYER CONVECTION IN LIQUID ENCAPSULATED CRYSTAL GROWTH

R. Narayanan, A.X. Zhao
Department of Chemical Engineering, Univ. of Florida
Gainesville, Fl 32611, USA

and

C. Wagner
Lehrstuhl für Strömungsmechanik
T.U.München, D 8000, München, Germany

ABSTRACT. Bilayer convection occurs during the encapsulated crystal growth of certain compound semiconductors. Interfacial tension gradients respond to the temperature field and cause convection of the 'Marangoni' type and this in turn is affected by buoyancy driven convection or Rayleigh convection. We calculate the onset conditions for flow in the presence of a lower solidifying interface and consider regions of low gravity. We show that the stability decreases with a lowering of gravity and with a lowering of liquid depths. The coupling is in one direction as the solid thickness has virtually no effect on the convection characteristics of the bilayer while the convection leaves it's signature on the solid. This is markedly different than the results of Davis et. al (1984) who considered the freezing solid at the top.

1. Introduction

Liquid encapsulated growth of semiconductor crystals from the melt offers a way to contain high volatiles and maintain appropriate stoichiometry. Fig.1 is a depiction of the vertical Bridgman method wherein Boron Oxide is used as a liquid encapsulant in the growth of Gallium Arsenide from it's melt. The essential fluid mechanics of convection lies in the Rayleigh-Marangoni interaction through the presence of a liquid-liquid bilayer. This is shown schematically in Fig.2. The main focus of this study is to determine the critical conditions and wavelength of the motion due to the convective flow in the presence of the solidifying interface. This is accomplished by linearized stability calculations.

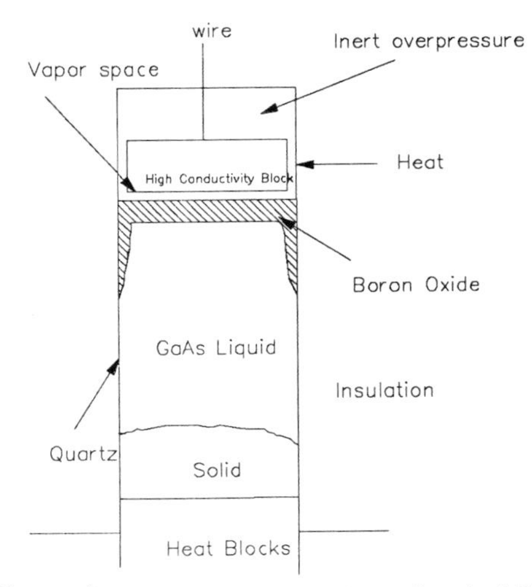

Figure 1 Liquid Encapsulated Bridgman Growth of GaAs

S. H. Davis et al. (eds.), Interactive Dynamics of Convection and Solidification, 15–18.

2. The Model

The basic derivation of the modeling equations and notation may be found in Ferm and Wollkind (1982) and the principle of exchange of stability is assumed. A spatial Fourier transform in the unbounded directions leads to ordinary differential equations. We get the momentum and energy equations in the upper(superscript'+') phase as:

$$-s(D^2-\omega^2)^2 W^+_0 + arR\,\omega^2\theta^+_0 = 0 \tag{1}$$

$$(D^2-\omega^2)\theta^+_0 = -\frac{W^+_0}{mn} \tag{2}$$

Here D is the total derivative, W_0 and θ_0 are the perturbed and transformed vertical component of velocity and temperature fields. ω is the disturbance wave number and R is the Rayleigh number based on the lower phase. The governing equations in the lower phase are of the same form with superscripts '+' replaced by '-' and the constants a,m,n,r and s each replaced by unity. These are the ratios of expansion coefficients, thermal diffusivity, conductivities,densities and viscosities respectively.

For the sake of brevity we only give the interfacial conditions here. The kinematic, no slip and momentum equations at the liquid-liquid interface are:

$$W_0^+ = W_0^- = 0, \qquad DW_0^+ = DW_0^- \tag{3}$$

$$-(D^2-3\omega^2)(sDW_0^+ - DW_0^-) = (R + (G+\omega^2)/C)\,\omega^2\eta_0 \tag{4}$$

$$-(D^2+\omega^2)(sW_0^+ - W_0^-) = M\,\omega^2(\eta_0 - \theta^-_0) \tag{5}$$

The continuity of temperature and heat flux at the liquid-liquid interface are:

$$\theta_0^+ - \theta_0^- = \eta_0(1-m)/m \tag{6}$$

$$D\theta_0^+ = D\theta_0^- \tag{7}$$

C, G and M are the Capillary, Bond and Marangoni numbers. η_0 is the perturbed transformed position of the liquid-liquid interface. The energy balance and continuity of temperature at the freezing deflecting interface give the following equation in a manner similar to Davis et. al (1984).

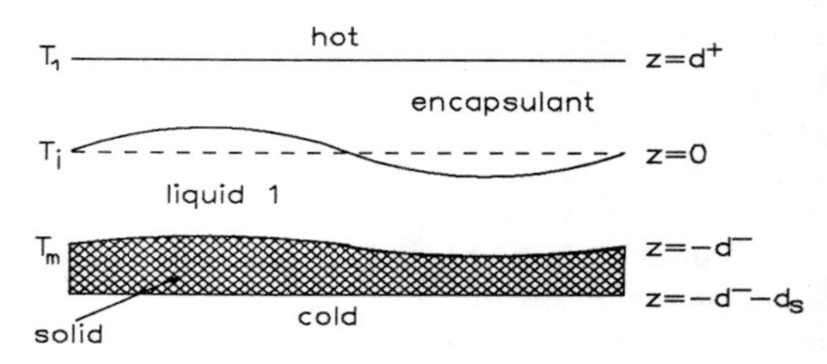

Figure 2 Schematic of physical problem

$$D\theta_0^- = \omega\theta_0^- \coth(\omega\Gamma) \tag{8}$$

Here the solidification rate is assumed to be negligible and Γ is the dimensionless ratio of freezing layer thickness to lower layer depth. We calculate the critical Marangoni number over a range of wave numbers for a given set of system properties. The calculated Marangoni number must be compatible to the input value of the Rayleigh number since they are related. The parameters that were varied were gravity level, depths and solid thickness. Changes in the thermo physical properties will naturally affect the quantitative results. However, we do believe that the basic mechanisms and the qualitative features will remain the same.

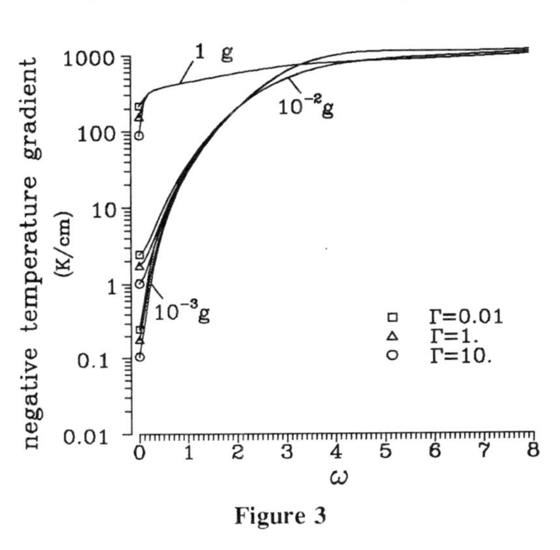

Figure 3

3. Results

The calculations for determination of the critical conditions of convective onset and the corresponding wavelength of the critical disturbance were performed for the Gallium Arsenide / Boron Oxide system. We note that the Boussinesq approximation has been verified to be valid in all of these calculations. The Gallium Arsenide is heavier than the Boron Oxide and the presence of a solidifying interface in this bilayer problem implies that we can only consider the 'heated from above' configuration. Gravity serves the purpose of only delaying the onset of the Marangoni driven convection. Thus it comes as no surprise that the instability is enhanced at lower gravity levels. This is seen in Figure 3. Note that the ratio of upper to lower phase depths is taken to be 1/3 and d^-=1cm. We also have checked variation of the instability limit with fluid depths. The Marangoni and Rayleigh numbers as well as Capillary and Bond numbers are dependent on this parameter. There is, naturally, a stabilizing effect of the rigid horizontal walls. However, this can be offset by the relative increase in the instability of interfacial tension

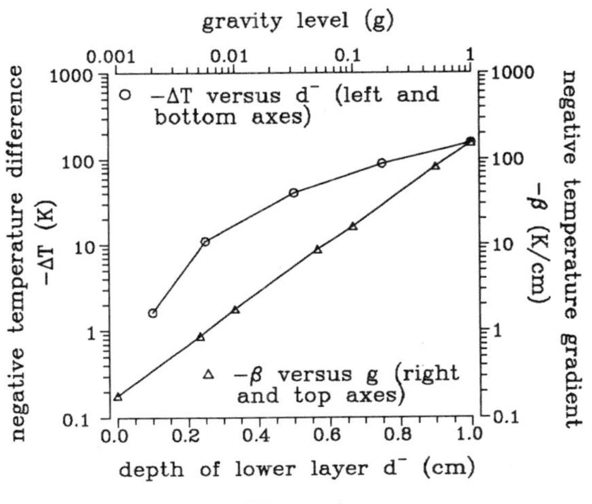

Figure 4

gradient in comparison to the stabilizing buoyancy mechanism. In fact our calculations for this system indicate that instability increases as we lower the depths (see Figure 4. Here Γ=1). The upper curve is calculated at '1 g' and the lower curve for d^{-}=1cm. This implies that the effect of convection on the solid-liquid interface morphology would be dominant towards the end of the solidifying process.

The variation of gradient with wave number and Γ indicate little sensitivity with respect to the latter variable. This is quite unlike the results of Sarma (1988) or Davis et al.(1984), who considered the single layer problem with an **upper cool solidifying surface**. In the present problem we have the freezing surface below. Recall that gravity is a considerable stabilizing force in our problem and so the effect of the gain in stability as the solid thickness is lowered is very small in '1 g'. In fact from Fig. 4 we can see that the effect of the solid thickness is only slightly more pronounced at lower 'g' levels. Finally, we observe that the instability does not vary monotonically with gravity level for a fixed wave number and this has implication on convection in bounded containers. This is a subject for a future study.

Table of Thermophysical Properties

Species	Density gr./cc	viscosity cm^2/s	conductivity 10^4 ergs/cm/s/°K	diffusivity 10^{-3} cm^2/s	negative expansion coefficient 10^{-4}/°K
Boron Oxide	1.648	23.665	20	25.2	0.9
Gallium Arsenide	5.72	.0049	178	71.7	1.87

Interfacial tension is taken as 455 dynes/cm and it's gradient with temperature as - 1.20dynes/cm/°K. (see Shetty (1989))

Acknowledgment

Support from NASA (Langley) NAG 1- 609 and NSF CBT890016P (PSC) is acknowledged.

References

1. Davis, S.H., Müller, U and Dietsche, C. (1984), 'Pattern selection in single-component systems coupling Bénard convection and solidification', J. Fluid Mech., **144,** pp.133-151.

2. Ferm, E.N. and Wollkind, D. (1982), 'Onset of Rayleigh-Bénard-Marangoni instability: Comparişon between theory and experiment', J. Non-Equilib. Thermodyn., **7**, N.3, pp.169-190

3. Sarma G.S.R., (1988), 'Interfacial effects on the onset of convection', in Physicochemical Hydrodynamics ed. M.G. Velarde, Plenum. Publ. Co., pp.271-289

4. Shetty, R. (1989), 'Surface tension and contact angle of Cadmium Telluride and Gallium Arsenide melts', M.S. Thesis, Clarkson Univ., Potsdam, New York

5. Wagner, C., Narayanan, R. and Friedrich, R. (1992), 'Gravitational effects and some extended results for bilayer convection of the Rayleigh-Marangoni type', accepted by Phys. of Fluids.

EFFECT OF MODULATED TAYLOR-VORTEX FLOWS ON CRYSTAL-MELT INTERFACES

G. B. McFadden, R. J. Braun, B. T. Murray, and S. R. Coriell
National Institute of Standards and Technology
Gaithersburg, MD 20899
USA

M. E. Glicksman
Department of Materials Engineering
Rensselaer Polytechnic Institute
Troy, NY 12180-3590
USA

ABSTRACT. During crystal growth from the melt phase, the interaction of the crystal-melt interface with convection in the melt is of fundamental importance. For growth of a metallic alloy or doped semiconductor, the distribution of solute in the growing crystal is determined by the prevailing conditions near the crystal-melt interface; convection in the melt and interface instability may both produce solute inhomogeneities that result in an inferior crystal. A specific example of strong interaction between the crystal-melt interface and flow in the melt is provided by a time-modulated Taylor-Couette flow of a single-component fluid with a cylindrical crystal-melt interface.

1. Introduction

Crystal growth from the liquid or melt phase is often accompanied by hydrodynamic flows in the melt [1]. These flows can have an important impact on the quality of the crystal produced, so that studies of the basic flow-interface interaction are of fundamental importance. In many technologically important techniques, such as Bridgman growth or Czochralski growth, strong electromagnetic fields are frequently used in efforts to control the flow effects. (Some descriptions of common crystal growth techniques are given by Hurle and Jakeman in [2] and references therein.) Avoiding natural convection is one of the main motivations for developing the capability of crystal growth under the microgravity conditions available in low earth orbit, where the driving force for natural convection is lower by orders of magnitude. In addition, such an environment allows more precise fundamental experiments on interface dynamics to be performed without the complicating effects of buoyancy-driven convection (see, e.g. [3]).

The study of the interaction of fluid flow with a crystal-melt interface is an area of fundamental importance in materials science, but despite much recent research [4, 5, 6]

S. H. Davis et al. (eds.), Interactive Dynamics of Convection and Solidification, 19–21.

the understanding of such interactions is fragmentary. The general problem combines the complexities of the Navier-Stokes equations for the fluid flow in the melt with the nonlinear behavior of the free boundary representing the crystal-melt interface. Some progress has been made by studying explicit flows that allow a base state corresponding to a one-dimensional crystal-melt interface with solute and/or temperature fields that depend only on the distance from the interface. This allows the strength of the interaction between the flow and the interface to be assessed by a linear stability analysis of the simple base state.

For example, one can examine changes in the *morphological stability* [7, 8] of the interface in the presence of flow in the melt. Specific flows that have been considered in this way include plane Couette flow, thermosolutal convection, plane stagnation flow, rotating disk flow, and the asymptotic suction profile. One can also examine changes in the *hydrodynamic stability* of a given flow that occur when a rigid bounding surface is replaced by a crystal-melt interface. Examples here include the instabilities associated with Rayleigh–Bénard convection, thermosolutal convection, plane Poiseuille flow, the asymptotic suction profile, and thermally-driven flow in an annulus, and steady Taylor-Couette flow.

2. Current Research

In our recent work we consider the influence of the crystal-melt interface on another form of the classical Taylor-Couette instability [9, 10, 11] of the flow between rotating concentric cylinders. The flow we consider is for a time-dependent sinusoidal rotation of the system, which gives rise to an unsteady, time-periodic base state, whose stability may be assessed by applying Floquet theory [12]. Glicksman and colleagues at RPI are conducting experiments using the transparent organic material succinonitrile (SCN) in a cylindrical geometry [13]. A radial temperature gradient produces an initially cylindrical crystal-melt interface, which forms the inner surface for an annular region of molten SCN. The entire system is subjected to a sinusoidal, torsional oscillation with mean zero about the cylinder axis.

An expression for the strength of the driving force is given by the dimensionless Taylor number, which is proportional to the square of the amplitude of the angular velocity. For small Taylor numbers, the cylindrical base flow is stable. As the driving force is increased, this base state becomes unstable to an axisymmetric perturbation, which takes the form of Taylor vortices in the fluid, accompanied by an axial sinusoidal distortion of the crystal-melt interface. Although the direction of the azimuthal shear force at the walls reverses direction during each period, the deformation of the interface shape appears to be time-independent.

The results of a linear stability analysis of this system indicate that the effect of the crystal-melt interface is to destabilize the system by an order of magnitude relative to a single-phase system with rigid, non-deforming walls. Comparisons with preliminary experimental results [13] confirms the size of this effect for SCN. More generally, the destabilization is found to be a strong function of the thermal properties of the material, as measured by the Prandtl number, given by the ratio of the liquid's kinematic viscosity and thermal diffusivity. The destabilization is strongest for materials having a large Prandtl number, and the effect of the crystal-melt interface becomes insignificant for materials with small Prandtl numbers.

3. Acknowledgements

This work was conducted with the support of the Microgravity Science and Applications

Division of the National Aeronautics and Space Administration. One of the authors (RJB) was supported by an NRC Postdoctoral Research Fellowship.

References

[1] Brown, R. A. (1988) 'Theory of transport processes in single crystal growth from the melt', AIChE J. 34, 881–911.

[2] Hurle, D. T. J. and Jakeman, E. (1981) 'Introduction to the techniques of crystal growth', PCH PhysicoChemical Hydrodynamics 2, 237–244.

[3] Glicksman, M. E., Winsa, E., Hahn, R. C., Lograsso, T. A., Tirmizi, S. H. and Selleck, M. E. (1988) 'Isothermal dendritic growth- a proposed microgravity experiment', Metall. Trans. 19A, 1945–1953.

[4] Coriell, S. R. and Sekerka, R. F. (1981) 'Effect of convective flow on morphological stability', PCH PhysicoChem. Hydrodyn. 2, 281–293.

[5] Glicksman, M. E., Coriell, S. R. and G. B. McFadden (1986) 'Interaction of flows with the crystal-melt interface', Annu. Rev. Fluid Mech. 18, 307–335.

[6] Davis, S. H. (1990) 'Hydrodynamic interactions in directional solidification', J. Fluid. Mech. 212, 241–262.

[7] Mullins, W. W. and Sekerka, R. F. (1964) 'Stability of a planar interface during solidification of a dilute binary alloy', J. Appl. Phys. 35, 444–451.

[8] Coriell, S. R., McFadden, G. B. and Sekerka, R. F. (1985) 'Cellular growth during directional solidification', Annu. Rev. Mater. Sci. 15, 119–145.

[9] Taylor, G. I. (1923) 'Stability of a viscous liquid contained between two rotating cylinders', Phil. Trans. Roy. Soc. A 223, 289–343.

[10] McFadden, G. B., Coriell, S. R., Glicksman, M. E. and Selleck, M. E. (1989) 'Instability of a Taylor-Couette flow interacting with a crystal-melt interface', PCH PhysicoChem. Hydrodyn. 11, 387–409.

[11] McFadden, G. B., Coriell, S. R., Murray, B. T., Glicksman, M. E. and Selleck, M. E. (1990) 'Effect of a crystal-melt interface on Taylor-vortex flow', Phys. Fluids A 2, 700–705.

[12] Murray, B. T., McFadden, G. B. and Coriell, S. R. (1990) 'Stabilization of Taylor-Couette flow due to time-periodic outer cylinder oscillation', Phys. Fluids A 2, 2147–2156.

[13] McFadden, G. B., Murray, B. T., Coriell, S. R., Glicksman, M. E. and M. E. Selleck (1992) 'Effect of modulated Taylor-Couette flow on crystal-melt interfaces: theory and initial experiments', in M. E. Gurtin and G. B. McFadden (eds.), On the Evolution of Phase Boundaries, The IMA Series in Mathematics and Its Applications, Vol. 43, Springer-Verlag, New York.

STIMULATED CONVECTION AND MORPHOLOGICAL INSTABILITY

K. BRATTKUS
Applied Mathematics
Caltech
Pasadena, CA 91125
USA

ABSTRACT. The directional solidification of a stably stratified binary mixture is examined in the limit of large Rayleigh number. Three distinct modes that correspond to either internal waves, buoyancy edge waves, or interfacial instabilities are identified. Buoyancy is found to localize interfacial disturbances and destabilize the morphological instability. We find that when internal waves that are normally damped are driven by gravity modulation, their interaction with morphological modes is weak and that this stimulated convection does not significantly alter the onset of interfacial instability.

Solute buoyancy influences the morphological stability [1] of a solidification front by either altering the depth of the disturbance boundary layer or inducing a lateral transport of solute [2]. Although sufficiently strong buoyancy *always* destabilizes interfacial instabilities by steepening solute gradients, transport due to buoyancy is stabilizing. For instance, if a light solute is rejected during upward solidification the fluid is susceptible to natural convection and one imagines that an enhanced transport of solute may mix the solute gradients at the front and stabilize interfacial instability. However, it is now understood [3,4] that this effect is weak since convection is rapidly dissipated at the lengthscales associated with a morphological instability.

In this article we suggest that if convection can be made to operate on scales comparable to the morphological instability then solute transport will significantly stabilize the front. Convection might be used to stabilize the solidification front if only the scales of convection were controllable. Fortunately, during solidification into a stably stratified fluid they are; convection can be produced at arbitrary length scales. The fluid in this case is convectively stable at all Rayleigh numbers but if it is subjected to an oscillatory body force of large enough amplitude, convective waves are generated with wavelengths that are inversely proportional to the frequency of the forcing. We examine here the coupling between this stimulated convection and the morphological instability in the large Rayleigh number limit.

We employ a standard model that contains the essential elements needed to test our proposition. To begin, we assume that an incompressible and viscous binary mixture rejects a light solute as it is directionally solidified downward through a temperature field of constant gradient. The mixture is subject to a uniform oscillatory body force which

S. H. Davis et al. (eds.), Interactive Dynamics of Convection and Solidification, 23–25.

we absorb into an effective acceleration of gravity. The equations governing the diffusion of solute and the motion of fluid ahead of the propagating front are first scaled and then linearized about a solution representing planar solidification into a quiescent melt. The resulting linear system controls the development of infinitesimal disturbances in vertical velocity, concentration, and interface deflection.

We will not discuss the initial value problem but analyze instead the normal modes of the linearized disturbance equations. Our aim is to first examine solutions to the undriven case and then track the potential for modal coupling when the forcing is small but nonzero. The undriven system is convectively stable for all Rayleigh numbers so we simplify the description of the eigenfunctions by finding their leading-order asymptotic behaviour in the limit of large Rayleigh number, R, and investigate their potential for coupling in this case.

Arguments of dominant balance reveal three possible distinguished limits for large Rayleigh numbers. The scalings described below have been confirmed numerically.

The first set of eigenfunctions occupy the entire boundary layer and are approximated to leading order by inviscid internal waves in an exponentially stratified, nondiffusive and semi-infinite fluid [5]. Away from the front, closed-form solutions for a discrete set of modes are well known and these internal waves propagate at frequencies proportional to the Brunt-Väisälä frequency, directly related to $R^{1/2}$.

Viscosity and diffusion become dominant effects in a Stokes layer of thickness $R^{-1/4}$ located at the solidification front. Applying the method of matched asymptotic expansions we complete the description of the eigenfunctions near the front, require that our approximations remain uniformly valid in time and find an equation for the amplitude of the internal waves as a function of a "slow" time $T = R^{1/4}t$. In the undriven case all internal waves eventually decay. The time scale for decay in typical solidification experiments is approximately 1/10 second.

A second set of modes is confined to a layer near the interface with thickness $R^{-1/6}$. Although the stratification is weak in this layer, the balance includes viscous, diffusive, inertial, and buoyancy effects all of which complicate the leading-order eigenvalue problem. It can be shown that if σ is the growth rate of a disturbance then $\mathrm{Re}(\sigma)$ is negative definite and these modes, which that we have labeled buoyancy edge waves, are stable and decay on a faster time scale than the decay of the internal waves. These solutions will not resonate with the modulated gravity and are ignored for remainder our discussion.

Both of the previous sets of eigenfunctions were characterized by small interfacial deflection. Deflection in the next class is substantial and solutions in this third distinguished limit are the interfacial instability modes that lead to morphological instability. Disturbances here do not occupy the entire boundary layer but are, as with the edge waves, confined to a layer near the interface with a thickness $R^{-1/6}$. In this case buoyancy is balanced by an inertialess viscous drag, the diffusion of an advected solute is quasi-steady, and the velocity of the interface is balanced by solute gradients at the front. From the leading-order characteristic equation it is clear that buoyancy has *destabilized* the morphological instability to the point of constitutional supercooling. We mention that this is *also* true for unstable stratification but since the convective modes for large Rayleigh numbers are of course unstable, this fact is of little importance. Finally we note that the morphological istability in a strongly and stably stratified fluid sets in as a long-wave instability [6]. This has possible application for the practical control of microstructural scale and leads to the development of an interesting nonlinear theory.

The parametrically forced problem is nontrivial even in the large Rayleigh number limit. Since there are three separate vertical length scales $(z, R^{-1/6}z, R^{-1/4}z)$ and three distinct time scales $(R^{1/2}t, R^{1/4}t, R^{1/6}t)$ we must effectively solve a triple-deck problem in both space and time. Our goal is to select the parametric forcing so that internal waves decay on the time scale of a developing interfacial instability and search for a convective stabilization of the morphological instability.

To solve the coupled problem we adopt a multi-scale procedure in time and employ matched asymptotics in space. The frequency of the parametric forcing is chosen to produce a first-order resonance with the internal gravity waves while the magnitude of this forcing is just strong enough to overcome their natural decay. Parametric forcing modifies the amplitude equation for the internal waves and we find that if the magnitude of forcing is less than some critical value, $A_0 < A_c$, then gravity waves rapidly decay on the time scale T, if $A_0 = A_c$ they relax on a time scale of the interfacial instability and if $A_0 > A_c$, disturbances grow exponentially in time until our original linearization breaks down.

Setting $A_0 = A_c$ and continuing the expansion to higher order, we find that the interfacial modes which are confined to the middle and lower decks are virtually decoupled from the convective waves for the weak parametric forcing which we chose consistent with the linearization. The leading order characteristic equation for the morphological instability is unchanged from the result (1) and the parametrically excited convection is not able to overcome the destabilizing influence of buoyancy when Rayleigh numbers are large. It is unknown at present whether this coupling remains weak for smaller values of Rayleigh number; it may be interesting to examine the driven system beyond our restrictions.

[1] Mullins, W.W. and Sekerka, R.F. (1964) 'Stability of a planar interface during solidification of a dilute binary alloy', Journal of Applied Physics 35, 444-451.

[2] Davis, S. H. (1990) 'Hydrodynamic interactions in directional solidification', Journal of Fluid Mechanics 212, 241-262.

[3] Coriell, S.R., Cordes, M.R., Boettinger, W.S. and Sekerka, R.F. (1980) 'Convective and interfacial instabilities during unidirectional solidification of a binary alloy', Journal of Crystal Growth 49, 15.

[4] Hurle, D.T.J., Jakeman, E. and Wheeler, A.A. (1982) 'Effect of solutal convection on the morphological stability of a binary alloy', Journal of Crystal Growth 58, 163.

[5] Turner, J.S. (1973) Buoyancy effects in fluids, Cambridge University Press, Cambridge.

[6] Coriell, S.R. and McFadden, G.B. (1989) 'Buoyancy effects on morphological instability during directional solidification', Journal of Crystal Growth 94, 513-521.

DYNAMICS AND STRUCTURE OF AN AGGREGATION GROWING FROM A DIFFUSION FIELD

Yukio Saito
Department of Physics, Keio University,
Yokohama 223, Japan
and
Makio Uwaha and Susumu Seki
Institute for Materials Research, Tohoku University,
Sendai 980, Japan

ABSTRACT. The growth velocity of a unidirectional aggregation in a gas of finite density is determined by matching two length scales: the thickness of the gas density depression region and the crossover length of the solid structure from the fractal to the compact one. This theoretical prediction is confirmed by Monte Carlo simulations of a lattice gas model. Drift flow of the gas affects the growth dynamics and the structure. With a flow parallel to the growth direction, the aggregate shows the first-order dynamical transition from the steady growth to the nongrowth mode. With a perpendicular gas flow, dendrites incline to the upstream direction. The lattice anistropy suppresses the inclination of main stems.

1. Aggregation Growth from a Gas without Drift Flow

In an aggregation growth such as solidification from a gas with a density n_g, the local density of diffusing gas $n(\vec{x},t)$ obeys the time-dependent diffusion equation: $\partial n/\partial t = D\Delta n$. Its static and low density-limit, $n_g \to 0$, is the Laplace equation, $0 = \Delta n$, and an adhesive growth therein produces the diffusion-limited aggregation(DLA).[1] The DLA aggregate is fractal without a characteristic length, since the Laplace equation does not involve any length scale. The density n of the solid atoms decays as $n \propto R^{-(d-D_f)}$ as the distance R from the central solid atom increases. Here the fractal dimension D_f characterizes the fractal structure of the aggregate, and is smaller than the spatial dimension d.

The Laplace equation, however, is merely an approximation for the diffusion equation when the growth is slow. We studied previously the aggregation growth from a time-dependent diffusion field, which is simulated by a lattice gas model with a finite gas density.[2] Our model is a natural generalization of the DLA model and, in the limit of high gas density, $n_g \to 1$, it becomes another well-studied growth model, the Eden model.[3] In a unidirectional diffusion growth, the spatial variation of the gas density is characterized by a diffusion length $\xi \sim D/V$, where V is the growth velocity. Introduction of this characteristic length changes the structure of our aggregate drastically. It is a DLA fractal up to the length scale ξ, whereas it is uniform and compact for scales larger than ξ. The characteristic length ξ is determined by the fractal dimension of the DLA as $\xi \approx n_g^{-\nu}$, where $\nu = 1/(d - D_f)$. This relation is an analogue of that between the correlation length and the temperature difference from the critical point value in second order phase transitions. Consequently, dynamics of the growth is determined by the fractal structure of the aggregate, and the growth velocity changes in a power of the gas density as $V \sim n_g^{\nu}$.

S. H. Davis et al. (eds.), Interactive Dynamics of Convection and Solidification, 27–29.

Monte Carlo simulations confirmed this power law behavior as is shown in Fig.1, and for the fractal dimension the value $D_f = 1.71$ is obtained, which agrees with the previously obtained value.[4]

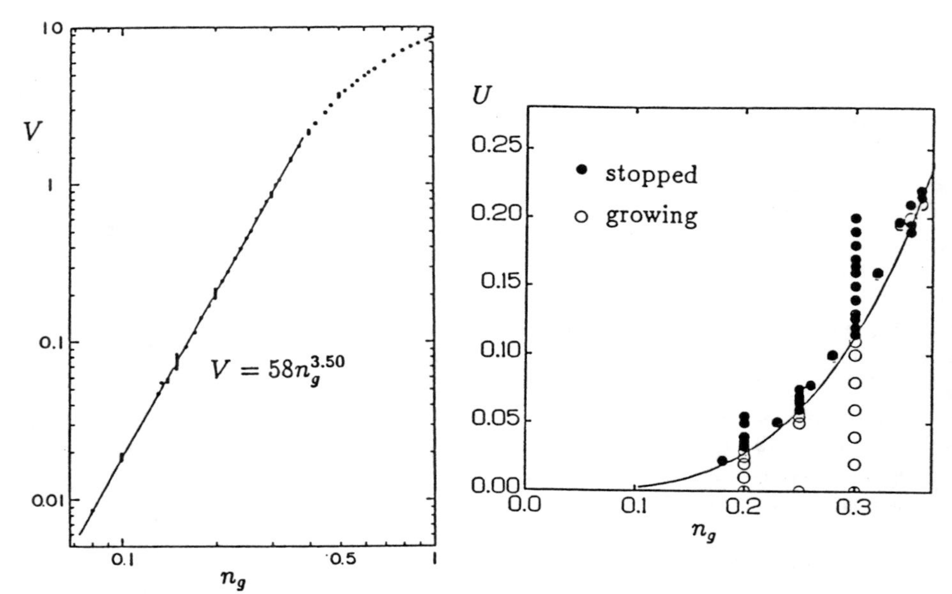

Fig.1 Growth velocity V versus gas density n_g without drift flow, $U = 0$.

Fig.2 Phase diagram in the space of flow velocity U and gas density n_g.

2. Aggregation Growth from a Gas with Parallel Flow

Homogeneous flow of the gas with a drift velocity $\vec{U}$ also introduces a characteristic length in the system via the drift term, $\vec{U} \cdot \vec{\nabla} n$, in the diffusion equation.

We studied coupled effect of the drift flow parallel to the growth direction and the finite gas density in the aggregation growth, and found a dynamical phase transition of the first order.[5] For a drift velocity smaller than a critical value U_c, the aggregate grows steadily, whereas for $U > U_c$ the aggregate stops growing. At $U = U_c$, the growth velocity and other physical quantities remain finite. The ratio of the critical drift velocity U_c to the growth velocity V_0 without flow takes a constant value, determined only by the fractal dimension of the DLA. Our Monte Carlo simulation confirms this theoretical prediction, and in two dimensions the value of fractal dimension $D_f = 1.70$ is obtained by fitting the phase boundary between the steadily growing phase and the nongrowing phase. The value is quite close to $D_f = 1.71$ obtained previously.

3. Aggregation Growth from a Gas with Perpendicular Flow

If the flow is perpendicular to the growth direction, it does not affect the growth velocity much, although it changes the growth direction. The aggregates incline to the upstream of the perpendicular flow. The inclination is affected by the anisotropy of the dendrite. The variation of the anisotropy is effectively simulated by allowing solidification only on a large grid mesh.[6] For the original model with mesh size m=1, main branches of the dendrite incline towards the upstream of the flow, and the inclination angle is determined approximately as the vector sum of the growth velocity and the drift flow velocity(Fig.3a). For systems with a larger mesh size as m=8, the main branches keep growing in its crystallographic direction, independent of the drift flow, as is shown in Fig.3b. The effect of the drift flow is reflected in the asymmetry of the sidebranch activities, which is nearly independent of m.

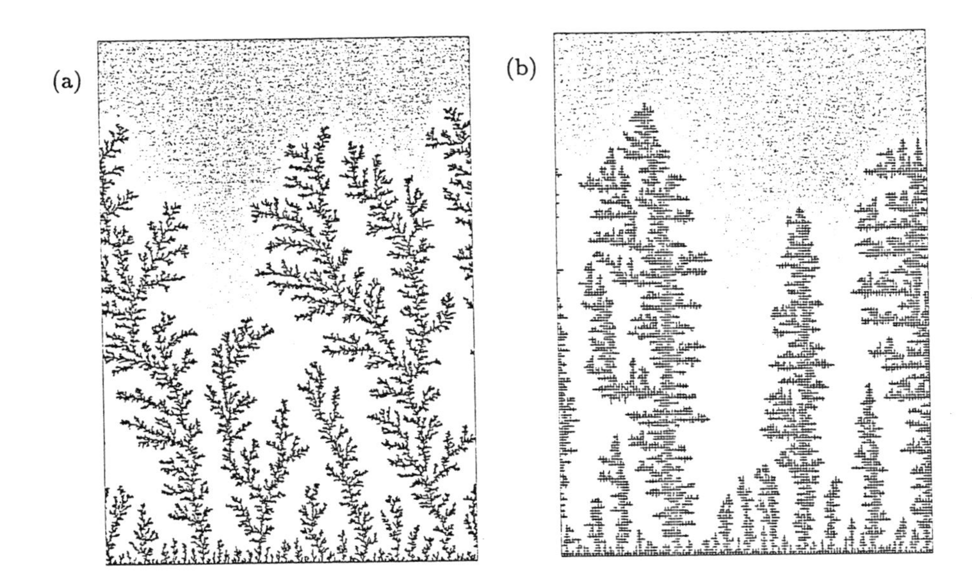

Fig.3 Aggregation growing in a gas with a perpendicular drift velocity U=0.01. The mesh size m and the density n_g are (a) m=1, n_g=0.1, and (b) m=8, n_g=0.065.

Part of this work is supported by the Grant-in-Aid for Scientific Research on Priority Areas by the Ministry of Education, Science and Culture with No.03243101, and the Visitor's Program of the IMR, Tohoku University.

References

[1] Witten, T. A. and Sander, L. M. (1983) Phys. Rev. B **27**, 5686.
[2] Uwaha, M. and Saito, Y. (1988) J. Phys. Soc. Jpn. **57**, 3285.
[3] Eden, M. (1961) in *Proc. 4-th Berkley Symp. on Math. Statistics and Probability*, Vol.4, Ed. F. Neyman (Berkley).
[4] Meakin, P. (1986) Phys. Rev. A **33**, 3371.
[5] Seki, S. Uwaha, M. and Saito, Y. (1991) Europhys. Lett. **14**, 397.
[6] Uwaha, M. and Saito, Y. (1990) J. Crystal Growth **99**, 175.

MICROSCALE COUPLING OF SOLIDIFICATION AND FLOW

S. H. DAVIS
Northwestern University
Dept. of Engineering Sciences and Applied Mathematics
McCormick School of Engineering and Applied Science
Evanston, IL 60208

ABSTRACT. Morphological instability is discussed via linear and nonlinear theories. Couplings with forced flows and convecting melts are discussed in terms of physical mechanisms and mathematical descriptions. Microscale evolutions of coupled phenomena involve the development of cells, and travelling waves, and the selection of patterns and scales.

1. INTRODUCTION

The microstructure of a solidified material largely depends on two interacting physical processes. Firstly, a solid-liquid interface is subject to morphological changes that are controlled by the conduction of heat and the diffusion of solute. These conspire to determine the interface shape be it be a simple conic surface, cellular or dendritic. Secondly, fluid flow, be it externally driven or the result of buoyancy-driven, volume-change or thermo-solute-capillary convection, can dramatically affect the microsegregation. The flow field can redistribute heat and solute so as to alter the intrinsic morphological development. Alternately, it can create interactions that produce new morphologies.

On the microscale one can envision three types of interactions between solidification and flow: (i) flow over existing morphology, (ii) morphological growth into a flow field and (iii) competing instabilities. The object of this paper is to discuss the interactions of the microscale processes principally from the point of view of instability theory. The aim is to delve into morphological instabilities of planar fronts and their interactions with flow fields. Although linear stability theory plays a central role in the subject, emphasis will be given to nonlinear studies, ones that have been rather recently given and others that do not exist, but are necessary for the understanding of microscale evolutions.

There are two categories of problems that have received recent quantitative study. The first area, which we call PROTOTYPE FLOWS, involves simple flow geometries that correspond to well-studied hydrodynamic instabilities. If one (or more) of the rigid boundaries

S. H. Davis et al. (eds.), Interactive Dynamics of Convection and Solidification, 31–52.

of such a geometry is replaced by a crystal interface, and if temperature and/or concentration gradients are posed to support this, then one can investigate the influences of fluid flow, heat and/or mass transfer on the conditions for instability and the morphologies of the interfaces. Typically, the interfaces in their basic states are supposed stationary. Thus, even though morphological changes are important and these studies involve morphological instabilities, these should not be confused with <u>the</u> morphological instability of moving fronts.

The second area involves <u>the</u> morphological instability in binary mixtures in DIRECTIONAL SOLIDIFICATION. Here the primary instability is driven by solute rejection and the solute distribution near the propagating interface. Clearly, fluid flows have the potential of homogenizing the solute distribution and hence delaying the onset of morphological change. However, the interaction of flows with interfaces can also generate new instabilities that promote morphological changes.

There are several surveys that would be useful to consult, e.g. Coriell, McFadden and Sekerka (1985), Glicksman, Coriell and McFadden (1986), and Davis (1990). There is a soon to appear three volume set called the Handbook of Crystal Growth, edited by D. T. J. Hurle. The articles, Coriell and McFadden (1993) on Morphological Stability and Davis (1993) on Effects of Flow on Morphological Stability are especially relevant here; these greatly expand on the topics given herein.

2. MORPHOLOGICAL INSTABILITY

Before discussing coupling problems let us consider "pure" morphological instability of a planar front. In the simplest version of the directional solidification of a binary alloy, there are three control parameters: the average solute concentration c_∞, the imposed temperature gradient G and the pulling speed V. Here the frozen-temperature approximation has been made in which the release of latent heat L is ignored, the thermal properties of the solute and liquid are taken to be identical, and the diffusion of solute is much slower than the diffusion of heat. Further, the diffusion of solute in the solid is neglected. The dimensional quantities can be gathered into two non-dimensional numbers: the morphological number M

$$M = \frac{m_E(k_E-1)c_\infty V}{DGk_E} \quad , \tag{1}$$

and the surface-energy parameter Γ,

$$\Gamma = \frac{T_M \gamma V k_E}{LDm_E(k_E-1)c_\infty} \quad . \tag{2}$$

Here γ is the surface energy, D is the solute diffusivity in the liquid, and m_E is the (equilibrium) liquidus slope. An independent parameter, as well, is the (equilibrium) segregation coefficient k_E.

Typically, one fixes k_E and Γ, and raises M until the critical value M_c where by the linearized theory of Mullins and Sekerka (1964) the planar interface becomes unstable. A typical neutral curve is shown in Figure 1a. In dimensional terms if one fixes G, the neutral stability curve is shown in Figure 1b. Inside the loop the planar front is unstable. The asymptote to the lower branch is the constitutional undercooling limit, and the asymptote to the upper branch is the absolute-stability limit.

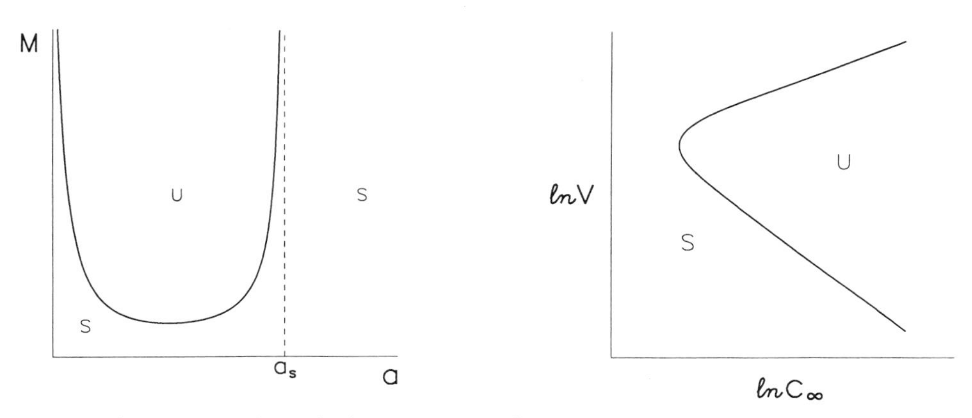

Figure 1. Neutral stability curves for morphological instability. (a) M versus a, where M is the morphological number and a is the non-dimensional wave number; (b) V versus c_∞ at fixed G.

Bifurcation theory is a means of examining the weakly-nonlinear behavior near $M = M_c$ and determining when the system will undergo smooth or jump transitions from the planar state to cells. The transitions are governed by a Landau equation for the amplitude A,

$$\frac{dA}{dt} = \sigma A - a_1 A|A|^2 \quad , \tag{3}$$

that gives the size of the interfacial corrugations. Here σ is linear-theory growth rate; the Landau constant a_1 is computable and its sign gives the required information: when $a_1 > 0$ (< 0) there is a smooth (jump) transition to steady cells. Figure 2 shows the results of Wollkind and Segel (1970) superimposed on the dimensional neutral curve. The upper branch is typically supercritical while most of the lower is subcritical. Alexander, Wollkind and Sekerka (1986) included

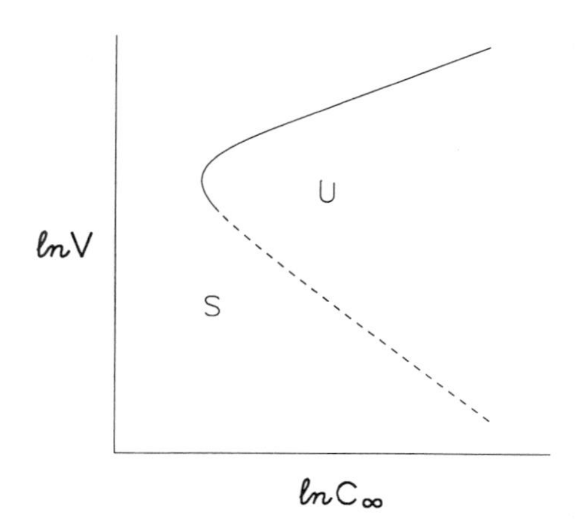

Figure 2. Neutral stability curves for morphological instability. Bifurcation results of Wollkind and Segel (1970) for zero latent heat indicate supercritical (subcritical) bifurcation by solid (dashed) curve.

the effects of latent heat, and hence calculated a modified value of a_1. Caroli, Caroli and Roulet (1982) showed that when $k > 0.45$, the entire branch is supercritical. Sriranganathan, Wollkind and Oulton (1984) and Wollkind, Sriranganathan and Oulten (1984) described weakly-nonlinear three-dimensional states. Ungar and Brown (1984a,b), Ungar, Bennett and Brown (1985) and Wheeler and Winters (1989) studied deep cells and their two-dimensional secondary bifurcations in spatially-periodic boxes using finite-element simulations. McFadden and Coriell (1984) studied Al-Ag and used finite differences to examine shallow two-dimensional cells and their secondary bifurcations. McFadden, Boisvert and Coriell (1987) used finite-difference simulation for the Al-Cr system, $k > 1$, and found stable hexagonal nodes.

Merchant and Davis (1989a) examined bifurcation theory near the nose of the curve of Figure 1. They found an amplitude equation of the form

$$\frac{dA}{dt} = (\alpha_1 \Delta c_\infty - \alpha_2 \Delta V)A - [\beta_1(\Delta n) + \beta_2(\Delta V)]A|A|^2 - \alpha_3 A|A|^4 . \quad (4)$$

Here n is the conductivity ratio k_S/k_L and n is manipulated so that the transition point is with Δn of the nose. This analysis simultaneously describes the upper and lower branches near the nose when latent heat is negligible. It is predicted that isolas, solutions detached from the neutral curve of Figure 1, may occur to the left of the nose.

Near the absolute stability boundary the preferred wave lengths of the cells, while short, are very long compared to the solute boundary-layer thickness $\delta_c = D/V$. Thus, one can use long-wave theory and derive a strongly nonlinear evolution equation for the interface shape $z = h(x,y,t)$. Brattkus and Davis (1988a) found that

$$h_{tt} - \nabla^2 h_t + \frac{1}{4}(1-\nu^2)\nabla^4 h + \nabla^2 h + \mu^{-1} h = h_t \nabla^2 h + |\nabla h|_t^2$$

$$- \frac{1}{2}(1-\nu)\nabla^2 |\nabla h|^2 - \nu\nabla\cdot[\nabla h \nabla^2 h] - \frac{1}{2}\nabla\cdot[|\nabla h|^2 \nabla h] \qquad (5)$$

where $\nabla^2 = \frac{\partial^2}{\partial x^2} + \frac{\partial^2}{\partial y^2}$, $\nu = (1 + 2k_E)^{-1}$ and $\mu \propto V$.

When equation (5) is specialized to weakly nonlinear disturbances, Brattkus and Davis find that (i) there exist stable, steady two-dimensional cells that bifurcate supercritically, and (ii) in three dimensions there exist stable hexagonal nodes near $k_E = 1$.

Brattkus (1990) has reconsidered the above analysis for the symmetric model in which solute diffusion is equally effective in the solid and liquid. He found that the evolution equation (in two dimensions) is equation (5) in two dimensions with only a single coefficient modified. He then examined the large μ regime and found secondary instability that has double the wave length of a basic cell and is oscillatory in time. This is a so called "optical mode".

Long-wave descriptions of solidification fronts can be very rewarding since one thereby reduces a free-boundary problem to a single equation. The first of these was Sivashinsky (1983) for $\Gamma = O(1)$ and $k_E \to 0$ with $k_E \ll a^2 \ll 1$ in which the weakly-nonlinear interface is given by

$$h_t + \nabla^2 h + \nabla^4 h + Kh = \nabla\cdot(h\nabla h) \quad , \qquad (6)$$

where K is a scaled version of k_E. Here a is the wave number of the spatially periodic structure. Novick-Cohen and Sivashinsky (1986) have included latent heat effects and Young and Davis (1989) have included end effects.

Riley and Davis (1990) described "all" such equations and found a new one intermediate between that of Sivashinsky and of Brattkus and Davis, a strongly nonlinear one that applies to the limit $k_E \to 0$, $\Gamma \to \infty$, with $a^2 = O(k_E)$. They find that

$$h_t - M\nabla^2 h_t + M\nabla^4 h + (M-1-\kappa M)\nabla^2 h + \kappa h =$$

$$\nabla\cdot(h\nabla h) - M\nabla\cdot(\nabla h \nabla^2 h) \quad , \qquad (7)$$

where κ is a scaled version of k_E. Impey, Riley and Wheeler (1992) have studied the bifurcation structure of equation (7).

When systems operate near the absolute stability boundary, the speeds V are so large that thermodynamic equilibrium at the interface is likely not to hold. Baker and Cahn (1971), Boettinger and Perepezko

(1985), Boettinger and Coriell (1986), Aziz (1982), and Jackson, Gilmer and Leamy (1980) have contributed to an altered model of the interface in which $k_E \to k(v_n)$ and $m_E \to m(v_n)$. A disequilibrium parameter β

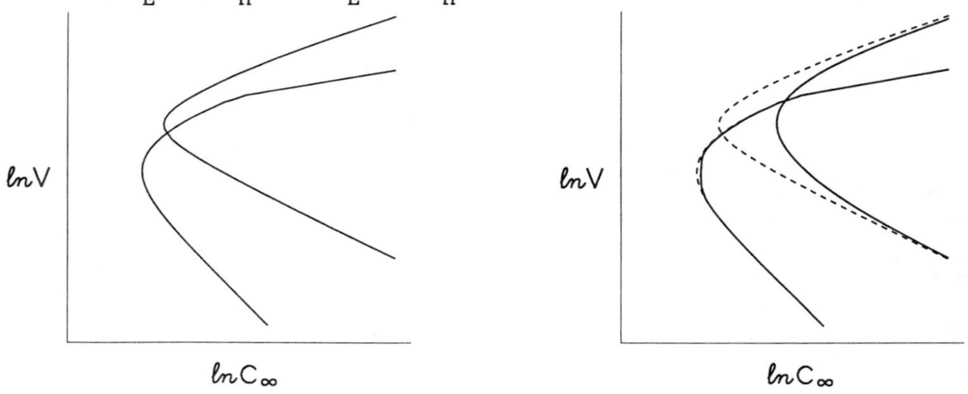

Figure 3. Neutral curves, V versus c_∞, for fixed G for stability in rapid solidification. (a) Lower (upper) curves describe cellular (pulsatile) modes according to frozen-temperature approximation results of Merchant and Davis (1990). Increasing β stabilizes upper branch of cellular mode. Increasing μ stabilizes pulsatile mode. (b) Similar curves as for (a) with $\beta, \mu \neq 0$ (solid curve) with and without (dashed curve) full thermal field and latent heat included. Latent heat slightly stabilizes the cellular mode and greatly affects the oscillatory mode though both absolute stability boundaries are unaffected by latent heat.

relates k and m to the local speed v_n of propagation of the interface. In addition, there can be a kinetic delay at the interface in which the attachment kinetics causes the temperature of the interface to become depressed via $T_e \to T_e - \mu v_n$.

Coriell and Sekerka (1983) showed that a similarly-altered model contains an oscillatory instability in the system. Merchant and Davis (1990) delved further into the characteristics of the system specifically for the model suggested by Baker and Cahn, Boettinger and Perepezko, Boettinger and Coriell, Aziz and Jackson, Gilmer and Leamy. Under the frozen-temperature approximation they found two distinct modes of instability. There is the cellular mode in which disequilibrium somewhat stabilizes the absolute boundary from that given by Mullins and Sekerka (1964); this mode is independent of attachment kinetics. There is a second distinct mode which is oscillatory in time, and in the frozen-temperature approximation, occurs at zero wave number; it represents a pulsatile modulation of the interface speed. This mode is independent of surface energy and has an absolute stability boundary determined by attachment kinetics. These modes are shown in Figure 3a. Merchant and Davis (1990) have conjectured that where the two modes are simultaneously present, the banding phenomenon should be present. Bands are alternate layers of cells and seemingly segregation-free material periodic in the pulling direction. Braun, Merchant, Brattkus and Davis (1992) have analyzed the strongly-nonlinear relaxation oscillations that develop from the pulsatile mode. Brattkus and Meiron (1992) have simulated these numerically. Braun and Davis (1992a) have examined the sideband instabilities of the oscillatory mode and shown that the zero-wave-number mode is very sharply selected at $V \to \infty$. Braun and Davis (1991) have examined the weakly nonlinear development of the cellular mode. Braun, Merchant and Davis (1992) have examined the coupling of the two modes via a co-dimensional two bifurcation analysis.

Huntley and Davis (1992) have reexamined the linear theory including latent heat and the full effects of thermal perturbations. They made the thermal conductivity and diffusivity of the two phases equal. Their results, as given in Figure 3b, show that thermal effects slightly stabilize the cellular branch near the nose and the absolute stability boundary is unchanged from before. However, the oscillatory mode is strongly stabilized by thermal effects though its absolute stability boundary is also unchanged by thermal effects. Further, the inclusion of latent heat may alter the preferred wave number of the mode. When $L = 0$, the critical wave number is zero. When $L \neq 0$, the critical wave number is generally non-zero (though quite small numerically) but is again zero if the diffusivity ratio D/κ is not too small. Karma and Sarkissian have simulated numerically the $L \neq 0$ problem for $a = 0$ (though $a \neq 0$ is preferred) and found relaxation oscillations.

3. FLOW OVER EXISTING MORPHOLOGY

Consider an isolated cell or dendrite of width ℓ growing into an undisturbed environment as shown in Figure 4a. The natural growth direction is **V**. When, as is shown in Figure 4b, a flow is imposed, say, orthogonal to **V**, the solidification process is altered and the cell tilts into the flow. Dantzig and Chao (1986) argued that if the

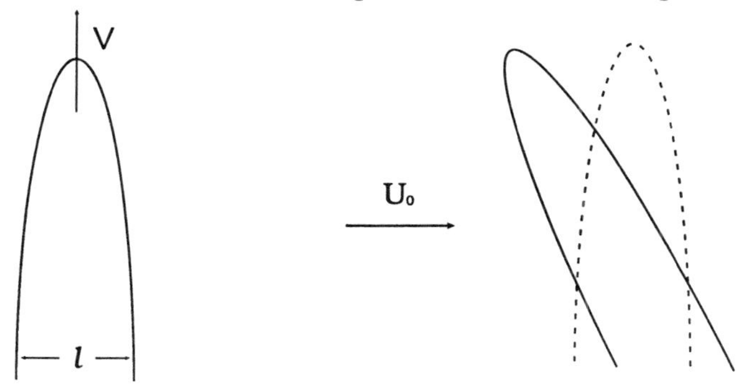

Figure 4. An isolated cell growing into an undercooled melt (a) with no flow, and (b) with flow causing a tilt into the shear according to Dantzig and Chao (1986).

Reynolds number Re = $U_0\ell/\nu$ is small so that the flow field is free of inertia, the solute field can still have significant convective contributions. The solute Peclet number $U_0\ell/D$ = Re Sc, where Sc = ν/D, may be significant since the kinematic viscosity ν of the fluid is 10-100 times the solute diffusivity D. Thus, on the upstream side of the cell there is a solute boundary layer that may have large concentration gradients which induce the cell to grow into the flow as shown in Figure 4b. This morphological change is due to the alteration of the solute distribution by the flow. A similar effect may occur with the thermal field. Here, the thermal Peclet number $U_0\ell/\kappa$ = Re Pr, where Pr = ν/κ. Now, a morphological alteration will occur if ν is much larger than the thermal diffusivity κ, i.e. only if the Prandtl number Pr is large. This will be the case with organic liquids but not with liquid metals.

4. MORPHOLOGICAL GROWTH INTO FLOW FIELDS

The effect of thermal convection has been seen in the now-classical experiments of Glicksman and Huang (1982), which grow isolated thermal dendrites of succinonitrile in various directions with respect to the gravity vector. The dendrites develop asymmetries in their sidearm developments that reflect the preferred direction of gravity. Theoretical analysis of the effect on an isolated dendrite of forced or convective flow are discussed in several papers in this volume.

There are several PROTOTYPE FLOWS that have received attention. These are as follows: (i) shear flows (parallel and non-parallel), (ii) free convection, (iii) Bénard convection and (iv) Taylor-Couette flow. Herein only shear flows imposed on a <u>propagating</u> interface will be discussed.

5. SHEAR FLOWS OVER PROPAGATING FRONTS

5.1 Parallel Shear Flows

In a frame of reference moving with the propagating front a parallel shear flow is imposed along the planar interface. Coriell, McFadden, Boisvert and Sekerka (1984) and Forth and Wheeler (1989) examined two-dimensional disturbances (periodic in the flow direction) and found that the flow stabilizes the interface in that the presence of the flow increases M_c. However, if one considers longitudinal-roll disturbances (periodic cross-stream and independent of the stream direction), the flow decouples from the problem and M_c is unchanged. Thus, under "normal" circumstances, the flow leaves M_c unchanged but selects the cellular state (longitudinal rolls) that appears.

Forth and Wheeler (1991) have examined the weakly-nonlinear development of two-dimensional waves and found that the flow promotes supercritical bifurcation and narrows the bandwidth of two-dimensional structures allowable in the nonlinear regime.

Forth and Wheeler (1989) have looked at a wide range of conditions and found for long waves that the flow destabilizes the interface for two-dimensional disturbances that propagate against the flow. However, they do not determine when this mode is preferred.

Hobbs and Metzener (1991) have examined the effect of a shear flow on the interface near the absolute stability boundary. Here the wave

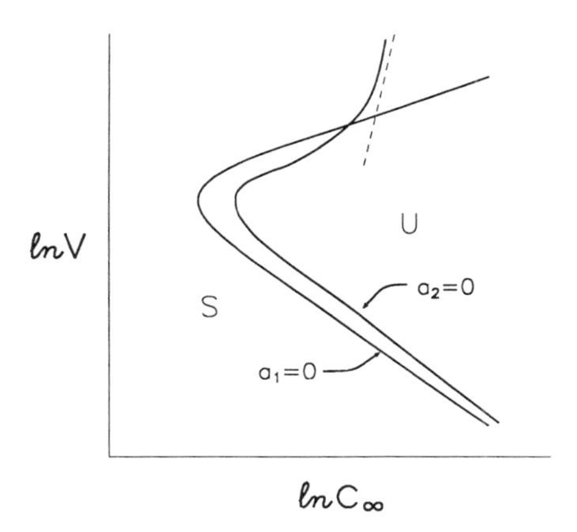

Figure 5. Neutral curves, V and c_∞, at fixed G for an interface with an imposed parallel flow. $a_1 = 0$ and $a_2 = 0$ corresponds to longitudinal roll and two-dimensional modes, respectively, according to Hobbs and Metzener (1991).

lengths are long compared to δ_c and the two-dimensional waves lower M_c and hence are preferred. They establish via linear stability theory conditions that determine neutral stability of the interface. Figure 5 shows that the neutral curve on the lower branch and beyond the nose corresponds to the flow-free values, but that a substantial destabilization occurs on the upper branch.

Hobbs and Metzener (1992) have extended their work into the nonlinear regime. Here the long-wave structure of the solution induces a flow correction to the equation of Brattkus and Davis. The resulting interface equation contains an extra linear term that accounts for the destabilization, but otherwise the equation is left unchanged. When they specialize their equation to the weakly-nonlinear regime, they obtain a modified Newell-Whitehead-Segel equation. Further, when this equation is examined for phase evolution, a modified Kuramoto-Sivashinsky equation is obtained, showing various sequences of "normal", as well as chaotic, solutions.

Remark: The phenomena uncovered for high speeds suggest a host of potentially interesting experiments. However, near the lower branch of the morphological instability, linear theory selects the morphology but nothing more is known. One should at least delve into the weakly nonlinear competition between longitudinal rolls and nearby three-dimensional states.

5.2 Non-Parallel Flows

Most flows are not parallel. These range from the von Kàrmàn swirl flow generated by the rotation of the crystal to the locally hyperbolic flows present when cellular convection exists at the interface.

Brattkus and Davis (1988b,c) studied flows with hyperbolic streamlines directed upon a solidifying interface. These were, respectively, a von Kàrmàn swirl flow and stagnation-point flows. We discuss here the simplest of these, two-dimensional stagnation-point flow with (x,z) coordinates corresponding to velocity (u,w). A Hiemenz flow has the form for $z \to \infty$, $u \sim (K\nu)^{1/2} x F'(z)$, $w \sim -(K\nu)^{1/2} F(z)$ where F is a function that is obtained numerically and K measures the strength of the flow. The linear-stability problem is made tractable by assuming that the viscous-boundary thickness δ_V is much larger than the concentration-boundary-layer thickness δ_c and that the Schmidt number Sc is very large. Explicitly, it is assumed that $\delta_c/\delta_V = (\frac{D}{V})(\frac{\nu}{K})^{-1/2} = O(Sc^{-1/3})$ as $Sc \to \infty$. Given the smallness of δ_c, the interface senses only the local forms of the imposed flow and the flow senses a flat interface; thus $u \sim \beta x \zeta$, $w \sim -\frac{1}{2}\beta\zeta^2$ where $\zeta = z/\delta_c$, and β measures F"(0), the local shear. Finally, Brattkus and Davis considered waves long compared to δ_c though smaller than δ_V. They solved the modified diffusion problem in which the lateral diffusion of solute is neglected. This neglect is a device that allows one to solve the stability equation in terms of the quasi-normal modes $\exp[\sigma t + ia \ell n\, x]$.

The flow produces a long-wave instability that creates waves that travel inward, toward the stagnation point. It is locally periodic in x but by the structure of the quasi-normal modes is not globally periodic. The instability exists for long waves, in a region where the Mullins and Sekerka condition gives only stability. Thus, it is called _flow-induced morphological instability_. The largest growth rate Re σ occurs for $a \to \infty$ where the long-wave theory is invalid and where surface energy should help stabilize the interface. Thus, "longish" waves would be preferred and these would grow for M just above unity, i.e. for any degree of constitutional undercooling; thus, it is a morphological instability. The conjectured neutral stability curve, valid for all wave numbers, would be as shown in Figure 6. When the wave numbers are large, the disturbances see only the local velocities and the flow appears to be locally parallel. Figure 6 shows the analog of the results of Coriell, McFadden, Boisvert and McFadden (1984) appropriate to locally-parallel flows. When the wave numbers are small, the disturbances see the curvature of the streamlines and hence the non-parallel-flow effects found by Brattkus and Davis (1988c), as shown in Figure 6. The connection between the small -|a| loop and the ordinary morphological instability loop is conjectured.

The destabilization by non-parallel flow depends on both velocity components. The component normal to the mean position of the interface is directed inward. Its presence causes boundary-layer alteration. The concentration boundary layer is compressed, steepening the local gradient. The lateral component of velocity (linear in x) varies with distance from the stagnation point and produces horizontal concentration gradients that drive the travelling cells that propagate into the oncoming flow. Brattkus and Davis (1988b) argued that these instabilities may be responsible for the "rotational striations"

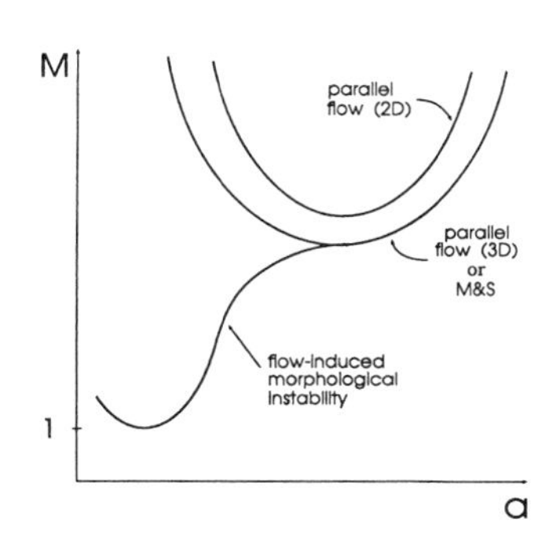

Figure 6. Conjectured neutral curve, M versus |a|, shown as the solid curve for general two-dimensional disturbances on a Hiemenz flow against a solidifying interface according to Brattkus and Davis (1988c). For small |a|, non-parallel effects dominate; for other a the curve coincides with the locally-parallel theory. Coriell, McFadden, Boisvert and Sekerka (1984) as shown by the dashed curve.

present in crystals produced in devices using crystal rotation.

The destabilization of long waves in the x-direction may be negated by "end" effects that disallow the "fitting" of such waves in the system. In this case the results of Coriell, McFadden, Boisvert and Sekerka (1984) would be regained. One could then allow disturbances of the form $\exp[\sigma t+iby]$ for cross-stream periodic waves that are x-independent (a=0). The full stagnation-point flow linear-stability problem has been examined in this case by McFadden, Coriell and Alexander (1988). They found that the flow would then slightly stabilize morphological instability.

The effects of unsteadiness in the melt flow have been investigated by Merchant and Davis (1989b). They considered plane stagnation-point flow against the interface but allowed the flow at infinity to be time periodic, where the strength K of the flow is replaced by a time-periodic function as follows: $K \rightarrow K[1 + \delta \cos \omega t]$. They considered long-wave two-dimensional disturbances and found that modulation at low frequency $\omega D/V^2$ stabilizes the interface against flow-induced morphological instabilities while high frequency modulation promotes the instabilities. The response of the system to instability is quite complex with a disturbance being composed of two independent frequencies, the imposed frequency and the travelling-wave frequency modified by the modulation.

<u>Remark</u>: The phenomena uncovered by linear theory with some non-parallel flows should be examined in the nonlinear range. Further, one should examine a broad range of non-parallel flows and attempt to quantify the interactions for use in diagnosing microstructures developed in complex flows.

6. COMPETING INSTABILITIES: CONVECTION AND SOLIDIFICATION

6.1 Volume-Change Convection

The interface is the site of a phase transformation in which the liquid of density ρ_L changes to solid of density ρ_S. Morphological changes are driven by solute rejection there and in addition if $\rho_L \neq \rho_S$, the interface drives a convective flow. For example the lead-tin alloy shrinks upon solidification, $\rho_S/\rho_L \approx 1.05$, and a weak flow from "infinity" is generated in order to conserve mass. On the other hand, silicon alloys expand upon solidification, $\rho_S/\rho_L \approx 0.91$, and the generated flow is from the interface to "infinity".

Since the interface is solid, the no slip condition still holds. When $\rho_L = \rho_S$, the interface is impermeable to flow, but when $\rho_L \neq \rho_S$, it acts as a porous surface that produces or consumes liquid at the rate required by the volume changes.

The linear theory for morphological instability with volume-change convection was given by Caroli, Caroli, Misbah and Roulet (1985b). The following explanation is a variant of theirs. Consider the case of shrinkage, $\rho_S > \rho_L$. There are two competing effects that are present.

On the one hand the flow from infinity will cause the concentration boundary layer of thickness $\delta_c = D/V$ to be compressed so that the local gradient is increased. This will enhance the morphological instability through what we can term boundary-layer alteration. Thus, shrinkage is destabilizing while expansion is stabilizing.

On the other hand consider the result of an initial corrugation of the interface. Since the interface is a no slip surface, all streamlines cross the interface normally. At the crest or trough the streamlines are vertical but elsewhere they are curved. The curvature is accompanied by transverse velocities that for $\rho_S > \rho_L$ carry the solute from the trough to the crests which homogenizes the solute and decreases the concentration gradient. Thus, for the case of shrinkage, the induced lateral transport of solute is stabilizing.

Caroli, Misbah and Roulet (1985b) found that low solidification speeds V, which correspond to thick concentration boundary layers, promote destabilization by shrinkage since the boundary-layer alteration is more effective than is lateral transport. At high speeds V, the opposite is the case. Brattkus (1988) showed that systems with small k are destabilized when compared to the constant-density case; the reverse is true in materials with moderate k. This switch-over of effect with material properties is important to recognize since for convenience in many experiments transparent organics are substituted for metallics; such a switch may also reverse the influence of volume-change convection.

Figure 7 shows our sketch of Caroli, Misbah and Roulet's results. When $\rho_S > \rho_L$ and k is moderate, the neutral curve is shifted downward as shown. If k is small, the shift is upward.

Wheeler (1992) has examined the coupling of morphological instability and volume-change convection by deriving a strongly-nonlinear evolution equation for the interface.

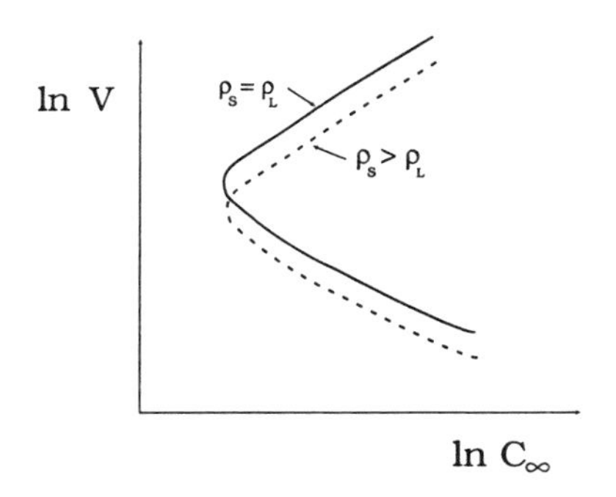

Figure 7. Neutral stability curves, V versus c_∞, for fixed G for $\rho_S = \rho_L$ and $\rho_S > \rho_L$ for moderate k_E according to Caroli, Caroli, Misbah and Roulet (1986).

Clearly, volume-change convection has small effects if $(\rho_L - \rho_S)/\rho_L$ is small. It should be of importance when other modes of convection are absent or very small in magnitude. Such would be the case in a microgravity environment in Space. It might also be important in the deep roots between cells where dimensions are small and where diffusion is very slow.

6.2 Coupled Morphological/Convective Instabilities

Consider the directional solidification of a binary liquid in which the front moves vertically upward. The rejected-solute profile gives not only the possibility of morphological instability, but also the possibility of buoyancy-driven convective instability. If the solute is of low density, the steady basic state consists of a "heated-from-above" temperature field and an unstably-stratified concentration field. Thus, there can be a double-diffusive Bénard instability on a semi-infinite domain containing an exponentially decaying concentration profile and a (locally) linear temperature field.

If the Boussinesq approximation is used, the variations in density with concentration and temperature are considered only in the buoyancy term. There is an assumed equation of state $\rho = \rho_0\{1 - \alpha(T-T_0) - \beta(c-c_0)\}$, where T_0 and c_0 are reference quantities, and α and β are the thermal and solutal expansion coefficients, respectively. Double-diffusive convection is described using two parameters, the solutal Rayleigh number $R_S = \frac{g\beta c_\infty \delta_c^3}{D\nu}$, and the thermal Rayleigh number

$R_T = \dfrac{g\alpha G\delta_c^4}{\kappa_L \nu}$. In addition there are the Prandtl and Schmidt numbers.
Note that the length scale of both Rayleigh numbers is δ_c, given that the solute field sets the scale for the instability.

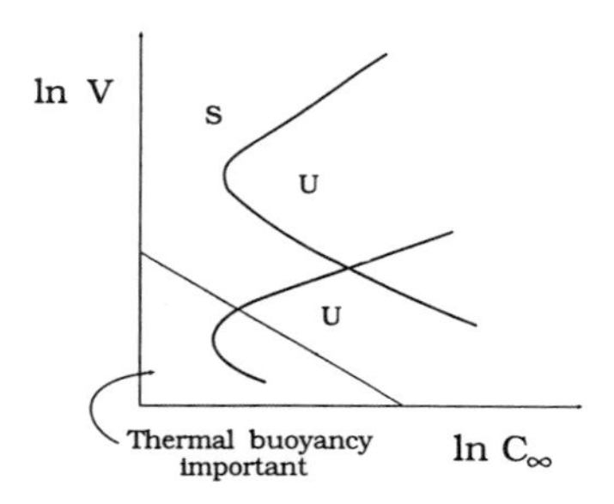

Figure 8. Neutral stability curve, V versus c_∞, for fixed G. The upper (lower) curve corresponds to morphological (convective) instability. Thermal buoyancy is negligible except in the shaded area.

Under general conditions the convection is driven by the two buoyancy effects jointly. However, under certain conditions typical of crystal growing, the thermal buoyancy is negligible. Consider now conditions under which thermal buoyancy is negligible. We evaluate the R_T/R_S ratio to obtain

$$\frac{R_T}{R_S} = \frac{\alpha}{\beta} \left| \frac{m_E(1-k_E)}{k_E} \right| \frac{D}{\kappa_L} M^{-1} \ll 1 \quad . \tag{8}$$

In relation (8), obtained jointly with D. S. Riley, the ratio of expansion coefficients, and the augment of the absolute value are numbers near unity.

In Figure 8 we show a typical morphological instability curve for fixed G, $\ell n\ V$ versus $\ell n\ c_\infty$. The upper branch asymptotes to slope unity, which represents the absolute stability boundary. The lower branch asymptotes to the constitutional undercooling limit, $M \to 1$, which has slope negative one. Plotted here as well is a typical convective instability curve. The line R_T/R_S-is-constant is shown, given inequality (8), the typical $D/\kappa_L \ll 1$, and the estimate $M \approx 1$. The line has slope negative one and is always well below the lower branch of the morphological instability. Below this line, thermal buoyancy is important while above this the convection is driven solely by solutal buoyancy.

The linear stability theory for soluto-convective instability with $\rho_L = \rho_S$ has been examined in great detail (Coriell, Cordes, Boettinger and Sekerka 1980; Hurle, Jakeman and Wheeler 1982,1983; Forth and Wheeler 1992). Such theories give results typified by Figure 9 showing, for fixed G and V, the mean solute concentration c_∞ versus (dimensional) wave number a; here c_∞ is used as the bifurcation parameter. There are two coexisting neutral curves, one for the "pure" soluto-convective mode and one for the "pure" morphological.

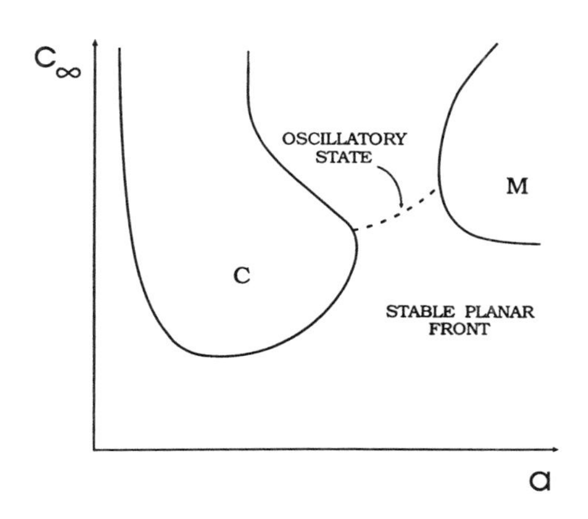

Figure 9. Neutral stability curves, M versus a, according to Coriell, Cordes, Boettinger and Sekerka (1989). The morphological (convective) mode is subject to the right-hand (left-hand) loop. The dashed curve, which occurs near the intersection of the loops in Figure 8, corresponds to an oscillatory mode.

Typically, the critical wave numbers $a_c^{(C)}$ and $a_c^{(M)}$ for the two are widely separated, with the convective mode of much longer scale. In the figure, $a_c^{(M)}/a_c^{(C)} \approx 6.5$ so that there is a large-scale convective flow with a small-scale morphology mode. As V is increased, the length scale δ_c in the solutal Rayleigh number decreases and the convective curve rises while the morphological curve falls.

At a specific value of V, the critical c_∞ of both modes coincide, giving the possibility of a coupled instability. The bifurcation theory (Jenkins 1985a,b; Caroli, Caroli, Misbah and Roulet 1985a), however, shows that the wave-length disparity mentioned above leads to only a very weak coupling between the instabilities. This conclusion, however, is based upon the examination of only a small set of material systems, mainly lead-tin and SCN-ethanol (Schaefer and Coriell 1982).

Clearly, if the linear-theory critical values of c_∞ are common, and if $a_c^{(M)} = a_c^{(c)}$, then the two-dimensional problem has a codimension-two-bifurcation structure and there is the possibility that strong interactions of the two modes would be possible; the appearance of secondary oscillations might also occur. More general interactions could give rise to oscillations; for example if $a_c^{(M)}/a_c^{(c)} = 2$, then strong interactions would also be possible. Riley and Davis (1989b) have undertaken a systematic study of such possibilities for both these cases over wide classes of solutes, solvents and their material properties. They do find that such interactions occur but only at temperature gradients and speeds too small to be of physical interest. Generally, each instability stabilizes the other mode albeit slightly (Caroli, Caroli, Misbah and Roulet 1985a; Riley and Davis 1991).

The soluto-convective mode can be sketched on a V-versus-c_∞ plot; this is shown in Figure 8. The curve is non-monotonic with the nose and lower branch lying in the shaded region where thermal buoyancy is unimportant.

Let us return to the results shown in Figure 9. We see that in addition to the weakly-coupled convective and morphological instabilities predicted, Coriell, Cordes, Boettinger and Sekerka (1980) found a new mode, indicated by the dashed curve, which represents a time-periodic instability generated by the coupling of the two pure modes. This periodic branch occurs near the point in Figure 8 where the two modes cross. For the lead-tin system, this time-periodic instability is not the first one to appear under ordinary operating conditions, though it can under extreme conditions of low G. Coriell (private communication) finds this to be the case for G = 0.1 K/cm and V = 10 μm/s. Schaefer and Coriell (1982) show for SCN-ethanol that the mode is present and can under ordinary conditions be preferred theoretically (though it was not observed in their experiment). Jenkins (1990) found that the oscillatory mode is preferred for high enough surface energy. His weakly-nonlinear theory showed that this mode can be sub- or supercritical and correspond to either travelling or standing waves. These theoretical results are very sobering since they caution the crystal grower. The "attempt" to impress convective flow on a growing crystal may give rise to unwanted oscillations, rather than the homogenization that one sought.

The physical mechanism responsible for this oscillatory "mixed" mode is not explained. A partial explanation in terms of lateral transport is given by Davis (1990).

For most common materials $a_c^{(M)} \gg a_c^{(C)}$. However, there is the possibility that limiting cases exist which display a stronger coupling. Sivashinsky (1983) showed for the pure-morphological problem that for $k \to 0$, $a_c^{(M)} \sim k^{1/4}$. Riley and Davis (1989a) showed for the pure-convective problem that for $k \to 0$, $a_c^{(C)} \sim k^{1/4}$, as well. Young and Davis (1986) examined the small-k limit for the coupled convective/morphological problem, and found the effect. Above the line Rayleigh number $R = R_c = 2(1 + S^{-1})$, there is convective instability as shown by Hurle, Jakeman and Wheeler (1982,1983). However, there is another curve R = R(M) to the right of which lies the region of morphological instability as modified by buoyancy. Thus, for $R < R_c$, the morphological instability is delayed by buoyancy. This is an effect of lateral transport as discussed above. When the basic state is perturbed by an interface corrugation, the rejected solute, if it is light, moves via a baroclinic motion, from troughs to crests, lowering the local concentration gradient. Riley and Davis (1990) examined coupling between morphology and convection in all the evolution equations (5), (6) and (7). Wheeler (1992) generalizes equation (7) to include solutal buoyancy and volume-change effects. The discussion below of the thermally-induced convective instability of Coriell and McFadden (1989) is relevant here as well.

The convective instability can be delayed by the imposition of solid-body rotation on the system. Oztekin and Pearlstein (1992) show that the convective curve of Figure 8 is shifted to the right as the rotation rate is increased, while the morphological instability is unaffected.

Coriell and McFadden (1989) have found a strong coupling between thermal convection and morphological instability. This occurs at wave lengths much larger than δ_c so that the estimate (8) is invalid. When solutal buoyancy is absent (R_S=0), they find a greatly-promoted morphological instability driven by thermal buoyancy. In this case the conductivity ratio $n = k_S/k_L$ is a crucial parameter. Moreover, when the rejected fluid is heavy (light) and solutal buoyancy is present, the destabilization is retarded (promoted).

Say that, typical of metals, $n > 1$, there is a convective flow driven laterally from crest to trough. (The solid acts as a cold, conductive body in a warm fluid.) The flow drives lateral transport that augments the concentration at the troughs, depletes that at the crests, and thus promotes morphological instability. Coriell and McFadden (1989) find for $R_S = 0$ that M_c can decrease by orders of magnitude. When the rejected solute is heavy, the excess in the troughs creates a stably-stratified fluid that retards the thermally-driven flow and hence the modifications of M_c. When the rejected solute is light, it should augment the lateral flow, consistent with the mechanism studied by Young and Davis (1986) at small k and in the frozen-temperature approximation.

When $n < 1$, as in semi-conductors, the flow is reversed, and morphological instability is retarded. However, a new, time-periodic mode is now present, Coriell and McFadden (1989), and this mode is unstable below M_c as well.

Remark: Linear theories show that the morphological and convective instabilities coexist and conspire to produce an oscillatory instability. Weakly nonlinear theories show only weak couplings. Hence, what is required is an entry into the strongly nonlinear regime via a new approach. This approach might be a truncation method analogous to the derivation of Lorenz equations in Bénard convection. It might be the development of phase equations valid a distance from the bifurcation point, or it might be a direct numerical simulation. This last approach is probably beyond reach at the moment.

7. DISCUSSION

In this survey only a sketch of the microscale couplings could be given. A number of phenomena have been illustrated and a few open areas have been mentioned.

Historically, a motivation for this work has been the seeking of flows that might stabilize a front susceptible to morphological instability. Much insight has been obtained though one could easily conclude that flow will at least as often destabilize the front and create oscillations that produce solute segregation.

There is another motivation for the studies outlined herein, especially those that examine nonlinear morphologies. This is the description of deep cells with the aim of describing the cellular-dendritic transition. Given such a transition one would have the ability to estimate the scale of dendritic microstructures and hence the mechanical properties of solids produced from the melt. This takes the investigation from the microscale range to the "gates" of the mesoscale structures.

Acknowledgement

This work was sponsored by the National Aeronautics Space Administration through the Microgravity Science and Applications Program. The author is grateful to D. A. Huntley for his aid with the preparation of this manuscript.

References

Alexander, J. I. D., Wollkind, D. J. and Sekerka, R. F. (1986) 'The effect of latent heat on weakly nonlinear morphological instability', J. Cryst. Growth 79, 849-865.

Aziz, M. J. (1982) 'Model for solute redistribution during rapid solidification', J. Appl. Phys. 53, 1158-1168.

Baker, J. C. and Cahn, J. W. (1971) 'Thermodynamics of solidification', in Solidification, (Amer. Soc. Metals, Metals Park, OH), 23-58.

Boettinger, W. J. and Perepezko, J. H. (1985) 'Fundamentals of rapid solidification' in S. K. Das, B. H. Keer and C. M. Adam (eds.) Rapidly Solidified Crystalline Alloys (Proc. TMS-AIME Northeast Regional Meeting, Morristown, NJ), 21-58.

Boettinger, W. J. and Coriell, S. R. (1986) 'Microstructure formation in rapidly solidified alloys'in Rapid Solidification Materials and Technologies, ed. P. R. Sahm, H. Jones and C. M. Adam, Martinus Nijhoff, Dordrecht, 81-108.

Brattkus, K. (1988) 'Directional solidification of dilute binary allows', Ph.D. Dissertation, Northwestern University.

Brattkus, K. (1990) 'Oscillatory instabilities in cellular solidification', in C. F. Chen (ed.), Mechanics USA 1990 Proc. of the 11th U.S. Nat. Cong. of Appl. Mech., Tucson, S56-S58.

Brattkus, K. and Davis, S. H. (1988a) 'Cellular growth near absolute stability', Phys. Rev. B38, 11452-11460.

Brattkus, K. and Davis, S. H. (1988b) 'Flow induced morphological instabilities: The rotating disc', J. Cryst. Growth 87, 385-396.
Brattkus, K. and Davis, S. H. (1988c) 'Flow induced morphological instabilities: Stagnation-point flows', J. Cryst. Growth 89, 423-427.
Brattkus, K. and Meiron, D. I. (1992) 'Numerical simulation of unsteady crystal growth', SIAM J. Appl. Math. (in press).
Braun, R. J. and Davis, S. H. (1991) 'Oscillatory instabilities in rapid directional solidification: Bifurcation theory', J. Cryst. Growth 112, 670-690.
Braun, R. J. and Davis, S. H. (1992) 'Cellular instability in rapid directional solidification', Acta metall. mater. (in press).
Braun, R. J, Merchant, G. J., Brattkus, K. and Davis, S. H. (1992) 'Pulsatile instability in rapid directional solidification: Strongly nonlinear analysis', SIAM J. Appl. Math. (in press).
Braun, R. J., Merchant, G. J. and Davis, S. H. (1992) 'Pulsatile and cellular mode interaction in rapid directional solidification', Phys. Rev. B. (in press).
Caroli, B., Caroli, C. and Roulet, B. (1982) 'On the emergence of one-dimensional front instabilities in directional solidification', J. Phys. (Paris) 43, 1767-1780.
Caroli, B., Caroli, C., Misbah, C. and Roulet, B. (1985a) 'Solutal convection and morphological instability in directional solidification of binary alloys', J. Phys. (Paris) 46, 401-413.
Caroli, B., Caroli, C., Misbah, C. and Roulet, B. (1985b) 'Solutal convection and morphological instability in directional solidification of binary alloys. II. Effect of the density differences between the two phases', J. Phys. (Paris) 46, 1657-1665.
Coriell, S. R., Cordes, M. R., Boettinger, W. S. and Sekerka, R. F. (1980) 'Convective and interfacial instabilities during unidirectional solidification of a binary alloy', J. Cryst. Growth 49, 13-28.
Coriell, S. R. and Sekerka, R. F. (1983) 'Oscillatory morphological instabilities due to non-equilibrium segregation', J. Cryst. Growth 61, 499-508.
Coriell, S. R. and McFadden, G. B. (1989) 'Buoyancy effects on morphological instability during directional solidification', J. Cryst. Growth 94, 513-521.
Coriell, S. R., McFadden, G. B., Boisvert, R.F. and Sekerka, R. F. (1984) 'The effect of forced Couette flow on coupled convective and morphological instabilities during directional solidification', J. Cryst. Growth 69, 15-22.
Coriell, S. R., McFadden, G. B. and Sekerka, R. F. (1985) 'Cellular growth during directional solidification', Ann. Rev. Mater. Sci. 15, 119-145.
Coriell, S. F. and McFadden, G. B. (1993) 'Morphological stability' in Handbook of Crystal Growth (in press).
Dantzig, J. A. and Chao, L. S. (1986) 'The effect of shear flow on solidification microstructures', (Proc. Nat. Cong. Appl. Mech., Austin, TX), 249-256.
Davis, S. H. (1990) 'Hydrodynamic interactions in directional solidification', J. Fluid Mech. 212, 241-262.

Davis, S. H. (1993) 'Effects of flow on morphological stability' in Handbook of Crystal Growth (in press).
Forth, S. A. and Wheeler, A. A. (1989) 'Hydrodynamic and morphological stability of the unidirectional solidification of a freezing binary alloy: a simple model', J. Fluid Mech. 202, 339-366.
Forth, S. A. and Wheeler, A. A. (1992) 'The effect of shear flow on coupled convective and morphological instability in a simple model of solidification of a binary alloy', J. Fluid Mech. 236, 61-94.
Glicksman, M. E. and Huang, S. E. (1982) 'Convective heat transfer during dendritic growth', in J. Zierep, H. Oertl, Jr. (eds.), Convective Transport and Instability Phenomena, Karlsruhe, 557-574.
Glicksman, M. E., Coriell, S. R. and McFadden, G. B. (1986) 'Interaction of flows with the crystal-melt interface', Ann. Rev. Fluid Mech. 18, 307-335.
Hobbs, A. K. and Metzener, P. (1991) 'Long-wave instabilities in directional solidification with remote flow', J. Cryst. Growth 112, 539-553.
Hobbs, A. K. and Metzener, P. (1992) 'Directional solidification: Interface dynamics and weak remote flow' (pending publication).
Huntley, D. A. and Davis, S. H. (1992) 'Thermal effects in rapid directional solidification', (pending publication).
Hurle, D. T. J., Jakeman, E. and Wheeler, A. A. (1982) 'Effects of solutal convection on the morphological stability of a binary alloy", J. Cryst. Growth 58, 163-179.
Hurle, D. T. J., Jakeman, E. and Wheeler, A. A. (1983) 'Hydrodynamic stability of the melt during solidification of a binary alloy', Phys. Fluids 26, 624-626.
Impey, M. D., Riley, D. S. and Wheeler, A. A. (1992) 'Bifurcation analysis of cellular interfaces in unidirectional solidification of a binary mixture' (pending publication).
Jackson, K. A., Gilmer, G. H. and Leamy, H. J. (1980) 'Solute trapping', in Laser and Electron Beam Processing of Materials ed. C. W. White and P. S. Peercy, (Proc. Symp. of Mat. Res. Soc., Academic Press, New York, 104-110.
Jenkins, D. R. (1985a) 'Nonlinear interaction of morphological and convective instabilities during solidification of a dilute binary alloy', IMA J. Appl. Math. 35, 145-157,
Jenkins, D. R. (1985b) 'Nonlinear analysis of morphological and convective instabilities during solidification of a dilute binary alloy', Physicochem. Hydrodyn. 6, 521-537.
Jenkins, D. R. (1990) 'Oscillatory instability in a model of directional solidification', J. Cryst. Growth 102, 481-490.
Karma, A. and Sarkissian, A. (1992) 'Dynamics of banded structure formation in rapid solidification', (pending publication).
McFadden, G. B. and Coriell, S. R. (1984) 'Nonplanar interface morphologies during unidirectional solidification of a binary alloy', Physica 12D, 253-261.
McFadden, G. B., Boisvert, R. F. and Coriell, S. R. (1987) 'Nonplanar interface morphologies during unidirectional solidification of a binary alloy: II. Three dimensional computations', J. Cryst. Growth 84, 371-388.

McFadden, G. B., Coriell, S. R. and Alexander, J. I. D. (1988) 'Hydrodynamic and free boundary instabilities during crystal growth: The effect of a plane stagnation flow', Comm. Pure Appl. Math. 41, 683-706.
Merchant, G. J. and Davis, S. H. (1989a) 'Directional solidification near minimum c_∞: Two-dimensional isolas and multiple solutions', Phys. Rev. B40, 11140-11152.
Merchant, G. J. and Davis, S. H. (1989b) 'Flow-induced morphological instabilities due to a temporally-modulated stagnation-point flow', J. Cryst. Growth 96, 737-746.
Merchant, G. J. and Davis, S. H. (1989c) 'Shallow cells in directional solidification', Phys. Rev. Lett. 63, 573-575.
Merchant, G. J. and Davis, S. H. (1990) 'Morphological instability in rapid directional solidification', Acta Metall. mater. 38, 2683-2693.
Mullins, W. W. and Sekerka, R. F. (1964) 'Stability of a planar interface during solidification of a binary alloy', J. Appl. Phys. 35, 444-451.
Novick-Cohen, A. and Sivashinsky, G. I. (1986) 'On the solidification front of a dilute binary alloy: Thermal diffusivity effects and breathing solutions', Physica 20D, 237-258.
Özetkin, A. and Pearlstein, A. J. (1992) 'Effect of rotation on the stability of plane-front solidification of dilute Pb-Sn binary alloys' (pending publication).
Riley, D. S. and Davis, S. H. (1989a) 'Hydrodynamic stability of the melt during the solidification of a binary alloy with small segregation coefficient', Physica D 39, 231-238.
Riley, D. S. and Davis, S. H. (1989b), Applied Mathematics Technical Report No. 8838, Northwestern University.
Riley, D. S. and Davis, S. H. (1990) 'Long-wave morphological instabilities in the directional solidification of a dilute binary mixture', SIAM J. Appl. Math. 50, 420-436.
Riley, D. S. and Davis, S. H. (1991) 'Long-wave interactions in morphological and convective instabilities', IMA J. Appl. Math. 45, 267-285.
Schaefer, R. J. and Coriell, S. R. (1982) 'Convective and interfacial instabilities during solidification of succinonitrile containing ethanol', in G. E. Rindone (ed.) Materials Processing in the Reduced Gravity Environment of Space, (Mat. Res. Soc. Symp. Proc., North-Holland, New York), 479-489.
Sivashinsky, G. I. (1983) 'On cellular instability on the solidification of a dilute binary alloy', Physics 8D, 243-248.
Sriranganathan, R., Wollkind, D. J. and Oulton, D. B. (1983) 'A theoretical investigation of the development of interfacial cells during the solidification of a dilute binary alloy: Comparison with experiments of Morris and Winegard', J. Cryst. Growth 62, 265-283.
Ungar, L. H. and Brown, R. A. (1984a) 'Cellular interface morphologies in directional solidification. The one-sided model', Phys. Rev. B29, 1367-1380.
Ungar, L. H. and Brown, R. A. (1984b) 'Cellular interface morphologies in directional solidification II. The effect of grain boundaries', Phys. Rev. B30, 3993-3999.

Ungar, L. H., Bennett, M. J. and Brown, R. A. (1985) 'Cellular interface morphologies in directional solidification III. The effects of heat transfer and solid diffusivity', Phys. Rev. B31, 5923-5930.
Wheeler, A. A. and Winters, K. H. (1989) 'On a finite element method for the calculation of steady cellular interfaces in the one-sided model of solidification', Comm. Appl. Num. Meth. 5, 309-320.
Wheeler, A. A. (1992) 'A strongly nonlinear analysis of the morphological instability of a freezing binary alloy: Solutal convection, density changes and non-equilibrium effects', IMA J. Appl. Math. (in press).
Wollkind, D. J. and Segel, L. A. (1970) 'A nonlinear stability analysis of the freezing of a dilute binary alloy', Philos. Trans. Roy. Soc. Lond. A 268, 351-380.
Wollkind, D. J., Sriranganathan, R. and Oulton, D. B. (1984) 'Interfacial patterns during plane front alloy solidification', Physica 12D, 215-240.
Young, G. W. and Davis, S. H. (1986) 'Directional solidification with buoyancy in systems with small segregation coefficient', Phys. Rev. B34, 3388-3396.
Young, G. W. and Davis, S. H. (1989) 'Morphological instabilities in directional solidification of a binary alloy: End effects', SIAM J. Appl. Math. 49, 152-164.

WEAKLY NONLINEAR MORPHOLOGICAL INSTABILITY ANALYSIS OF A SPHERE CRYSTALLIZING FROM AN UNDERCOOLED MELT

ROBERT F. SEKERKA and PARTHA P. DEBROY
Departments of Physics and Mathematics
Carnegie Mellon University
Pittsburgh, Pennsylvania 15213
United States of America

ABSTRACT. Over the last twenty-five years, linear morphological stability analysis has become a standard methodology that has been applied to a wide variety of situations involving crystal growth and the motion of phase boundaries. Nonlinear analyses, based mostly on weakly nonlinear expansion techniques, are much less well-understood. These techniques focus attention on the time evolution of a specific planform. We describe briefly a weakly nonlinear stability analysis of a sphere crystallizing from an undercooled melt. We perturb the sphere by a single spherical harmonic and carry out a nonlinear expansion up to third order in the amplitude of the perturbation. Results depend strongly on the symmetry properties of the harmonic under consideration. For harmonics such that the shapes produced by perturbations with positive and negative amplitudes are not related by symmetry, the bifurcation is transcritical and a second order perturbation expansion is sufficient. For harmonics such that shapes having perturbations with positive and negative amplitudes are related by symmetry, the bifurcation is either subcritical or supercritical and a third order perturbation expansion is necessary.

1. Introduction

The general problem of morphological stability is concerned with the stability of the shape of a moving interface that separates the phases during a first order phase transformation in which the rate of transformation is governed primarily by the transport of heat and/or chemical species. Linear morphological stability analyses, based on first order perturbation expansions, were first performed by Mullins and Sekerka (1963) for a spherical crystal growing by either heat flow or diffusion, by Mullins and Sekerka (1964) for a planar interface considering both heat flow and diffusion, and by Coriell and Parker (1965) for a circular cylinder growing by diffusion. Since then, linear morphological stability theory has become a standard method of analysis of new effects and has been extended to include the effects of such things as anisotropy of surface tension and bulk properties, multicomponent solutions, departures from local equilibrium at the interface (interface kinetics) and coupling with fluid dynamical instabilities. Details may be found in review articles such as those by Langer (1980) and Sekerka (1986), a forthcoming review by Coriell and McFadden [6], and references therein.

Wollkind and Segel (1970) pioneered the use of weakly nonlinear expansion techniques, originally developed to analyze hydrodynamic instabilities, to study the nonlinear instability of a planar interface near the onset of instability, considering both heat flow and diffusion. Such techniques are used to study the nonlinear interaction, brought about through coupling via higher harmonics, of a specific planform with itself by using perturbation expansions, usually to second or third order in the amplitude of the planform. Thus, one begins to learn something about the nature of the bifurcation (transcritical, subcritical or supercritical) as well as

S. H. Davis et al. (eds.), Interactive Dynamics of Convection and Solidification, 53–55.

about the stabilities of the non-planar states (such as rolls, nodes, and hexagonal cells) corresponding to various planforms. See McFadden et. al. (1988) for a treatment that includes the effect of anisotropy of surface tension, as well as references to the earlier literature.

Subsequently, Brush et. al. (1990) extended the weakly nonlinear analysis to a non-planar geometry, in particular to the case of a two dimensional circular crystal growing into an undercooled melt. Their analytical results predicted subcritical bifurcations (except for perturbations having low wavenumber and favorable values of the thermal conductivity ratio of crystal and melt) and were in agreement with numerical computations based on the boundary integral technique. The present paper gives results for an extension of this analysis to a sphere in three dimensions.

2. Problem Statement and Methodology

We analyze the weakly nonlinear stability of a pure spherical crystal growing into an undercooled melt. Details will be given in the doctoral dissertation of Debroy [10] and in forthcoming publications in the *Journal of Crystal Growth*. Here, we briefly state the problem and outline the methodology.

We study shapes of the form

$$r = f(\theta,\phi) = R + A\, H(\theta,\phi) \tag{1}$$

where

$$H(\theta,\phi) = \begin{cases} Y_{L,0}(\theta,\phi) & for\ M = 0 \\ [Y_{L,M}(\theta,\phi) + Y_{L,-M}(\theta,\phi)]/\sqrt{2} & for\ M \neq 0. \end{cases} \tag{2}$$

Here, r, θ and ϕ are spherical coordinates, $Y_{L,M}$ are spherical harmonics, R is the radius of an unperturbed sphere, and A is the amplitude of a perturbation having the indicated planform. For R near R_0, the critical radius for linear stability, we solve this problem by carrying out a weakly nonlinear perturbation expansion of the governing PDE's and boundary conditions up to third order in A. For example, the radius of the unperturbed sphere itself is given by an expansion of the form

$$R = R_0 + A\, R_1 + A^2\, R_2 + A^3\, R_3 + \ldots \tag{3}$$

Temperature fields in crystal and melt, temperature gradients at the crystal melt interface, and interface curvature are similarly expanded and and substituted into the boundary conditions. The equations at each order corresponding to each spherical harmonic generated by the nonlinearities are analyzed. Such a calculation is tractable because the nonlinearities involve products of spherical harmonics which can be expanded in terms of a small set of neighboring spherical harmonics because of the "selection rules" that apply to the Clebsch-Gordon coefficients in such expansions. Depending on the symmetry of the planform, as discussed below, we obtain from a solvability condition, either at second or third order, a determination of the value of R_1 or R_2 for which the fundamental planform neither grows nor decays.

3. Results and Conclusions

The nature of the bifurcation is dictated by the symmetry of the planform as follows:

For $M = 0$ and L even, the shapes produced by perturbations with positive and negative amplitudes are not related by symmetry. There is therefore no reason for the coefficient R_1 in Eq(3) to be zero, and a second order perturbation expansion is sufficient to determine it. The bifurcation is therefore transcritical, i.e., $R - R_0 = R_1 A$. R_1 may have either sign depending on material parameters.

For $M \neq 0$ or for $M = 0$ and L odd, shapes having perturbations with positive and negative amplitudes are related by symmetry. The coefficient R_1 in Eq(3) must be zero, and a third order perturbation expansion is necessary to determine R_2. The bifurcation is either subcritical or supercritical, i.e., $R - R_0 = R_2 A^2$. Except for low values of L and extreme values of the thermal conductivity ratio of crystal and melt, we find $R_2 < 0$, so most bifurcations are subcritical.

The case $L = 1$ is anomalous since it is neutrally stable in linear theory. We can now calculate its nonlinear instability, and find that it excites the $L = 2$ mode at second order and couples with itself at third order.

Acknowledgement This work was supported by the Division of Materials Research of the National Science Foundation under grant DMR9043322.

References

1. Mullins, W. W. and Sekerka, R. F. (1963) 'Morphological Stability of a Sphere Growing by Diffusion or Heat Flow', J. Applied Physics 34, 323-329
2. Mullins, W. W. and Sekerka, R. F. (1964) 'Stability of a Planar Interface During Solidification of a Dilute Binary Alloy', J. Applied Physics 35, 441-451
3. Coriell, S. R. and Parker, R. L. (1965) 'Stability of the Shape of a Solid Cylinder Growing in a Diffusion Field', J. Applied Physics 36, 632-637
4. Langer, J. S. (1980) 'Instabilities and Pattern Formation in Crystal Growth', Reviews of Modern Physics 52, 1-28
5. Sekerka, R. F. (1986) 'Phase Interfaces: Morphological Stability', in Michael E. Bever (ed.), Encyclopedia of Materials Science and Engineering, Pergamon Press, Oxford, pp. 3486-3493
6. Coriell, S. R. and McFadden, J. B. 'Morphological Stability', in D. T. J. Hurle (ed.), Handbook of Crystal Growth, to be published
7. Wollkind, D. J. and Segel, L. A. (1970) 'A Nonlinear Stability Analysis of the Freezing of a Dilute Binary Alloy', Philosophical Transactions of the Royal Society of London 268, 351-380
8. McFadden, G. B., Coriell, S. R. and Sekerka, R.F. (1988) 'Effect of Surface Tension Anisotropy on Cellular Morphologies', J. Crystal Growth 91, 180-198
9. Brush, L. N., Sekerka, R.F and McFadden, G. B. (1990) 'A Numerical and Analytical Study of Nonlinear Bifurcations Associated with the Morphological Stability of Two-dimensional Single Crystals', J. Crystal Growth 100, 89-108
10. Debroy, P. P. (1992) Doctoral Thesis, Carnegie Mellon University

FLOW-INDUCED SPATIO-TEMPORAL DYNAMICS IN DIRECTIONAL SOLIDIFICATION

A. K. Hobbs and P. Metzener
Swiss Federal Institute of Technology at Lausanne
Department of Mathematics
CH-1015 Lausanne, Switzerland

ABSTRACT. When a dilute binary alloy is directionally solidified into a velocity field which is constant far from the solid-liquid interface, long-wave morphological instabilities are still present, depending upon the parameter range. Linear theory shows that the flow selects a unique neutrally stable state. This state consists of travelling waves when the flow destabilizes the morphology. Nonlinear theory and numerical simulations demonstrate that the flow induces rich spatio-temporal dynamics. The effects of nonequilibrium thermodynamics and attachment kinetics are discussed.

1 Introduction

The presence of flow in the melt has an effect on the dynamics of a solidifying interface. To better understand this coupling, we study the directional solidification of a dilute binary alloy with a flow field imposed in the melt which is constant far from the interface [1]. We specialize to the case where this remote flow is weak, i.e. $R/S \ll 1$, where R is the ratio of flow speed to pulling speed and S is the Schmidt number. Of particular interest in rapid solidification is the absolute stability limit, characterized by the critical pulling velocity above which the planar interface is always stable. As developed in [2], the effects of attachment kinetics μ and nonequilibrium thermodynamics β are important in this regime. To carry out our analysis for this case, we consider the morphological number $M \gg 1$, the surface energy coefficient Γ near its absolute stability value Γ_0, and μ near its absolute stability value μ_0, where $\Gamma_0 = \Gamma_s$ and $\mu_0 = \hat{\mu}_{T_s}$ as defined in [2].

2 Formulation

We consider a one-sided frozen temperature model and seek an asymptotic solution to the system of equation for conservation of mass, momentum, and solute in the melt. The conditions taken at the interface are no slip, mass balance, and conservation of solute. First, we will discuss the case where local thermodynamic equilibrium is assumed. Then, we include the effects of nonequilibrium thermodynamics and attachment kinetics, using the model of Boettinger *et al*, Aziz , and Jackson *et al* as referenced in [2]. The system

S. H. Davis et al. (eds.), *Interactive Dynamics of Convection and Solidification*, 57–59.

described above has a simple solution, or basic state, given by an exponential flow profile and a planar interface. To investigate long-wavelength instabilities of this basic state, we define our small parameter $\varepsilon \ll 1$ by the size of the Schmidt number $S \equiv s\varepsilon^{-2}$, $s \sim O(1)$. Then we scale lengths $L \sim O(\varepsilon^{-1})$, time $t \sim O(\varepsilon^{-2})$, $M^{-1} \equiv m\varepsilon^4$, $\Gamma_0 - \Gamma \equiv \varepsilon^2\Gamma_2$, $\mu - \mu_0 \equiv \varepsilon^2\mu_2$, and take $R \sim O(1)$. In this scaling, the velocity problem is singularly perturbed and is solved by the method of matched asymptotic expansions. The solutal problem is solved by a regular perturbation expansion. From the solvability condition, we obtain a strongly nonlinear equation for the interface position $h = h(x, y, t)$

$$\begin{aligned} &a_1 h_{tt} + (a_2\mu_2 + a_3\nabla^2)h_t + (a_4 m + a_5\Gamma_2\nabla^2 + a_6\nabla^4)h + a_7 R\partial_x\mathcal{L}(h) \\ &= c_1 h_t^2 + c_2\mu_2(\nabla h)^2 + c_3(\nabla h)_t^2 + c_4 h_t\nabla^2 h + c_5\nabla^2(\nabla h)^2 + c_6\nabla\cdot[(\nabla h)^2\nabla h] \\ &\quad + c_7 h_t(\nabla h)^2 + c_8\nabla^2 h(\nabla h)^2 + c_9\nabla h\nabla(\nabla h)^2 + c_{10}(\nabla h)^4, \end{aligned}$$

where $[\mathcal{L}(h)]^* \equiv |\vec{\kappa}|h^*$. The asterisk denotes Fourier transform, and $\vec{\kappa}$ is the wavenumber in Fourier space. This linear nonlocal operator comes from the matching condition in the velocity problem and can be interpreted physically as the horizontal shear stress resulting from the remote flow [3]. The coefficients a_i and c_i depend on β and the equilibrium segregation coefficient k_e as given in [5]. When β and μ are set to zero in this equation, we recover our interface equation derived in [3], and if furthermore R is set to zero, we recover the interface equation derived by Brattkus and Davis [6].

3 Thermodynamic equilibrium and infinite attachment kinetics

In the case where $\beta = \mu = 0$, a linear stability analysis of the basic solution $h \equiv 0$ leads to the following marginal surface in the space (κ_1, κ_2, m),

$$m = k_e^{-1}[-\Gamma_2 k_e\sigma^2 - (1 + k_e^{-1})\sigma^4 + R^2(2 + k_e^{-1})^{-2}\kappa_1^2\sigma^{-2}]$$

where κ_1 and κ_2 are the wavenumbers of disturbances in the x and y directions, $\sigma^2 = \kappa_1^2 + \kappa_2^2$, and $\Gamma_2 < 0$. This is one of the results analyzed in [4]. The marginal state, at the maximum of $m \equiv m_c$, consists of a backward travelling wave of the form $\exp[i\omega t + i\sigma_c x]$. The critical wavenumber σ_c does not depend on R, but m_c does, increasing as R^2. The flow has destabilized the system and has selected a unique pattern. In addition, for a given melt concentration, the absolute stability limit is postponed to a higher pulling velocity.

If m is slightly below m_c, one can investigate the nonlinear development of the interface by a weakly nonlinear analysis. Here, however, we shall suggest a special type of solution. Suppose that $m_c - m = \varepsilon^2$ and $R \sim O(\varepsilon^2)$. Take

$$h = \varepsilon h_1 \exp[i(\sigma_c + \varepsilon^2 a)x + i\varepsilon^2\omega t + i\Psi](1 + \varepsilon^2\rho) + c.c. + O(\varepsilon^2),$$

where $\Psi = \Psi(T, Y)$ and $\rho = \rho(T, Y)$ satisfy

$$\begin{aligned} \Psi_T &+ \Psi_{YY} + \Psi_{YYYY} + r\Psi_Y^2 - q^2(\Psi_Y^3)_Y = 0 \\ \rho &= -\frac{1}{2}a^2 + f(\Psi), \end{aligned}$$

where $T = \varepsilon^{-4}b^4 t$, $Y = \varepsilon^{-1}by$, r is proportional to R, and the scaling factors h_1, a, b, and q can be found in [3]. The equation for Ψ (a damped Kuramoto- Sivashinsky (K-S) equation) shows phase instabilities developing along the cross-stream direction. Numerical simulations described in [3] exhibit various time-dependent stable solutions and show that an increase in the flow strength r implies that the K-S dynamics will be predominant.

4 Nonequilibrium thermodynamics and finite attachment kinetics

When the nonequilibrium parameter β and the attachment kinetics parameter μ are nonzero, we perform a linear stability analysis of the basic state. The resulting marginal equation serves to motivate interesting parameter ranges for further analysis. After rescaling m, μ_2, and Γ_2, we obtain

$$m = -\Gamma_2\sigma^2 - A\sigma^4 + (R\kappa_1\sigma)^2[\mu_2 + B\sigma^2]^{-2},$$

where $\mu_2 > 0$, $A = (1 + k^{-1})(1 + \beta(k_e + \beta)^{-1})$, $B = 2 + k^{-1} + \beta(k_e + \beta)^{-1}$, and $k = (k_e + \beta)(1 + \beta)^{-1}$. If $\Gamma_2 < 0$, the situation is very close to that described in the previous section, except that σ_c depends upon R^2. The basic features of this marginal surface are better illuminated if we consider a 6th order polynomial expansion in κ_1 and κ_2. This expansion suggests (at least) two different models for long-wave, small amplitude dynamics.

For the first model, $\Gamma_2 < 0$ and $(R/\mu_2)^2 < A$. We find, after appropriate scaling, the marginal surface which leads to following dynamical equation for h

$$h_T + [m - \Gamma_2\nabla^2 + A\nabla^4 - (R/\mu_2)^2\nabla^2\partial_X^2]h + \frac{\mu_2}{2}(\nabla h)^2 = 0.$$

Note that this is a type of K-S equation, with the nonlinear linear term governed by attachment kinetics, especially when $m \equiv 0$.

In the second model, $\Gamma_2 > 0$ and $0 < (R/\mu_2)^2 - A \ll 1$. We find, after scaling, the interface equation

$$h_T + [m - \Gamma_2\partial_X^2 - ((R/\mu_2)^2 - A)\partial_X^4 + A\partial_X^2\partial_Y^2 - 2R^2\mu_2^{-3}B\partial_X^6]h + \frac{\mu_2}{2}h_X^2 = 0.$$

Note that in the first model, the K-S equation describes the *interface* rather than the phase. In the second model, the presence of the sixth order derivative means that the interface equation is a higher order generalization of a K-S equation. The derivation and interpretation of these interface equations can be found in [5].

References

[1] S.A. Forth and A.A. Wheeler, J. Fluid Mech. **209** (1989) 339.

[2] G.J. Merchant and S.H. Davis, Acta. Metall. mater. **38** (1990) 2683.

[3] A.K. Hobbs and P. Metzener, J.Crystal Growth, to appear 1992.

[4] A.K. Hobbs and P. Metzener, J.Crystal Growth. **112** (1991) 539.

[5] A.K. Hobbs, P. Metzener, and S.H. Davis, "Interfacial dynamics in directional solidification", EPFL technical report (1992).

[6] K. Brattkus and S.H. Davis, Phys. Rev. B **88** (1988) 11452.

LONG-WAVE INTERACTIONS IN MORPHOLOGICAL AND CONVECTIVE INSTABILITIES

D.S. Riley
School of Mathematics
University of Bristol
Bristol BS8 1TW
UK

ABSTRACT. A binary liquid that undergoes directional solidification is susceptible to morphological and solutal-convective instabilities that cause the solid/liquid interface to change from a planar to a cellular state. This paper outlines results of an integrated analytical and numerical study of a particular long-wave equation that describes the weak effect of convection on the evolution of interface morphology.

Introduction

It is now well established that, in the unidirectional solidification of dilute binary mixtures, changes in the morphology of the solid/liquid interface and convection in the liquid phase are two of the main controlling factors in determining the homogeneity of single crystals grown from the melt.

These features occur because the equilibrium solute concentration in the liquid is different than that in the solid. This results in a region of light-solute-rich (or heavy-solute-deficient) melt near the solid/liquid interface. For growth of the crystal to proceed, diffusion must act to reduce the solute excess (or deficit). For a given temperature gradient and bulk solute concentration in the melt, there is a critical pulling speed above which diffusion acts too slowly, and the planar interface breaks down into a spatially-periodic structure. Moreover, if the process takes place in a gravitational field, baroclinic motions and/or convective instabilities may occur.

The degree of difficulty of the mathematical problem has necessitated the formulation of simplified models in order to gain theoretical understanding of interfacial-pattern formation. By further restricting attention to special distinguished limits, reductions in the dimensionality of the problem and, more importantly, its complexity may be achieved. This approach has been used, for example, by Sivashinsky [1], Brattkus and Davis [2], Young and Davis [3] and Riley and Davis [4,5], who all focused attention on parameter ranges in which morphological instabilities first develop to disturbances with asymptotically small wave numbers. Whether the underlying assumptions are met in practice is difficult to justify a priori: the use of the model for quantitative analysis needs to be examined for each specific material.

S. H. Davis et al. (eds.), Interactive Dynamics of Convection and Solidification, 61–63.

Results

We have studied a strongly nonlinear, time-dependent, two-space dimensional equation governing the evolution of long-wave interfacial displacements from planarity, h. This equation arises in a large-surface-energy, small-segregation-coefficient limit [5], where gravitational instabilities are absent and the flow is purely baroclinic:

$$\frac{\partial A}{\partial t} - \nabla . [(1 - rA)(\nabla A + A\nabla h)] - k(1 - A) = 0,$$

where

$$A = 1 - M^{-1}h + \nabla^2 h,$$

k is a scaled segregation coefficient, M is a morphological parameter, r is a solutal Rayleigh number (scaled by a factor involving the Schmidt number), and a surface- energy parameter has been scaled out. Given that regular interfacial structures having the form of rolls and hexagons have been found experimentally [6], it is natural in a theoretical study to work relative to a regular lattice that accommodates such structures. In [7], we did this and showed that the bifurcation structure of the equation near the onset of linear instability is equivalent to that of a codimension-one normal form. Examination of the bifurcation structure of this normal form indicated that three-dimensional solutions may correspond to interface patterns that are hexagonal in appearance, although not necessarily hexagonally symmetric, and led to an explanation for the development of hexagonal interface patterns. In common with problems having perfect hexagonal symmetry [8,9], local analysis revealed that there is no stable solution in the neighbourhood of the critical value for linear instability. We speculated, however, that a solution branch stabilized at a limit point (outside the range of validity of a local bifurcation analysis) giving rise to a physically realizable fully-nonlinear hexagonal interfacial pattern. This conjecture was consistent with the work of Brattkus and Davis, but required numerical validation.

To determine nonlinear steady solutions, we have used ENTWIFE, a state-of-the-art computer code which discretizes the system using a finite-element method and employs modern developments in numerical bifurcation and continuation theory. This work is continuing, but we have confirmed the existence of strongly nonlinear hexagonal patterns. The family of patterns is very similar to that found by McFadden et al. [10]. We are also systematically investigating models lacking the full hexagonal symmetry. A complete report of this work will be made in a later paper.

Acknowledgements

This work was carried out in collaboration with Mike Impey, Adam Wheeler and Keith Winters in the UK, and with Steve Davis and Jeff McFadden in the USA. The work was supported in part by the United Kingdom Science and Engineering Research Council, in part by the United States National Aeronautics and Space Administration, Microgravity Science and Applications Program, and in part by NATO.

References

[1] Sivashinsky, G.I. (1983) 'On cellular instability in the solidification of a dilute binary alloy', Physica D 8, 243-248.

[2] Brattkus, K. and Davis, S.H. (1988) 'Cellular growth near absolute stability', Phys. Rev. B 38, 11452-11460.

[3] Young, G.W. and Davis S.H. (1986) 'Directional solidification with buoyancy in systems with small segregation coefficient', Phys. Rev. B 34, 3388-3396.

[4] Riley D.S. and Davis S.H. (1990) 'Long-wave morphological instabilities in the directional solidification of a dilute binary mixture', SIAM J. Appl. Math 50, 420-436.

[5] Riley D.S. and Davis S.H. (1990) 'Long-wave interactions in morphological and convective instabilities', I.M.A. J. Appl. Math. 45, 267-285.

[6] de Cheveigne, S., Guthmann, C., Kurowski, P., Vincente E. and Biloni, H. (1988) 'Directional solidification of metallic alloys: the nature of the bifurcation from planar to cellular interface', J. Cryst. Growth 92, 616-628.

[7] Impey, M.D., Riley, D.S. and Wheeler, A.A. (1992) 'Bifurcation analysis of cellular interfaces in unidirectional solidification of a dilute binary mixture', SIAM J. Appl. Math (in the press).

[8] Buzano E. and Golubitsky, M. (1983) 'Bifurcation on the hexagonal lattice and the planar Benard problem', Phil. Trans. R. Soc. Lond. A308, 617-667.

[9] Golubitsky, M., Swift, J.W. and Knobloch, E. (1984) 'Symmetries and pattern selection in Rayleigh-Benard convection', Physica 10D, 249-276.

[10] McFadden, G.B., Boisvert, R.F. and Coriell, S.R. (1987) 'Nonplanar interface morphologies during unidirectional solidification of a binary alloy. II Three dimensional computations', J. Crystal Growth 84, 371-388.

NONLINEAR ANALYSIS OF MORPHOLOGICAL INTERACTIONS BETWEEN FLOW AND INTERFACE SHAPE IN THE DIRECTIONAL SOLIDIFICATION OF A BINARY ALLOY

REZA MEHRABI AND ROBERT A. BROWN
Department of Chemical Engineering
Massachusetts Institute of Technology
Cambridge, MA 02139

ABSTRACT. A Finite-element Newton algorithm is used to solve the coupled equations for thermosolutal convection and morphological structure for a two-dimensional model for binary alloy solidification. The calculations of the states that evolve from the convective branch demonstrate the development of large amplitude deflections in the melt/solid interface that are similar to the cellular interface shapes caused by morphological instability.

1. Introduction

If not complex enough in their own right, morphological instabilities in the melt/solid interface can interact with other hydrodynamic instabilities due to the coupling of solute and temperature fields. Coriell et al. (1980) first realized this possibility and presented a linear stability analysis that allowed simultaneously for morphological instability in a planar melt/solid interface growing at constant rate and for thermosolutal convection in the melt. They identified three modes of instability which are distinguished on the basis of the spatial wavenumber for the disturbance to the interface. Morphological modes are found at high wavenumber and correspond to the interfacial instability first discovered by Mullins and Sekerka (1964). Convective modes caused by thermosolutal instability in the density profile in the diffusion layer occur at relatively small wavenumbers. Coriell et al. (1980) uncovered oscillatory instabilities which occur at intermediate wavenumbers.

The purpose of this short contribution is to demonstrate the danger associated with considering the states created by the convective and morphological modes as lossely coupled. We consider the case of thermophysical parameters similar to those for the organic alloy succinnonitrile-acetone in which the convective instability occurs at a much lower acetone concentration c_0 than the morphological mode.We use a finite-element/Newton method to compute the finite amplitude steady-states in the convective family and show that the interface morphology evolves rapidly (for small increases in c_0 above critical) to resemble the deep cells seen in morphological calculations.The analysis is based on solution of the Boussinesq equations for thermosolutal convection, solute transport in the melt and heat transfer in the melt and solid. The interface is a free-boundary described by the condition for thermal equilibrium and by the Gibbs-Thomson equation. The equations and boundary conditions are discretized using a combination of finite element methods that we have developed for solving problems in macroscale convection and microscale morphology. The interface is represented as $y = h(x)$ and the mesh is fixed to the interface. Steady-state solutions are found by Newton's method.

2. Results

The calculations presented here are for the parameters given by Coriell et al. (1992) for the succinonitrile-acetone system with two exceptions; we set the latent heat to zero and the thermal conductivities of the melt and solid to the value for the melt. With these

S. H. Davis et al. (eds.), Interactive Dynamics of Convection and Solidification, 65–67.

approximations, the temperature field is fixed except for the contributions of convective heat transfer driven by the solidification velocity and by thermosolutal motion. The calculations are for V = 4.02 μm/s and an average temperature gradient of 20 K/cm. The steady-state portions of the neutral stability curves are shown in Fig. 1 for the critical acetone concentration as a function of the wavenumber ω; these curves were computed using the finite element discretization shown in Fig. 2. The curves show the features described by Coriell et al. (1980), i.e. the convective branch is dominant at low wavenumbers and the morphological branch is dominant at high wavenumbers. Moreover, the morphological branch approaches the results of analysis without thermosolutal convection (the curve for g = 0 in Fig. 1) at high ω. We have not searched for the mixed oscillatory modes that are expected for ω near the junction of the two curves.

Calculations of the nonlinear steady-states that evolve from the neutral stability curves at the points marked on Fig. 1 are shown on Fig. 2. These states is develop large interface deflections with only a small increase in c_0 above the critical value; the interface shapes appear to be very similar to the deep cells seen for a purely morphological instability. For the wavenumber $\omega = 25.13$ cm^{-1}, the critical concentration is $c_0 = 0.0128$, and the range of c_0 covered in Fig. 2 spans only to 1.05 of the critical value. States for higher values of c_0 could not be computed because of the steepness of the melt/solid interface in the groove. The flow concentration fields computed with each state also are shown in Fig. 2; the cellular structure of the flow has been emphasized by subtracting out the uniaxial streamlines that correspond to the growth velocity .

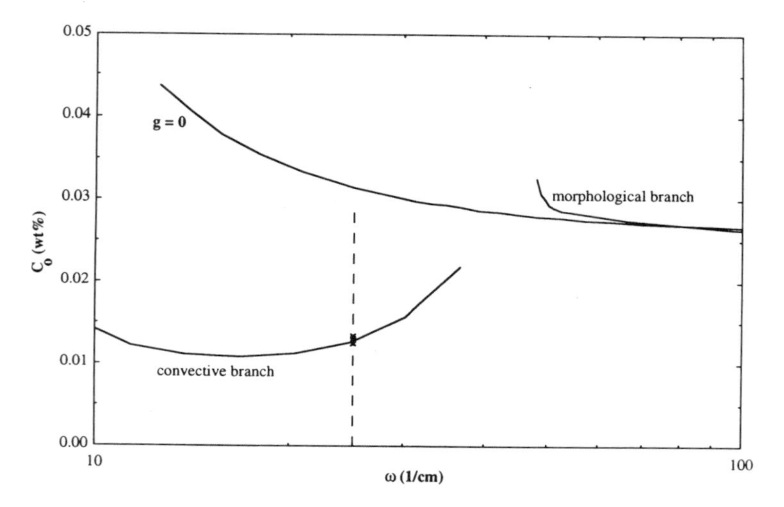

Figure 1.Neutral stability curves computed for the succinonitrile-acetone system.

3. Discussion

The calculations shown in Fig. 2 demonstrate that the thermosolutal induced convection in the melt cannot be thought of as being decoupled from the interface shape. In fact, the interface deformation shown there is as large as is traditionally associated with finite amplitude solidification cells. These calculations are not extensive enough to distinguish between the two candidates for the mechanism for the large deformation: nonlinear mode coupling between the the morphological and convective branches, and interface deformation caused by lateral segregation of solute across the interface induced by the flow. However, the calculations suggest the former hypothesis.

The lateral segregation of solute (acetone) increases rapidly with increasing c_0; the percentage difference in concentration across the interface for the four cases shown in Fig. 2 are 0, 49, 72, and 93 percent, respectively. It seems likely that a weak cellular flow causes uneven mixing of solute and leads to a high concentration where the flow moves away from the interface. This higher concentration makes the interface locally morphologically unstable and leads to large interface deformation, which enhances the lateral solute segregation. Accordingly, convection and morphology couple to cause large interface deflection and solute segregation. This effect can only be predicted by an analysis that accounts for both effects.

This research was supported by the Microgravitry Sciences and Application Program of NASA.

References

Coriell, S.R., Cordes, M.R., Boettinger, W.J., and Sekerka, R.F. (1980) "Convective and interfacial instabilities during unidirectional solidification of a binary alloy," *J. Crystal Growth* 49, 13-28.

Coriell,S.R., Murray,B.T., McFadden,G.B. and Leonartz,K. (1992) "Convective and morphological stability during directional solidification of a succinonitrile-acetone system, " *Proceedings Institute Mathematics and Analysis Meeting on Free-Boundary Problems*, to be published.

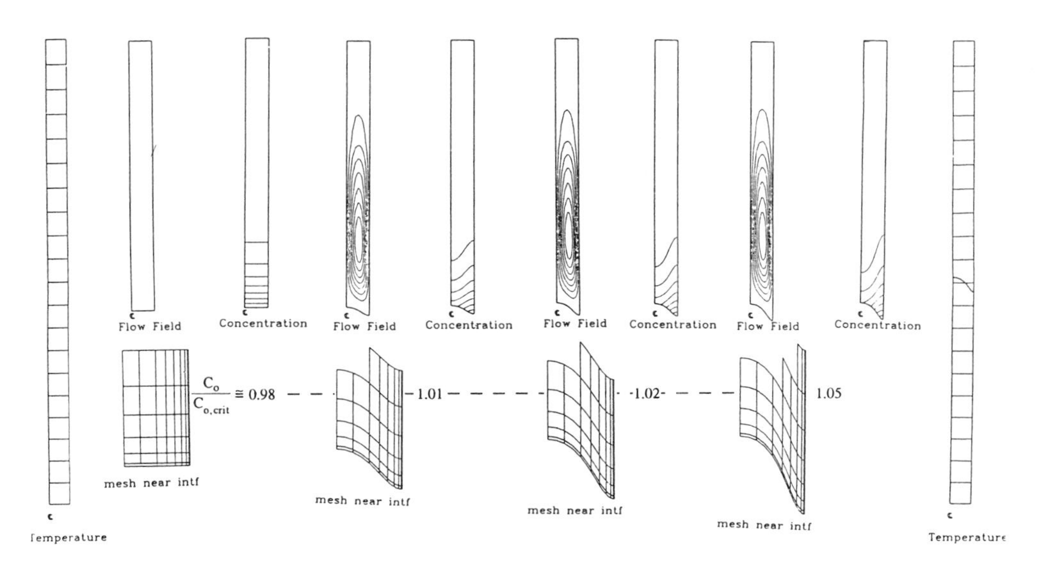

Figure 2. Interface shapes, concentration and flow fields computed at $\omega = 25.13$ cm^{-1}.

NONLINEAR DYNAMICS NEAR THE ONSET OF CELLULAR GROWTH DURING THIN-FILM SOLIDIFICATION OF A BINARY ALLOY: ON THE OBSERVABILITY OF WEAKLY NONLINEAR STATES

ROBERT A. BROWN, T.C. JOHN LEE AND KOSTAS TSIVERIOTIS
Department of Chemical Engineering
Massachusetts Institute of Technology
Cambridge, MA 02139

ABSTRACT. Results from a combination of thin-film directional solidification experiments and two-dimensional numerical simulations are described and suggest that steadily growing patterns of shallow solidification cells with a unique wavelength cannot be observed in the succinonitrile-acetone organic alloy. Thin-film directional solidification experiments performed on long time scales and for conditions very close to the onset of morphological instability show the evolution to complex patterns in space and time — possibly spatiotemporal chaos — with mean wavelength significantly below the critical value predicted by linear theory. Numerical simulations confirm many of the features of the cells observed in experiments and link the mechanisms for the birth and death of cells with the proximity of codimension-two bifurcations caused by the flatness of the neutral stability curve with spatial wavelength.

1. Introduction

The linear stability analysis of the onset of cellular crystal growth during the directional solidification of a binary alloy has spawned an enormous amount of experimental and theoretical research aimed at describing the transition from a planar interface morphology to the deep solidification cells and dendrites seen in most processing conditions. This theory describes the onset of cellular growth in terms of the conditions for growth of infinitesimal amplitude cells from a planar interface. For a sample held in a constant temperature gradient G and with a set alloy concentration c_o, morphological stability theory predicts the onset of small-amplitude cells with wavelength λ_c at a critical value of the growth rate $V_c = V(\lambda_c)$. An important and relatively unresolved problem has been the evolution of the melt/solid interface from the small-amplitude cells predicted by linear theory to the deep cells seen in experiments. A specific question is prediction of the evolution of the cellular wavelength from the onset value as the growth rate is increased.

We have addressed this question using a combination of experimental measurements and numerical simulations to present the hypothesis that the interface composed of a large collection of cells for conditions near onset is not composed of cells with a single wavelength, but is in a state of spatiotemporal chaos, exhibiting a range of spatial wavelengths in time. Thus, no unique wavelength is selected for steadily solidifying cells; this observation agrees with steady-state calculations of cellular crystal growth (Ramprasad et al., 1989). Moreover, the dynamics of the interface is driven by nonlinear interactions between neighboring cells along the front that cause cell birth, tip splitting, and cell death, by annihilation of a cell in the groove between two others, all occurring on a long time-scale. We have demonstrate this dynamics in long time-scale simulations of cellular crystal growth (Bennett and Brown, 1988; Bennett et al., 1992).

The most striking feature of the nonlinear dynamics in thin-film solidification of

S. H. Davis et al. (eds.), Interactive Dynamics of Convection and Solidification, 69–72.

organic alloys is that complex spatiotemporal behavior occurs at operating conditions very close to the critical value, i.e. for $|V-V_c| << 1$, so that neither the linear state, nor states predicted by weakly nonlinear analysis are easily observable in experiments. We have argued before (Bennett and Brown, 1989) that this behavior is a result of the extreme flatness of the neutral stability curve $V_c = V(\lambda_c)$, which allows for the linear instability of several decades of wavenumber with only a few percent increase in V. Experimental evidence for this nonlinear dynamics is presented here for thin-film solidification that is especially designed for careful analysis of long experiments performed at conditions close to the onset value $V = V_c$. The local interfacial dynamics seen in the experiments are compared to results of numerical simulations for the same alloy system and show the importance of cell birth by tip splitting and cell death by annihilation in grooves.

2. Experiments

Lee and Brown (1992a &b) have developed a new thin-film solidification system that has been designed especially to study the conditions near the onset of cellular solidification. The system is described in the above references and differs from previous thin-film solidification in size and in the degree of control of the translation rate. The solidification cell is 1000x45 mm with a gap of 75 μm. The experiments reported here are for a 30 K/cm temperature gradient. The sample is translated using a meter-long translation stage with an accuracy of ± 0.01 μm/s. The experiments use the organic alloy formed by dissolving a small amount of acetone in succinonitrile. The solidification system allows for 50-500 cells across the sample and for experiments with duration up to 10,000 diffusion time units defined as $t^* = \lambda_c^2/D$, where D is the solute diffusivity in the melt. Widths of up 1.2×10^4 μm of the interface are imaged to construct Fast Fourier Transforms.

Results for an experiment with a nominal acetone concentration of $c_o = 0.256$ mole percent and the temperature gradient G = 30 K/cm are shown in Fig.1 for the translation rate Vc = 0.75 μm/s, which is as close to the value for the onset of cellular growth as can be obtained in this system. The experiment was carried out by translating the sample at 0.67 μm/s for 1×10^4 s before increasing V to the higher value.Linear stability theory predicts that the most dangerous wavelength is $\lambda_c = 556$ μm. Cells with these long wavelengths are only seen at short times (Lee and Brown, 1992a). At long times the long wavelength modes decrease in importance and the microstructure is dominated by smaller wavelength cells, as shown in Fig. 1. The long time behavior of the cellular front appears to be spatially and temporally chaotic.The dominant dynamics along the front is the continual birth of new cells by tip splitting and cell annihilation in the grooves between others; this dynamics is qualitatively similar to the long time-scale numerical simulations.

3. Simulations

Numerical simulations of steady-state and time-dependent cellular growth of the SCN-ACE system has been carried out using the finite element analysis described in a series of papers (Tsiveriotis, 1992). The calculations use the Solutal Model for solidification in which the temperature field is assumed to be fixed with a constant gradient and solute diffusion in both the melt and crystal are included. The calculations for the SCN-ACE alloy demonstrate all the elements of the nonlinear dynamicsseen in the experiments; secondary bifurcations in steady-state cells that reduce the spatial wavelength by tip splitting, rapid reduction in the wavelength caused by the flatness of the neutral stability curve, and time-periodic states due to mode interactions and oscillation of the cell grooves. Each phenomena is seen in the experiments described above.There is not room here to describe these features. Instead, we reproduce as Fig 2. only the bifurcation diagram for steady-state cells computed using the most dangerous wavelength λ_c from linear stability theory. Within a small range of growth rate V, cellular states with spatial wavelengths of λ_c , λ_c /2, and λ_c /4. Sample cell shapes are shown as inserts; none of these cells can be thought of as the small-amplitude states predicted by weakly nonlinear theory. Also shown are cells taken from the above experiments at different times, but as the same position along the front. There is an amazing resemblance between the slowly evolving dynamical states in the experiments and the coexisting steady-state forms with the three wavelengths. Note that the transition to cells with wavelength corresponds to a reduction from 560 to 140 μm, which is well within the band observed in the experiment.

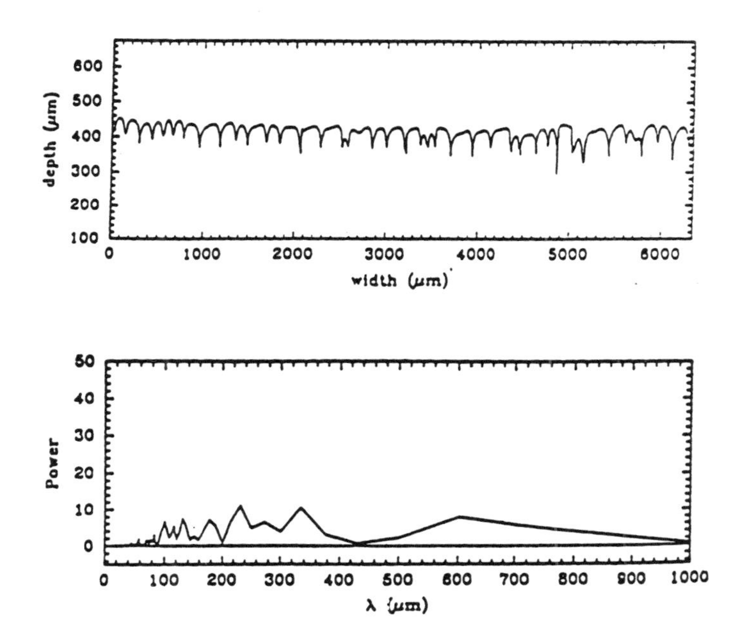

Figure 1. Power spectrum and digitized interface shape for the solidification experiment described in the text. Results are 15000 s after a step increase in the translation rate.

4. Conclusions

We believe that the simulations of large collections of cells reported by us (Bennett and Brown, 1989; Bennett et al., 1992) and the experiments described here and elsewhere strongly suggest that it may be impossible to observe the linear and weakly nonlinear states that are predicted to occur in thin-film directional solidification of model organic alloys. The difficulty is closely connected to the flatness of the neutral stability curve for these systems and the occurrence of spatially resonant mode interactions that lead to nonlinear dynamics via tip splitting and cell death for growth rates only very slightly above V_c. The unobservability of linear and weakly nonlinear states because these interactions makes the utility of any theory that neglects these states in doubt. Our experiments and calculations also suggest that large cellular fronts should be considered characterized as spatially and temporally chaotic states with a band of observable wavelengths that evolves with increasing growth rate.

This research was supported by the NSF and by the Microgravity Sciences and Application Program of NASA.

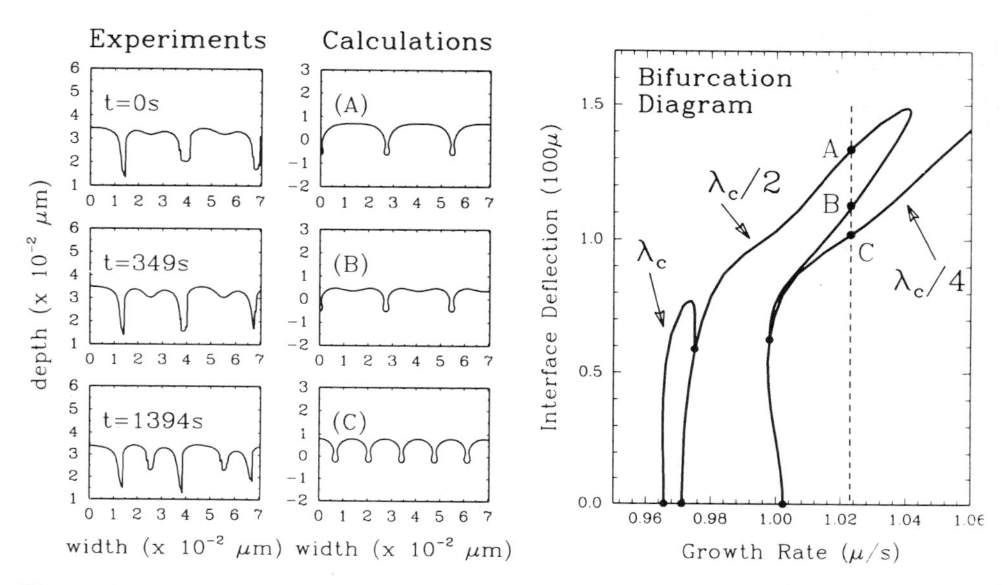

Figure 2. Evolution of cells for the SCN-ACE system under the conditions for the experiment described above. The cells shapes predicted to coexist for V = 1.02 μm/s are compared with the dynamical cell observed in the experiments with V = 0.832 μm/s.

References

Bennett, M.J. and Brown, R.A. (1989) 'Cellular dynamics during directional solidification: interactions of multiple cells', *Phys. Rev. B*39, 705-723.

Bennett, M.J., Tsiveriotis, K. and Brown, R.A. (1992) '. Nonlinear Dynamics in Periodically repeated sets of directional solidification cells', *Phys. Rev. B*45, xxx-xxx.

Lee, J.T.C., Tsiveriotis, K., and Brown, R.A. (1992) 'Spatiotemporal chaos neat the onset of cellular growth during thin-film solidification of a binary alloy,' *J. Crystal Growth*, in press.

Lee, J.T.C. and Brown, R.A. (1992b) ' Experimental study of the planar to cellular transition during thin-film directional solidification: observations of the long time-scale dynamics of microstructure formation', *Phys. Rev. B* submitted.

Ramprasad, N., Bennett, M.J. and Brown, R.A. (1988) ' Wavelength dependence of directional solidification cells with finite depth. *Physical Review* ', *Phys. Rev. B*38, 583-592.

Tsiveriotis, K. Doctoral thesis, Massachusetts Institute of Technology, 1992.

A STRONGLY NONLINEAR ANALYSIS OF MORPHOLOGICAL STABILITY OF A BINARY ALLOY: SOLUTAL CONVECTION AND THE EFFECT OF DENSITY MISMATCH

A. A. WHEELER
School Of Mathematics
University Walk, Bristol
BS8 1TW, UK.

ABSTRACT. Here we describe the results of a strongly nonlinear analysis of morphological stability of a binary alloy, in the presence of buoyancy driven convection. Despite providing an analysis of this situation in an distinguished limit that is not commonly encountered in directional solidification, we show how this approach is useful in providing a simplified description of the dynamics of the system in the fully nonlinear regime. In particular we emphasise the physical mechanisms that can be revealed by this approach.

1. Introduction

It has been recognised for some time that in real crystal growth processes convection in the melt has a profound effect. The simple linear stability theory of Mullins and Sekerka [1] has been extended to include thermal and solutal induced buoyancy first by Coriell et al [2] and subsequently by Hurle et al [3]. The inclusion of convection renders the eigenvalue problem that results from a linear stability analysis much less tractable, and numerical techniques are required for its solution. These indicate that there are two stationary modes of instability; one associated with a morphological mode, and one with a convective mode. In addition there are overstable modes which result from the coupling between the flow in the melt and the interface. This situation has, most recently, been investigated in more detail by Forth and Wheeler [4]. Because of the severe technical difficulties involved in analysing this situation, no clear picture has emerged. A different tack has been employed first by Sivashinsky [5] and most prolifically by Davis and his co-workers [6, 7, 8], whereby realistic material and growth parameters are replaced by values that ensure the critical wavenumber of the instability is small so that a long-wave analysis may be conducted. This has the advantage that the system may analysed more easily, particularly in the nonlinear regime.

The aim of the work reported here (which partly summarises that given by Wheeler [9]) is to use a strongly nonlinear long-wave analysis as a setting for investigating one particular mechanism, the effect of density mismatch between the solid and the liquid, with the aim

S. H. Davis et al. (eds.), Interactive Dynamics of Convection and Solidification, 73–76.

of providing some physical insight into the interaction between flow in the melt and the dynamics of the interface.

2. Model and Discussion

We adopt the one-sided model for solidification of a binary model, modified to include the effect of convection in the melt, whose density is assumed to depend only on the solute concentration. Thus the Navier-Stokes equations in the Boussinesq approximation are employed. The density of the solid and liquid phases at the interface are assumed different which ensures that a flow is always present in the melt as the crystal grows. At the interface boundary conditions are posed which represent conservation of mass and solute as well as thermodynamic equilibrium.

The base state is given by

$$\mathbf{u} = -V\Delta \mathbf{k}, c(z) = c_{\infty}\left[\frac{1}{k} + \frac{k-1}{k}(1 - e^{-V(1+\Delta)z/D})\right],$$

where $\mathbf{u}$ is the melt velocity and $c(z)$ is the melt solute concentration, V is the planar interface velocity, $\Delta = \frac{\rho_S - \rho_L}{\rho_L}$, and ρ_S, ρ_L denote the densities of the solid and liquid. It is clear that the flow induced by the density mismatch has the effect of attenuating the solute boundary layer thickness of the base state solute profile. Indeed this is reflected in the dimensionless governing equations (by employing a length scale of $D(1+\Delta)/V$) in which the solutal Rayleigh number Ra, Sekerka number (or morphological parameter) Sk and capillarity parameter A are given as

$$Ra = (1+\Delta)^3 Ra^*, Sk = Sk^*(1+\Delta), A = A^*(1+\Delta),$$

where superscript star denotes the value in the absence of density mismatch ($\Delta = 0$). This rescaling of the control parameters is the sole effect of the boundary layer attenuation. An additional effect of the density mismatch, not apparent in the base state is streamline curvature, as recognised by Caroli et al [10] in a linear stability analysis of this situation. This results from the no-slip condition at the interface which ensures that the streamlines are normal to it, and so induces a curvature in them. The direction of the flow (to or from the interface) is determined from conservation of mass at the interface, which is also evident in the base state flow given above.

A linear stability analysis of the morphological mode in the absence of any flow in the melt provides the dispersion relation

$$Sk^{*-1} = 1 - A^* a^2 + \frac{2k}{1 - 2k - \sqrt{1+4a^2}},$$

where a is the wavenumber. An analysis of this in the limit $a \to 0$ provides a variety of distinguished limits. We like Riley and Davis [7, 8] focus on the particular limit in which $a = (A^{-\frac{1}{2}}), k = (A^{-1})$ as $A \to \infty$; a large surface energy limit.

Motivated by these scalings from the linear stability theory we introduce the following scalings $x = \epsilon^{-1}X, t = \epsilon^{-2}T, k = \epsilon^2\bar{k}, A = \epsilon^{-2}$. The solute concentration, flow variables and interface are then expanded in powers of ϵ. At leading order this gives that

$c^{(0)}(z) = c_\infty \left(\frac{1}{k} + \frac{k-1}{k}[1 - \alpha(X,T)e^{-z}]\right)$, where $\alpha = \left(1 - Sk^{*-1}\chi^{(0)} + \chi^{(0)}_{XX}e^{\chi^{(0)}}\right)$, and at second order a solvability condition provides an amplitude equation for the leading order interface position $\chi^{(0)}$, which is too lengthy (and uninformative) to be given here. When $\Delta = 0$ it reduces to that given by Riley and Davis [8]. However in the absence of flow $\Delta = Ra = 0$ it may be written as

$$c^i_T = \frac{d}{dX}(c^i_X)_X - \bar{k}c^i,$$

where c^i is the leading-order interfacial concentration, and d/dX denotes a total derivative along the interface. This represents conservation of solute at $\mathcal{O}(\epsilon^2)$.

The result of a standard linear stability treatment of the ampitude equation shows that if Δ is sufficiently large the system is stabilised with increasing Ra^* and ultimately results in absolute stability. If however Δ is sufficiently negative increasing Ra^* destabilises the system, leading finally to convective instability. The effect of increasing Δ for a fixed Ra^* is to promote stability. Thus increasing ρ_S in favour of ρ_L is always stabilising, whereas the effect of buoyancy may be either stabilising or destabilising depending on the value of Δ.

The morphological instability is controlled by the local freezing temperature at the interface, which in turn is determined by the local solute concentration and interface curvature. Convection in the melt acts directly on the former by changing the solute transport to the interface. Here k is less than unity and so increasing the interface concentration decreases the freezing temperature and is thus stabilising. A peak in the interface encounters a region with lower concentration and so diffusion acts by causing the perturbed concentration to be maximum above the peaks and is therefore a stabilising agency. If we now allow for density mismatch ($\Delta \neq 0$) then the streamline curvature effect results in an up-flow or down-flow of the perturbed flow at the peaks depending on the sign of Δ and so the lateral solute transport to the peaks increases as Δ increases. Thus we expect the system to be stabilised by increasing Δ as observed above. Now if we allow for buoyancy, $Ra > 0$, in which case the lateral density gradient promotes up-flow above the peaks. If $\Delta > 0$ then the streamline curvature effect and diffusion conspire together to increase the perturbed concentration above the peaks, in which case buoyancy induces an up-flow there, resulting in enhanced stabilisation as Ra increases. If however $\Delta < 0$, the streamline curvature effect requires the lateral transport to be away from the peaks, above which there is now down-flow. This acts against diffusion and results in diminution of the concentration maxima above the peaks. For Δ sufficiently negative, the lateral concentration transport is so effective as to result in the perturbed concentration having minima above the peaks. Moreover, in this situation, buoyancy now enhances the lateral convective transport away from the peaks, resulting in further destabilisation.

References

[1] W. W. Mullins and R. F. Sekerka. (1964) 'Stability of a planar interface during solidification of a dilute binary alloy', J. Appl. Phys., 35, 444–451.

[2] S. R. Coriell, M. R. Cordes, W. J. Boettinger, and R. F. Sekerka. (1981) 'Convective and interfacial instabilities during unidirectional solidification of a binary alloy', J. Crystal Growth, 49, 13–28.

[3] D. T. J. Hurle, E. Jakeman, and A. A. Wheeler. (1982) 'Effect of solutal convection on the morphological stability of a binary alloy', J. Crystal Growth, 56, 67–76.

[4] Forth, S. A. and Wheeler, A. A. (1992) 'Coupled convective and morphological instability in a simple model of the solidification of a binary alloy, including a shear flow', J. Fluid Mech., 35, 61–94.

[5] Sivashinsky, G. I . (1983) 'On cellular instability in the solidification of a dilute binary alloy', Physica D, 8, 243–248.

[6] Young, G. W. and Davis, S. H. (1986) 'Directional solidification with buoyancy in systems with small segregation coefficient', Phys. Rev. B., 34, 3388–96.

[7] Riley, D. S. and Davis, S. H. (1990) 'Long-wave morphological instabilties in the directional solidification of a dilute binary mixture', S.I.A.M. J. Appl. Math, 50, 420–36.

[8] Riley, D. S. and Davis, S. H. (1990) 'Long-wave interactions in morphological and convective instabilities', I.M.A. J. Appl. Math., 45, 267–85.

[9] Wheeler, A. A. (1991) 'A strongly nonlinear analysis of the morphological instability of a freezing binary alloy: solutal convection, density change and non-equilibrium effects', I.M.A. J. Appl. Math., 47, 173–192.

[10] Caroli, B., Carol, C., Misbah, C., and Roulet, B. (1985) 'Solutal convection and morphological instability in directional solidification of binary alloys. ii. effect of density difference between the two phases', J. Phys. (Paris), 46, 1657–65.

BUOYANT CONVECTION NEAR A SOLIDIFYING DENDRITE

D. CANRIGHT
Mathematics, Code MA/Ca
Naval Postgraduate School
Monterey, CA 93943 USA

S. H. DAVIS
Engineering Sciences & Applied Mathematics
The McCormick School of Engineering and Applied Science
Northwestern University
Evanston, IL 60208 USA

ABSTRACT. An isolated, axisymmetric dendrite of pure material solidifies steadily downward into an undercooled melt. Surface energy and kinetic undercooling are negligible. Latent heat released at the solid-liquid interface drives buoyant convection. We construct an approximate solution for the flow and temperature in powers of a buoyancy parameter G. The results show that buoyant convection *enhances* growth at the tip for large Prandtl number P, but may *diminish* growth for small P. Buoyant convection also modifies the dendrite shape; the base widens more quickly than a paraboloid.

1. Introduction

We consider the effects of buoyant convection on the tip of an isolated axisymmetric dendrite growing steadily downward into a pure undercooled melt, neglecting surface energy and attachment kinetics. We construct a perturbation solution, linearized about the basic state given by Ivantsov (1947), valid near the tip when the buoyancy parameter G (defined below) is small. The results give an analytic description of the buoyant flow, temperature, and shape of the dendrite near the tip, showing the explicit dependence on the Stefan number S and the Prandtl number P. Because surface energy is neglected, this solution has the same "selection" problem as that of Ivantsov: the Peclet number $Pe \equiv RV/2\kappa$ is determined (where κ is the thermal diffusivity) without selecting which tip radius R and growth speed V are realized.

Ananth and Gill (1988) also examined buoyant flow near a dendrite. They used a coordinate expansion, assuming the dendrite was paraboloidal. They solved the resulting coupled nonlinear equations numerically for $P = 23.1$ and several choices of undercooling and buoyancy parameters. Our approach gives a different range of applicability than that of Ananth and Gill; while their method applies for arbitrarily large buoyancy G but only very near the dendrite tip, our approach applies throughout a much larger region about the

S. H. Davis et al. (eds.), Interactive Dynamics of Convection and Solidification, 77–80.

tip but only when buoyancy is small. Further, rather than presuming a paraboloidal tip, perturbations to the shape of the dendrite are determined as part of our solution.

Our results, augmented by a simple selection criterion based on tip stability, compare well with the experiments of Huang and Glicksman (1981) on succinonitrile when $G < 1000$, but overpredict the buoyancy effects for larger G. (The results of Ananth and Gill, using a selection criterion based on matching the experimental Peclet number, agree well with the experiments over the whole range.)

Our solution predicts two main effects of buoyant convection: both the growth rate and the dendrite shape are modified. The modified shape is no longer paraboloidal, but rather widens more quickly away from the tip. When the Prandtl number P is moderate to large, buoyant flow is found to enhance growth at the dendrite tip. Applying the selection criterion then shows that buoyancy both increases the growth speed and decreases the tip radius. However, when P is small, there is a range of undercooling S for which buoyant convection *diminishes* growth at the tip, giving a lower Peclet number, and, using the same selection criterion, results in *slower growth* with a *larger tip*. These results suggest that buoyancy effects for metals (low P) may be qualitatively different from those for organics (high P).

2. Method

Consider a smooth dendrite growing into pure, undercooled, quiescent melt. The dendrite is axisymmetric, and it grows downward along the gravity vector. The growth is steady in the reference frame of the solid-liquid interface. Surface energy is ignored and thermodynamic equilibrium holds on the interface so that the liquid solidifies at the equilibrium melting temperature T_m, hence the entire solid phase is isothermal at T_m. The material properties in each phase are assumed constant, with no density change upon solidification, and the Boussinesq approximation is applied in the melt. The latent heat released at the interface balances the heat removed by conduction into the melt, and the solid dendrite and the undisturbed melt are motionless in the laboratory reference frame.

We nondimensionalize the temperature difference by the overall undercooling $T_m - T_\infty$ (where T_∞ is the temperature of the undisturbed fluid), the velocity by the tip growth speed V, and lengths by the thermal boundary layer thickness κ/V. Then three dimensionless parameters govern the problem: the Stefan number (undercooling) $S \equiv (T_m - T_\infty)c_p/L$, the Prandtl number $P \equiv \nu/\kappa$, and the gravitational parameter $G \equiv g\alpha(T_m - T_\infty)\kappa/V^3$. ($G$ is the product of the Prandtl number and the Rayleigh number based on the thermal boundary layer thickness.) Here c_p is the specific heat, L is the latent heat of fusion, ν is the kinematic viscosity, g is the acceleration of gravity, and α is the thermal expansion coefficient.

When $G = 0$, there is no flow, and the thermal problem is separable in paraboloidal coordinates, giving the Ivantsov (1947) solution for a paraboloidal dendrite. For small G, we express the temperature, vorticity, stream function, and interface position as power series in G, with the Ivantsov solution at zeroth order. The linearized equations for the first-order perturbations are solved by separation of variables. The details are given in Canright and Davis (1991).

The flow solution is not uniformly valid, in two ways. (i) The perturbations grow linearly with distance along the dendrite, as buoyancy accumulates in a growing thermal boundary layer, and so at a distance $O(G^{-1})$ the nonlinear convective effects become important.

(ii) The separable solution cannot match the quiescent conditions far from the dendrite, so that far away the potential flow must depart from the separable form. Nonetheless, the solution satisfies all the interface conditions and is locally valid within a distance $O(G^{-1})$ from the tip in all directions.

3. Results and Discussion

The solution gives the dimensionless position of the solid-liquid interface in axisymmetric parabolic coordinates (ξ, η) as $\eta = H(\xi)$:

$$H(\xi; S, P, G) \sim H_0(S)\{1 + G[(\xi - 1)A_0(S, P) + A_1(S, P)]\} \quad , \tag{3.1}$$

where H_0 is the Ivantsov interface position, which increases monotonically with S, and the perturbations A_0 and A_1 are positive quantities (given in Canright and Davis, 1991). This result predicts two main effects of buoyant convection: enhanced or diminished tip growth, and a modified dendrite shape.

The tip growth is measured by the Peclet number $Pe = RV/2\kappa$, given by

$$Pe \sim H_0\{1 + G[(H_0 - 1)A_0 + A_1]\} \quad , \tag{3.2}$$

Thus, where the relative perturbation $Pe_1 \equiv (H_0 - 1)A_0 + A_1 > 0$, tip growth is enhanced by buoyancy effects; this occurs for moderate to large Prandtl number $P > 0.15$ or large undercooling $S > 0.6$. For smaller P, however, there is a range of undercooling where $Pe_1 < 0$, and tip growth is decreased. These two regions in parameter space are shown in the following figure (along with contours of Pe_1 at intervals of 0.005).

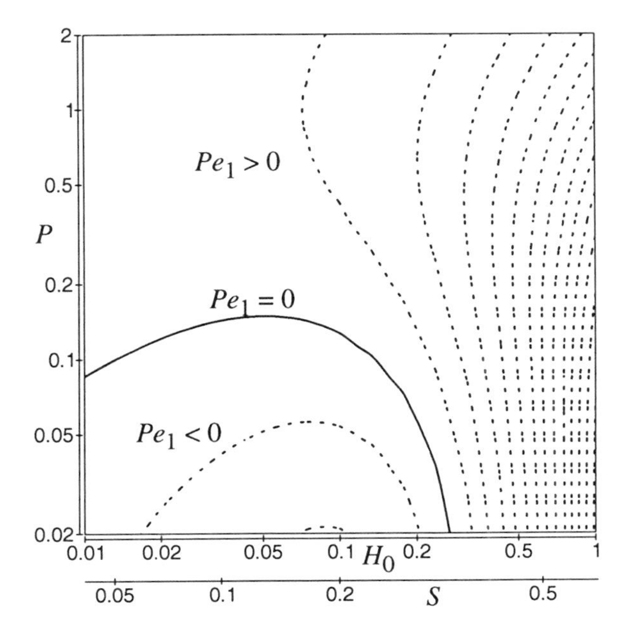

These differing effects at large and small P can be understood in terms of the interaction between the buoyant flow and the thermal field. The perturbation flow solution shows a layer of buoyant fluid, warmed by released latent heat, rising along the dendrite. Farther away, fluid is drawn inward and downward to supply the growing buoyant layer. (Fluid is not drawn upward from below, because ahead of the dendrite the fluid is still undisturbed.) As a consequence, a general feature of these flows is a *secondary stagnation point* on the axis ahead of the tip, below which the inward flow continues downward, and above which the flow turns upward toward the primary stagnation point on the tip.

There are *two competing thermal effects* of the secondary stagnation-point flow ahead of the tip. (i) Above the stagnation point the flow is upward, in the direction of increasing temperature, and locally convection tends to cool the fluid. (ii) Below the stagnation point the flow is downward, so convection tends to warm the fluid. These effects are *cumulative*; it is the *balance* that determines whether tip growth is enhanced or reduced. This balance depends on the relation of the stagnation point to the thermal layer: roughly, when the stagnation point is outside the thermal layer then growth is enhanced, while when the stagnation point is well inside the thermal layer, growth is diminished. The latter case requires that the viscous length scale be relatively small, which only occurs for small Prandtl number. Geometrical effects also play a role; for a wide dendrite the stagnation point is fairly far from the tip even for small P.

The dependence of H on ξ in (3.1) shows that the perturbed interface is no longer paraboloidal. The results show that A_0 is always positive, thus the perturbed shape widens more quickly than a paraboloid away from the tip, and approaches an asymptotic cone.

4. Acknowledgments

This work was supported in part by a grant from the National Aeronautics and Space Administration, Microgravity Science and Applications Program, and in part by the Naval Postgraduate School Research Council.

5. References

Ananth, R. and Gill, W. N. (1988) 'Dendritic growth with thermal convection', *J. Cryst. Growth* **91**, 587-598.

Canright, D. and Davis, S. H. (1991) 'Buoyancy effects of a growing, isolated dendrite', *J. Cryst. Growth* **114**, 153-185.

Huang, S. C. and Glicksman, M. E. (1981) 'Fundamentals of dendritic solidification - I. Steady-state tip growth', *Acta metall.* **29**, 701-715.

Ivantsov, G. P. (1947) *Dokl. Akad. Nauk, SSSR* **58**, 567-571.

FLOW INTERACTIONS WITH DENDRITIC MUSHY ZONES

C.J. PARADIES AND M.E. GLICKSMAN
Rensselaer Polytechnic Institute
Materials Research Center
Troy, New York 12180
USA

ABSTRACT. Cast shops have found that reduced grain size can be obtained in a finished casting if stirring velocity and melt temperature are carefully controlled. Forced convection in the melt causes fragmentation of dendrites in the mushy zone. A model system has been developed that allows independent control of concentration, temperature, and fluid flow velocity while viewing the mushy zone with an optical microscope. Initial results indicate an underlying mechanism associated with a decreasing interfacial area per unit volume of mushy zone combined with recalescence. This mechanism causes the dendrite side branches to pinch off at their bases, where they connect to the primary dendrite stem. Crystal fragments are then transported by convection in the melt and can either develop into independent equiaxed grains or remelt. Preliminary results indicate that the velocity of the flow dramatically alters the mushy zone causing both a change in the amount of fragmentation and the fragmentation rate.

1. INTRODUCTION

Investigations have shown a pronounced decrease in grain size with increasing vigor of forced convection.[1,2] Observations by Desnain, et al.,[2] indicate that a maximum flow velocity is reached above which the grain size either does not change or even increases. Reducing the grain size improves low temperature strength, toughness, heat treatment characteristics, and hot tearing tendency. Therefore, great interest has been generated in developing an understanding of the effects of stirring and melt flow during solidification and in optimizing grain refinement while avoiding the undesirable aspects of macrosegregation.[3] Redistribution of liquid and solid caused by stirring the melt can increase macrosegregation. The effects of macrosegregation on an Al-Cu alloy casting's ultimate tensile strength, yield stress, percent elongation, hardness and conductivity can offset the beneficial effects of reduced grain size. To optimize the benefits produced by imposing a forced flow, a complete understanding of how the melt flow rate affects grain size is necessary. The vigor of the melt flow, the addition of alloying elements, and the overall heat transfer characteristics have all been shown to influence the grain size.[3] Unfortunately, engineering alloys of interest are opaque to visible light; therefore, no direct observation of the process of fragmentation is possible. Both transparent single component materials and alloys have been observed to generate fragments during solidification. In the present study we used a

S. H. Davis et al. (eds.), Interactive Dynamics of Convection and Solidification, 81–91.

transparent model system to reveal the effects of the flow rate on the morphology of the mushy zone and the fragmentation of the dendrite side branches.

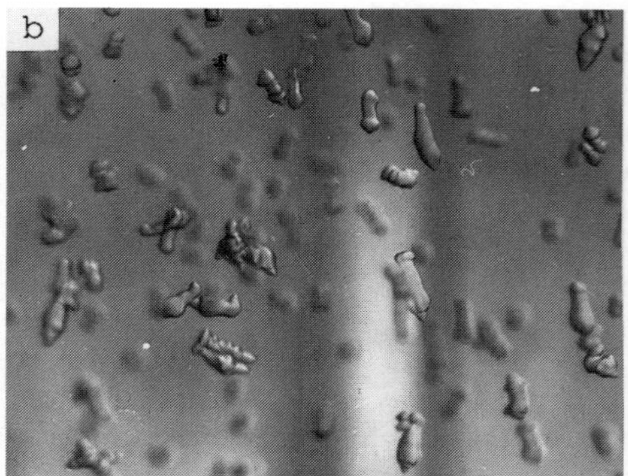

Figure 1. Photographs of (a) single side branch separating from the dendrite stem and (b) many crystallites detached from dendrites on the top surface of the growth chamber.

1.1 Pure materials

Rubinstein[4] observed the detachment of dendrite side branches during the growth of pure camphene dendrites. Figure 1(a) shows a side branch detached from a camphene dendrite growing in a pure 0.3 K supercooled melt. No significant bulk fluid convection or strong vibrations were apparent; therefore, mechanical shearing as a mechanism of detachment seems unlikely. Notice in figure 1(a) the tear drop shape of the free side branch. Video images of the process reveal that the side branches "pinch off" from the dendrite stem. Coarsening of the side branches occurs during dendritic growth often leading to remelting of selected side branches while other side branches continue to grow[5]. Coarsening occurs due to a redistribution of solid driven by a decrease in interfacial free energy. The Gibbs-Thomson equation[6],

$$\Delta T = T_\infty - T_{eq} = 2\,\gamma\,\Omega\,H_i\,/\,\Delta S_f, \tag{1}$$

gives the minimum undercooling necessary to continue solidification where T_∞ is a planar interface's melting temperature, T_{eq} is the equilibrium melting temperature at the curved interface, γ is the solid-melt interfacial energy at the melting point, Ω is the molar volume of the liquid at the melting point, H_i is the mean curvature defined as $1/2\ (\ 1/R_1 + 1/R_2\ )$, see figure 2, and ΔS_f is the entropy of fusion. As the temperature of the melt adjacent to the interface (T_{local}) increases due to the release of the latent heat of fusion, large side branches ($T_{eq} > T_{local}$) continue to grow while the small side branches ($T_{eq} < T_{local}$) remelt; therefore, the latent heat of fusion flows from large to small side branches. If the base of the side branch is narrow compared to the tip, then the side branch can "pinch off" during remelting. In fact, coarsening can also take place

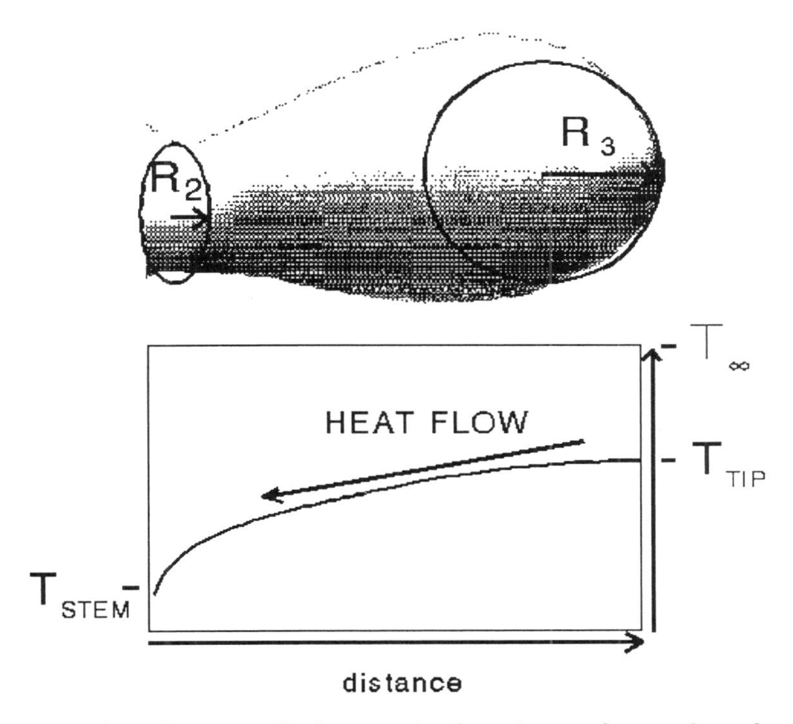

Figure 2. Illustration of heat transfer between the tip and stem of a camphene dendrite side branch. ΔT_{STEM} and ΔT_{TIP} are defined by T_{eq} in equation 1.

between the stem and the tip of the side branch hastening the remelting of the stem. Figure 2 illustrates the heat transfer from the stem to the tip of a side branch due to the temperature gradient corresponding to the difference in mean interfacial curvatures. The temperature gradient established by the difference in curvature between the bulbous tip and the narrow stem establishes the heat transfer sketched in figure 2; therefore, the tip will tend to grow at the expense of the stem.

It is also possible for natural convection in the melt to increase T_{local} as dendrites continue to grow. Rubinstein discovered a remarkable rate of side branch detachments during isothermal dendritic growth of camphene, figure 1(b). In this case the nucleation occurred on the upper, inside surface of a spherical growth chamber. As the dendrites grew downward from the top of the growth chamber, the latent heat of fusion was carried by thermal convection toward the top of the growth chamber. It is likely that an increase in the local melt temperature caused by the thermal convection contributed to the remelting of the dendrite side branches. Indeed, recalescence could occur if the increase in temperature due to the convection of the latent heat of fusion was sufficient to overcome the rate of cooling through the top of the growth chamber. However, both the dendritic growth and the fragmentation appeared to continue throughout solidification of the camphene, and remelting of entire dendrites was not evident. Therefore, it is more likely that coarsening combined with slight increases in T_{local} caused selective remelting of the smaller dendrite side branches.

1.2 Alloy systems

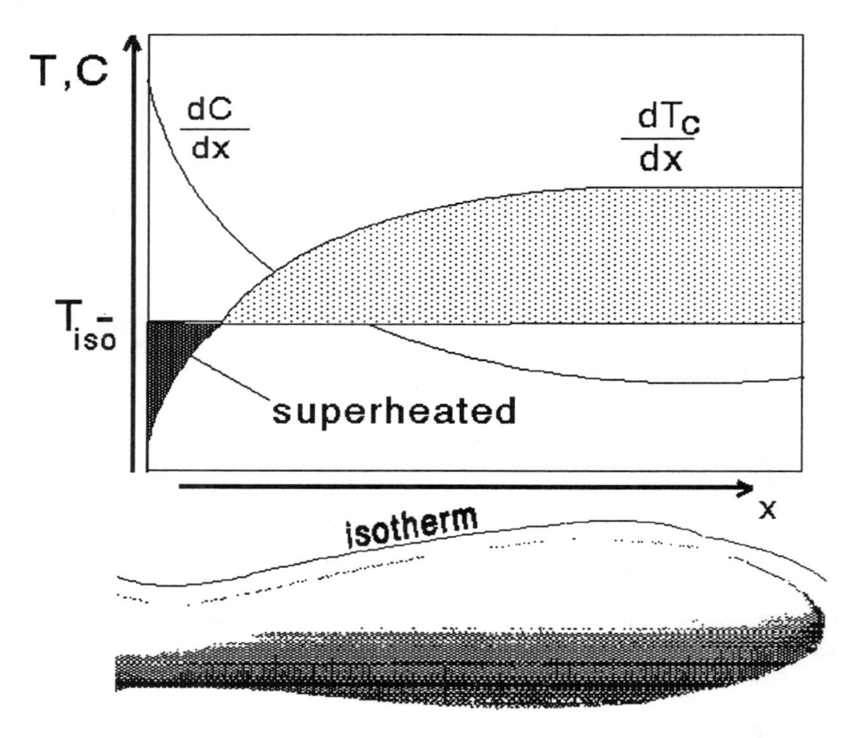

Figure 3. Illustration of "constitutional remelting" showing the effect of solute concentration along a temperature isotherm.

Jackson, et al.,[7] examined fragmentation of mushy zones in ammonium-chloride and water alloys and concluded that localized remelting was the process responsible for the formation of the equiaxed region in a casting. For most metal alloys an increase in the local solute concentration increases the solute concentration in the solid phase. A higher solute concentration in the solid phase means that the equilibrium melting temperature of the solid is lowered. Jackson, et al., explained that side branches begin growing in the region of high solute concentration around a growing dendrite. As the side branches extend into the bulk melt and the local solute concentration diminishes, the equilibrium melting temperature for the tip becomes higher than that of the stem. Therefore, growth at the tip is favored over growth of the stem. According to Jackson, et al., remelting of the stem, also called "pinch off", is a result of an increase in the local concentration and temperature at the stem attributed to continued growth of surrounding side branches.

The behavior of dendritic alloys can be related to the previous discussion on the remelting of pure materials. For pure materials the initially high local temperature experienced by the emerging side branch is, in some respects, similar to the high local solute concentration. As the pure dendrite grows the tip enters a region of lower melt temperature. Therefore, the tip can grow even if growth of the stem stops due to increasing local temperature. When the curvature of the stem becomes large compared to the curvature of the tip, the difference in equilibrium temperature due to curvature can become a significant driving force contributing to the remelting of the stem. In alloy systems differences in mean curvature between the stem and the tip might remain as a contributing factor during "pinch off", especially just prior to pinching when the large curvature of the stem creates a significant gradient. Initially, the difference between the solute concentration in the solid phase of the dendrite side branch tip and stem could be the primary reason that remelting of the stem occurs. Figure 3 illustrates the mechanism causing this "constitutional remelting." A growing dendrite rejects heat and solute which diffuse away at different rates. In metals heat conduction from the interface occurs very rapidly when compared to the diffusion of solute. The thermal field around the dendrite side branch can be approximated by a nearly isothermal field due to the relatively high thermal diffusivity. The accompanying graph shows a possible solute distribution of the solid phase taken along the isotherm. A specified concentration can be related to an equilibrium constitutional temperature on the phase diagram. The constitutional temperature profile is superimposed on the graph. The temperature of the isotherm and the equilibrium temperature profile intersect. The region to the left of the intersection can be described as the "constitutional remelting" zone since it is a superheated region. To the right of the intersection solidification continues.

1.3 The SCN-acetone alloy system

The SCN-acetone alloy system used in this experiment has been used previously [8-9] to study the microscopic length scales associated with equiaxed dendritic growth of alloy systems. SCN's properties are well established. SCN is a "plastic" crystal which freezes like a cubic metal; the dendrites are not faceted. The phase diagram exhibits a nearly linear solidus and liquidus in the region of interest, with acetone as the rejected component at the solid-liquid interface, creating a solute concentration profile typical of metal alloys. Also, the acetone concentration can be determined precisely by starting with 99.9999% pure SCN. The SCN-acetone system can be employed to model the phenomena of nucleation, mushy zone growth, and recalescence. Although the thermal diffusivity of SCN-acetone is more than an order-of-magnitude less than that of most metal alloys, the fundamental processes being investigated should be adequately represented by the model alloy since the thermal diffusivity of SCN is still two orders-of-magnitude greater than the diffusivity of acetone in the liquid phase.

2. EXPERIMENTAL

The forced convection system employed in our experiment is illustrated in figure 4 and includes a mushy zone growth chamber (inner radius = .7 cm & outer radius = 3.83 cm), a melt reservoir with an acetone reservoir attached, a variable speed vane pump, and an independently controlled chill loop. In the present study the growth chamber, seen on the left in figure 4, was oriented horizontally. Pure (99.99 %) SCN was introduced into the melt reservoir under vacuum. Acetone

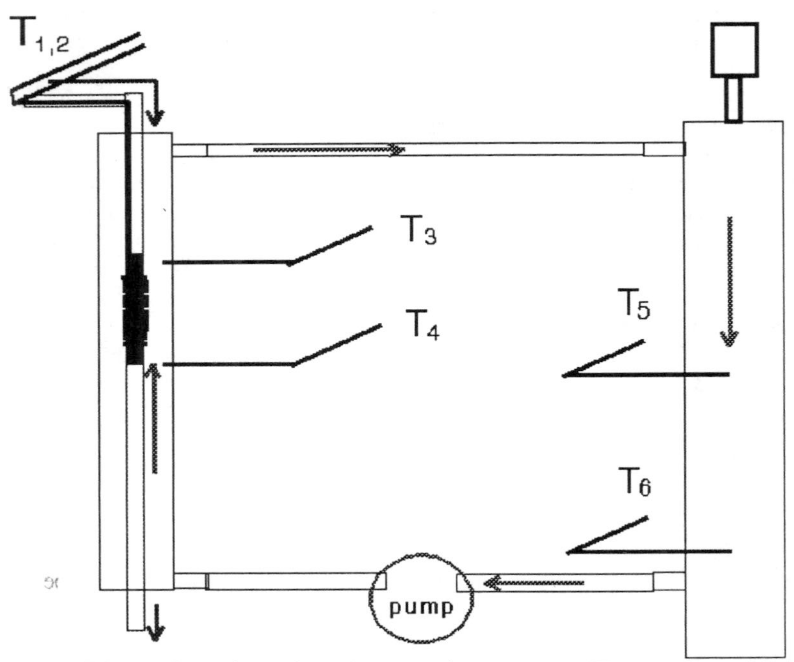

Figure 4. Illustration of the forced convection system. The results published here were obtained using the system with the growth chamber oriented horizontally.

was introduced from the acetone reservoir to a solute concentration of 0.8 mol%. The system was enclosed in a temperature controlled "hot box". The hot box provided an environment with a uniform temperature that was controlled to ±0.5 K. The cooling loop used a temperature controlled bath to maintain the chill tube at ±0.1 K, and the temperature of the chill tube was decreased by approximately 0.1 K per minute. The temperature at the chill tube and outer wall of the growth chamber were measured at the center of the field of view to within 0.1 K.

The velocity of the fluid within the annulus previous to the nucleation of the mushy zone is represented by figure 5. This is a three dimensional representation of the solution to the Navier-Stokes equations for fully developed laminar flow with no slip boundary conditions and an average flow rate of 2 cm/sec (maximum flow of 3.08 cm /sec) neglecting any natural convection effects. The solution has been verified by *in situ* measurements of the speed of tracer particles using laser doppler velocimetry.

The experimental observations were recorded using a CCD camera attached to a video recorder. Photographs presented in this paper were captured using a Polaroid Freezeframe recorder with a total magnification of 31 x. The data in figure 6 were obtained by counting, frame by frame, the crystallites entering and leaving a control volume bounded by an arbitrary length along the axis of the cylinder of 1.4 mm (4.3 cm on the screen), a depth equal to the depth of focus (approximately equal to 1 mm), and a height equal to the field of view. The height of the field

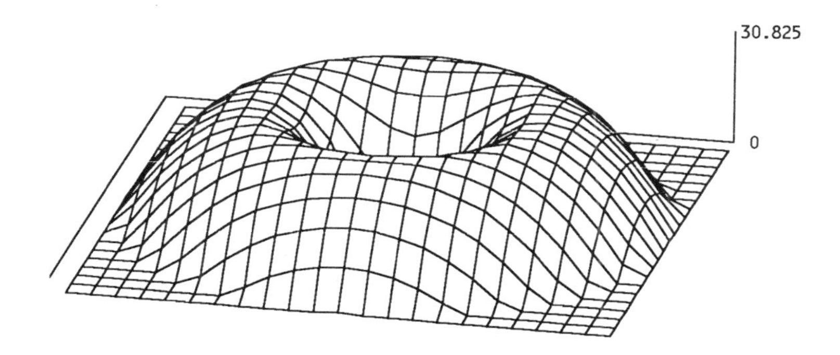

Figure 5. Three dimensional solution of the Navier-Stokes equations for fully developed flow through the growth chamber with no slip boundary conditions and ignoring natural convection.

of view does not alter the count since only those crystallites emerging from the mushy zone could be counted. By subtracting the number of fragments entering the control volume by the number exiting, a cumulative total of the number of free side branches emerging from the mushy zone bounded by the control volume was obtained over time. Side branches that may have detached but remained trapped in the mushy zone could not be counted. Crystallites visible in the control volume but not in focus were ignored. The criterion established to judge if a crystallite was in focus was quite arbitrary but was applied rigorously for all flow speeds by a single observer. Use of an image analysis engine to count the fragments based upon a defined algorithm is being investigated, but significant efforts have been made to insure the count currently obtained is reproducible. The presentation of the cumulative number of fragments per cm represents the number of crystallites detaching from the mushy zone in the control volume divided by 0.14 cm, the length along the axial direction of the mushy zone. The scale should be used primarily for comparison between the different flow speeds; however, the data are essentially proportional to the number of fragments created by any arbitrary area of mushy zone.

3. RESULTS

Flow rates of 1.8 ℓ/min (Re = 85.3), 3.3 ℓ/min (Re = 146) and 5.3 ℓ/min (Re = 244) were maintained during nucleation and growth of a SCN-8 mol% acetone alloy with ΔT_{an} = 8 K.

$$Re = 4\, R_H\, \rho\, V / \mu, \tag{2}$$

a dimensionless ratio of inertial to viscous forces which is frequently used to compare convective states between fluids, where R_H is the hydraulic radius $(R_{out}^2 - R_{in}^2) / (2\, R_{out})$, ρ is the density of the liquid phase, V is the mean velocity, and μ is the absolute viscosity. The results are shown in figure 6. The chill tube temperature was ramped at approximately 0.1 K/min with the box temperature held constant. The mushy zone nucleated on the chill tube and grew rapidly into the supercooled melt adjacent to the chill tube. The growth of the initial mushy zone was most rapid for the highest flow rate. Growth arrested with the onset of recalescence once the dendrites reached the extent of the supercooled region. Although the chill tube continued a downward temperature ramp, mushy zone growth remained very slow following recalescence.

The video images obtained during the experiment clearly show side branches detaching from the mushy zone. Nucleation of crystallites ahead of the mushy zone did not occur. It is probable that the remelting during recalescence, possibly combined with "constitutional remelting" and coarsening, caused the side branches to "pinch off". The side branches then became free crystallites subject to the influences of gravity and the fluid motion of the melt.

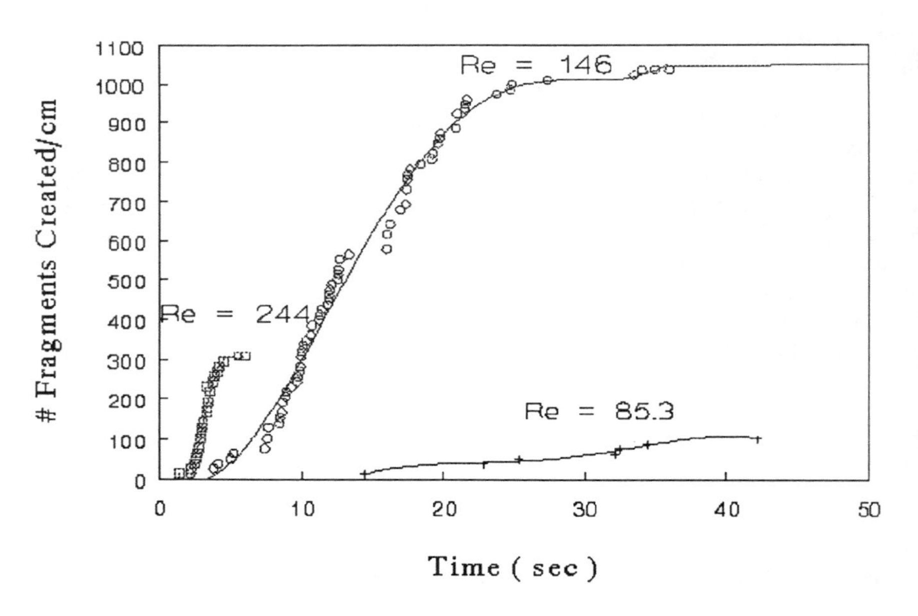

Figure 6. Results of the experiment. Average velocities of 0.7, 1.2, and 2.0 cm/sec in the growth chamber correspond to Re equal to 85.3, 146, and 244, respectively.

The fragmentation, in all cases, did not begin until after recalescence occurred and caused remelting of the dendrite side branches. As displayed by figure 6, the high velocity flow

(Re=244) caused the highest fragmentation rate, indicated by the slope of the fragmentation vs. time curve. However, the duration of the fragmentation events for the high flow rate was brief compared to either the moderate (Re=146) or slow (Re=85.3) flow rates. Therefore, the moderate flow rate was able to produce a far greater number of fragments. Due to an extremely low rate of fragmentation, the low flow rate produced considerably fewer fragments than either the moderate or fast flow rates. To produce a large number of free crystallites, a balance between the rate of fragmentation and the duration of the fragmentation events seems to be required.

Results of these experiments were recorded in real time on video tape. Photographs of the images on the video tape are shown in figure 7. The low flow rate has the coarsest and thickest mushy zone, and the high flow rate has the finest and narrowest mushy zone. It is clear from the photographs which were taken immediately after the initiation of fragmentation that the flow rate has a profound effect on the morphology of the mushy zone. Additionally, the size of the free

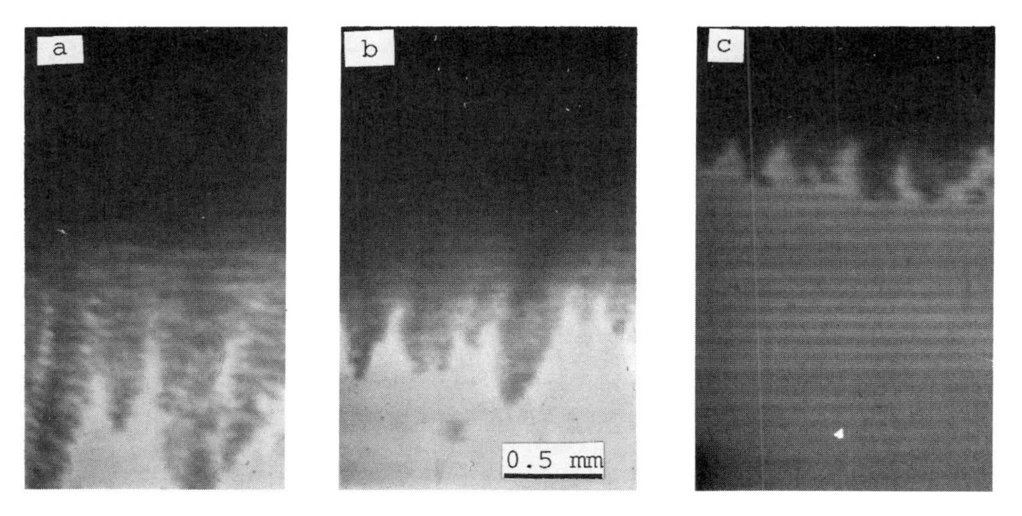

Figure 7. Photographs of the mushy zones immediately after the fragmentation begins for (a) Re = 85.3, (b) Re = 146, and (c) Re = 244.

crystallites decreases with increasing flow rate. After fragmentation ended, the dendrites in the mushy zone associated with the high flow rate appeared denuded of side branches. The mushy zone created by the moderate flow rate continued to produce detached side branches, although at a much reduced rate, even after recalescence appeared to be complete and slow growth of the mushy zone continued. The low flow rate produced such a coarse mushy zone that few side branches remelted at the stem. Instead, many side branches appeared to remelt while remaining attached to the dendrites.

4. CONCLUSIONS

The fact that fragments were not created until after recalescence began indicates that shear forces are not sufficient to cause direct, mechanical fragmentation.

Nucleation events, *per se*, are not required to produce the equiaxed zone in a casting since the present study indicates that a large number of free crystallites can be produced by recalescence. The crystallites created during recalescence are free to be carried by the flow to remote regions where either remelting or continued growth are possible. Thus, this work supports the early ideas of Jackson, et al., that the equiaxed region of a casting might be the result of fragmentation of the mushy zone due to localized remelting.

High flow rates are not required, and possibly not even desired, for the production of a large number of free crystallites.

A balance between the fragmentation rate and the duration of fragmentation events is required to produce the large number of crystallites necessary for significant grain refinement. We postulate that the morphology of the mushy zone and the establishment of favorable temperature and solute gradients by the flow are the critical factors contributing to the combination of a high rate of fragmentation and a long duration of dendrite side branch detachments. The flow speed influences the morphology by altering the thermal and solutal gradients in the annulus. Without a sufficiently robust dendritic mushy zone, few side branches will be available for remelting either by recalescence or by "constitutional remelting". The slow flow produces a coarse mushy zone that resists remelting of the side branches. The side branches eventually coarsen but do not "pinch off" in large numbers.

The free crystallites observed in these experiments were originally dendrite side branches. It is not surprising, then, that the size of the crystallites decreases with increasing flow, since the side branches themselves are finer in the mushy zones influenced by higher flow rates. It is likely that the reduction in the diameter of the side branches contributes to the increased fragmentation rate. Additionally, the fastest flow rate had the fastest initial growth. The fast initial growth increases the effect of recalescence, since less time is permitted for solute and heat to diffuse away from the mushy zone. Likewise, the slow growth of the low flow rate mushy zone permitted more time for the solute and latent heat to diffuse.

By creating an optimized mushy zone morphology and by producing a flow that will evenly distribute the free crystallites throughout the melt, one should be able to produce an alloy with limited macrosegregation and a fine, equiaxed structure.

The results reported here may be applicable to other alloys, but they are only indicative of situations where recalescence of the mushy zone is occurring. Further experimentation will be necessary to determine the effect of flow rates on mushy zones which continue to grow. Experiments involving additional solute concentrations, flow rates, and temperature gradients are also needed to establish more clearly the effect of forced flow on the morphology of mushy zones and the relationship of the flow rate to the production of detached crystallites.

5. ACKNOWLEDGEMENT

The authors appreciate the assistance, encouragement, and funding provided for these studies by the Alcoa Technical Center, Pittsburgh, Pennsylvania and the support provided by the New York State Science and Technology Foundation.

6. REFERENCES

1. Vives, Charles (1990) 'Hydrodynamic, thermal, and crystallographic effects of an electromagnetically driven flow in solidifying aluminum alloy melts', Int. J. Heat Mass Transfer, v.33, no. 12, 2585-2598.

2. Desnain, P., F. Durand, Y. Fautrelle, D. Bloch, J.L. Meyer, J.P. Riquet (1988) "Effects of the electromagnetic stirring on the grain size of industrial aluminum alloys: experiments and theoretical predictions', Light Metals 1988, 487-493.

3. Mehrabian, Robert (September 1984) 'A review of our present understanding of macrosegregation in ingots', NASA Conference Publications 2337, 169-186.

4. Rubinstein, E.R. (1989) 'Dendrite growth kinetics and structure' PhD Thesis, Rensselaer Polytechnic Institute.

5. Huang, S.C. and M.E. Glicksman (1981) 'Fundamentals of dendritic solidification II: development of sidebranch structure', Acta Metall., v. 29, 717-734.

6. Kurz, W. and D.J. Fisher (1986) Fundamentals of Solidification, Trans Tech Publications LTD, Aedermannsdorf, Switzerland.

7. Jackson, K.A., J.D. Hunt, D.R. Uhlmann, and T.P. Seward III (1966) 'On the origin of the equiaxed zone in castings', Transactions of the Metallurgical Society of AIME, v.236, February 1966, 149-157.

8. Chopra, M.A., M.E. Glicksman, and N.B. Singh (Dec 1988) 'Dendritic solidification in binary alloys', Metallurgical Transactions A, 19A, 3087-3096.

9. Lipton, J., M.E. Glicksman, and W. Kurz (1984) 'Dendritic growth into undercooled alloy melts', Materials Science and Engineering, 65, 57-63.

INTERACTION OF THERMAL AND FORCED CONVECTION WITH THE GROWTH OF DENDRITIC CRYSTALS

W. N. GILL, Y. W. LEE, K. K. KOO, R. ANANTH
Department of Chemical Engineering
Rensselaer Polytechnic Institute
Troy, NY 12180-3590 USA

ABSTRACT. New experiments with ultrapure succinonitrile, SCN, show that the selection parameter, σ^*, increases by about 50% above 0.0195 as the forced flow velocity increases from zero to 1 cm/sec. Existing theory does not predict well the forced convection experimental results for σ^*. The tips of the basal plane of ice crystals split in quiescent melts, at undercoolings less than 0.35°K. Then the tip regenerates itself, and the growth velocity remains invariant during this entire event. The causes of tip splitting are believed to be thermal convection and the lack of anisotropy in the basal plane.

1. INTRODUCTION

Dendritic growth has been studied recently by many investigators because of its practical importance in the crystallization of materials and as an example of spontaneous pattern formation. One can observe this phenomenon when crystals grow into undercooled melts of a pure single component system or into solutions containing two or more components one of which is rejected selectively at the solid-solution interface. Here we study experimentally two types of morphological instabilities in the growth of pure SCN and ice when convection, forced or thermal, is playing a significant role.

2. RESULTS AND DISCUSSION

We have measured the tip radii, R_1 and R_2, of the edge and basal planes of ice dendrites and the growth rate of ice dendrites in ultrapure undercooled water with and without air dissolved in it. The tip of the basal plane splits when the undercooling is 0.35°K or less and these results are

S. H. Davis et al. (eds.), Interactive Dynamics of Convection and Solidification, 93–96.

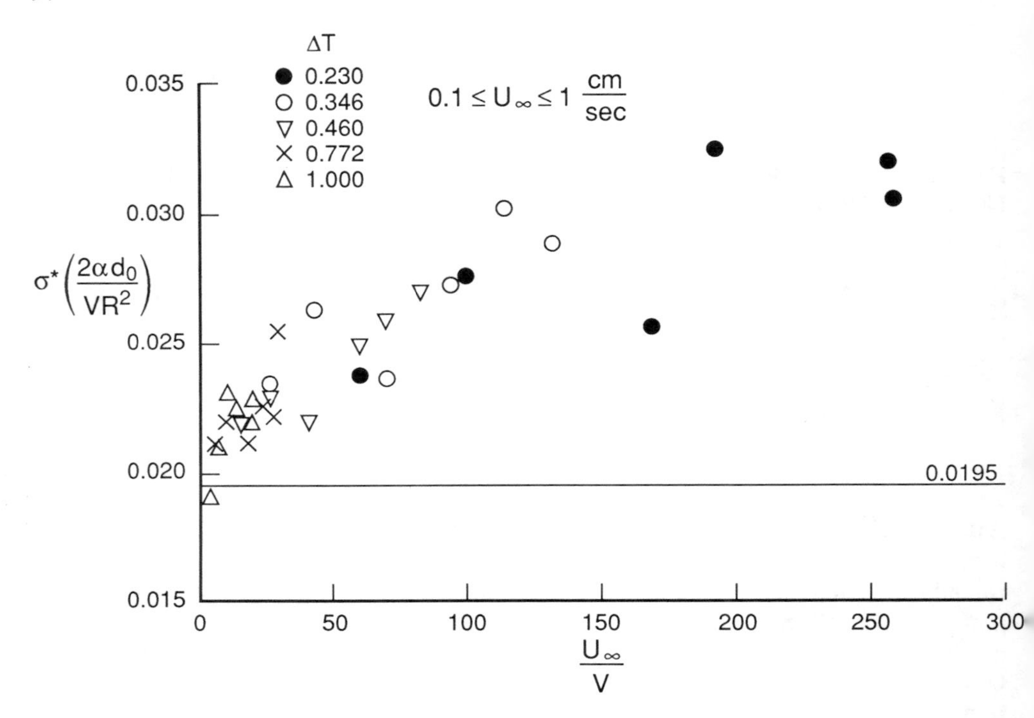

Figure 1. The selection parameter, σ^*, increases with the ratio of the forced velocity to the growth velocity, U_∞/v.

independent of whether the water is saturated with air or free of it. At the same time the growth velocity is constant throughout the tip splitting event. Furthermore, the selection parameter, σ^*, remains essentially constant at 0.075 over the entire range of undercooling, $0.035 \leq \Delta T \leq 1.0°K$. We speculate that tip splitting occurs because of the interaction of convection and the lack of anisotropy in the basal plane. Detailed experimental results on this phenomenon have been published recently by Koo et al, 1991.

Our forced convection experiments were conducted in a novel apparatus in which the entire melt was set in motion at a precisely controlled uniform velocity relative to the crystal. This velocity, U_∞, ranged up to 1 cm/sec which is about 35 to 40 times larger than the velocity due to thermal (natural) convection according to the analyses of Ananth and Gill, 1991. The ratio U_∞/v, where v is the growth velocity, varied from about 3 to 300 over which range σ^* increased by about 50%, as shown in Figure 1, which seems to be qualitatively different from the theory of Bouissou and Pelce (1989) which is based on a solvability condition. Our result for the increase of σ^* with U_∞ also is opposite to that in the experiments reported for the binary alloy of pivalic acid (PVA) and ethanol by Bouissou et al, 1989 in which the forced convection velocities were three orders of magnitude **smaller** (~10μm/sec) than those in our experiments.

Single component melts, such as ultrapure SCN and binary systems are different. For example, it is known that binary alloys of SCN-acetone and water-salt show maxima in the growth rate versus concentration curve at the same time the tip radius, R, exhibits what appears to be a minimum. Thus, the product vR^2 decreases significantly with concentration in SCN-acetone. Because the densities of PVA and ethanol are significantly different, 0.91 and 0.78 g/cm^3, and the Schmidt number is large, convection may be important. Furthermore, Dougherty, 1991 has reported that the anisotropy, ε, in PVA-ethanol is an order of magnitude smaller than that in pure PVA. Therefore, it seems that considerably more work is needed to understand the effect of convection in binary systems.

3. CONCLUSIONS

For SCN, the growth velocity, v, responds quickly to the introduction of flow, while R changes relatively slowly. Both the forced and natural convection data show a systematic increase in the stability parameter, σ^*, as U_∞ is increased. This effect is more pronounced at small ΔT. The tips of ice dendrites split when the undercooling is less than 0.35°K.

REFERENCES

Koo, K.K., Ananth, R. and Gill, W.N. (1991), "Tip Splitting in Dendritic Growth of Ice Crystals", Phys. Rev. A 44, 3782-3790.

Ananth, R. and Gill, W.N. (1991), "Self-Consistent Theory of Dendritic Growth with Convection", J. Crystal Growth 108, 173-189.

Bouissou, P., and Pelce, P. (1989) "Effect of a Forced Flow on Dendritic Growth", Phys. Rev. A 40, 6673-6680.

Bouissou, P., Perrin, B., and Tabeling, P. (1989), "Influence of an External Flow on Dendritic Crystal Growth", Phys. Rev. A 40, 509-512.

Dougherty, A. (1991), "Surface Tension Anisotropy and the Dendritic Growth of Pivalic Acid", J. Crystal Growth, 110, 501-508.

Acknowledgement

This work was supported in part by the New York State Energy and Research Development Authority and by NSF.

KEY WORDS

Dendritic Growth/Convection/Ice/Succinonitrile/Morphological Stability/Selection Parameter/Pattern Formation/Growth Rate/Tip Radius/Crystals/Tip Splitting/Anisotropy

ABSTRACT. New experiments with ultrapure succinonitrile, SCN, show that the selection parameter, σ^*, increases by about 50% above 0.0195 as the forced flow velocity increases from zero to 1 cm/sec. Existing theory does not predict well the forced convection experimental results for σ^*. The tips of the basal plane of ice crystals split in quiescent melts, at undercoolings less than 0.35°K. Then the tip regenerates itself, and the growth velocity remains invariant during this entire event. The causes of tip splitting are believed to be thermal convection and the lack of anisotropy in the basal plane.

EFFECT OF A FORCED FLOW ON DENDRITIC GROWTH

Ph. Bouissou and P.Pelcé

Labo physique E.N.S.
24 rue Lhomond
75231 Paris Cedex 05 France

Laboratoire de recherche en combustion
Universite de Provence-St Jerome
13397 Marseille Cedex 13 France.

ABSTRACT. The effects of a forced flow on dendritic growth rate are studied theoretically. By using a solvability condition one determines the eigenvalue $C = \rho^2 V/Dd_0$ as a function of the velocity of the forced flow in the two-dimensional model. The results are compared critically with recent experiments.

1.Introduction.

A basic problem in dendritic growth is to determine growth rates as a function of various control parameters such as undercooling, concentration of an impurity or velocity of a controlled external flow. When the effect of an axial external flow is considered, experiments show that for a given flow velocity U, $\rho^2 V$ is still constant, linear increasing function of the external velocity U (Bouissou et al.[1]) .Still in this case, the Ivantsov paraboloids are solutions of the problem when surface tension effects are neglected, either in the large Reynolds number limit (potential flow approximation (Dash and Gill [2], Ben Amar et al.[3])), or in the small Reynolds number limit (Oseen approximation) (Dash and Gill [2], Saville and Beaghton [4])).Only the Péclet number related to the crystal Pc is determined as a function of the undercooling and the Péclet number related to the flow $Pf = \rho U / 2D$. Our work will be devoted to the determination of the the constant $\rho^2 V$ as a function of the anisotropic factor and the external flow velocity U by using as in the purely diffusive case, a solvability condition [5].

S. H. Davis et al. (eds.), Interactive Dynamics of Convection and Solidification, 97–99.

2. The model.

The growth is limited by convection - diffusion in the pure liquid and surface tension. Kinetic effects are neglected. The temperature satisfies the advection diffusion in the liquid with the Gibbs-Thomson condition at the interface. Far ahead of the crystal, the fluid velocity field is uniform U in the opposite direction of the crystal velocity. In the following, one assumes that crystal and melt densities are equal so that there is no exchange of mass between the fluid and the crystal as it grows and the normal component of the flow vanishes at the interface. Furthermore, one assumes a no-slip condition at the interface so that the tangential component of the flow vanishes too (see Gliksman et al. [6]).

3. Results [7]

When surface tension effects are taken into account the Ivantsov parabolas are no longer solutions of the free boundary problem.Nevertheless, when surface tension effects are weak, in a sense that will be precised later, steady solutions can be found close to an Ivantsov parabola if a solvability condition is satisfied.In its most general form, derived previously by Pelcé and Bensimon (see ref.5) this condition appears as the vanishing of an oscillating integral as :

$$\int_{-\infty}^{+\infty} dl\, G\,[\,X_0(l)\,]\, \exp\left(i \int_0^l k_m(l')\, dl'\right) = 0 \qquad (1)$$

Here X_0 (l) is the Ivantsov parabola function of the curvilinear coordinate l ; G, the curvature operator ; and k_m , the local nonzero marginal mode of the conjugate dispersion relation, written in a frame moving with constant velocity U in the z direction (frame at rest with respect to the unperturbed solution).

One determines the growth rate of a perturbation of the planar interface with a wavelength small compared to the scale of the unperturbed solution.obtain the dispersion relation for the perturbations as :

$$\omega = iVk \sin\theta + \varepsilon Vk \cos\theta - 2d_0 D\varepsilon k^3$$
$$+ \frac{V^2}{D}\cos\theta\left(\frac{1}{2}\exp(-i\varepsilon\theta) - \cos\theta\right) + iVd_0k^2\varepsilon \sin\theta + \omega\left(\frac{5}{4}\varepsilon\frac{V}{Dk} - \varepsilon\, d_0k\right) \qquad (2)$$
$$+ d_0k\, ia\frac{U}{4\rho}\sin\theta\cos\theta$$

The first line of this relation corresponds to the usual Mullins-Sekerka growth rate.When all the terms of this line balance one obtains the usual order of magnitude $\omega \approx Vk$ and $k \approx (V/Dd_0)^{1/2}$.The second line of the relation corresponds to the effects of Péclet number related to the crystal, i.e. the correction to the Mullins-Sekerka growth rate due to the advective effects of the motion of the crystal . Then, the third line of the dispersion relation corresponds to the effect of the external flow. When the effects of the flow dominate the effects due to the advection of the crystal, the selected values of C are found from (1) and (2) as:

$$C = \beta^{-7/4} f\left(\frac{a(Re)}{\beta^{3/4}} \frac{d_0 U}{\rho V} \right) \quad (3)$$

where f is a function determined numerically.

When the argument of f in eqn.(3) is small the following selected values of C are: where b is a numerical constant.

$$C = \frac{n^2}{\beta^{7/4}} \left(1 + b \left(\frac{a(Re)}{\beta^{3/4}} \frac{d_0 U}{\rho V} \right)^{11/14} \right) \quad (4)$$

Comparisons with experiments are for the moment controversial. This theory is in agreement with an experiment of Bouissou et al. but in complete disagreement with experimental data presented by Gill and Glicksman at the conference. Such controversy must be resolved in future works.

Acknowlegments.

This work was supported by a contract CNES.

References.

[1] BOUISSOU, PH., PERRIN, B. and TABELING, P. Submitted to Phys.Rev.A , Rapid Communications.

[2] DASH, S.K. and GILL, W.N. J. of Mass and Heat Transfer 27, 1345 (1984).

[3] BEN AMAR, M.,BOUISSOU, PH. and PELCE, P. J.Crystal Growth, 92, 97 (1988).

[4] SAVILLE, D.A. and BEAGHTON, J.P. Phys.Rev.A 37, 3423 (1988).

[5] PELCE, P. " Dynamics of curved fronts", published by Acad.Press. (1988).

[6]GLICKSMAN,M.E., CORIELL,S.R. and McFADDEN,G.B., Ann.Rev.Fluid.Mech. 18, 307 (1986).

[7] BOUISSOU, P. and PELCE, P. Phys.Rev.A 40, 6673 (1989).

THE EFFECT OF CONVECTION MOTION ON DENDRITIC GROWTH

JIAN-JUN XU
Department of Mathematics and Statistics
McGill University
(For the conference at Chamonix, France, in March, 1992)

1. INTRODUCTION

In the past several years, we have extensively studied the problems of dendrite growth with no convection, from a pure melt, as well as from a binary mixture. An interfacial wave theory has been established for selecting the tip-velocity and determining the formation of micro-structure at the later stage of growth. The theoretical predictions agree with the available experimental data very well (see Figure 1). In the present work, we turn to investigate the effect of convection in melt. Assume that a single dendrite growing into an undercooled pure melt in the negative z-axis direction with a constant average velocity U. At the far field, a uniform external flow against the dendrite with the velocity $(U_\infty)_D$ may be applied. We assume that the mass density of the liquid phase is ρ, while the mass density of the solid phase is ρ_S. Due to the external flow and/or the change of density in solidification, a convective motion in melt is produced. The fluid motion will affect the heat transport process and change the temperature distribution. We consider the melt as an incompressible Newtonian fluid. Then system involves the following parameters:

$$T_\infty; \quad \varepsilon = \frac{\sqrt{\Gamma}}{\eta_0^2}; \quad \alpha = \frac{\rho_S - \rho}{\rho}; \quad U_\infty; \quad Pr = \frac{\nu}{\kappa_T}. \tag{1.1}$$

2. THE EFFECT OF CONVECTION INDUCED BY DENSITY CHANGE

1. The Steady Basic State for the case of zero surface tension ($\Gamma = 0$): a similarity solution is found under arbitrary undercooling (McFadden et. al., 1986).

$$\begin{cases} T_B &= T_B(\eta); \quad T_{SB} = T_B(1) \\ \eta_B &= 1 \\ \zeta_B &= 0; \quad \Psi_B = \frac{\xi^2}{2} f(\eta). \end{cases} \tag{2.1}$$

2. The Linear Perturbed State:
(1) the local dispersion relationship

$$\sigma = \Sigma(\xi, k_0) = \frac{k_0}{S^2}\left[1 - \frac{(2+\alpha)k_0^2}{S}\right] - i\frac{\xi}{S^2}\,k_0\,, \quad S = \sqrt{1+\xi^2}. \tag{2.2}$$

As $\alpha = 0$, the above local dispersion formula is reduced to that obtained for dendritic growth without density-change.

S. H. Davis et al. (eds.), Interactive Dynamics of Convection and Solidification, 101–103.

(2) the quantum condition for global trapped wave modes:

$$\frac{1}{\varepsilon}\int_0^{\xi_c}\left(k_0^{(1)}-k_0^{(3)}\right)d\xi = (2n+1+\frac{2}{3}+\frac{\theta_0}{2})\pi - \frac{i}{2}\log\alpha_0; \quad n = \left(0, \pm1, \pm2, \pm3, \cdots\right) \quad (2.3)$$

where

$$\alpha_0\, e^{i\theta_0\pi} = k_0^{(1)}(0)\Big/k_0^{(3)}(0)\,. \quad (2.4)$$

(3) the global neutrally stable state:

$$\sigma_R^*(\varepsilon_*) = 0, \quad (n=0).\,; \quad \varepsilon_* = \varepsilon_*(\alpha) \quad (2.5)$$

The critical number ε_* yields the tip velocity of dendrite. The effect of α on the tip velocity is shown in Figure 2.

3. THE EFFECT OF CONVECTION INDUCED BY EXTERNAL FLOW

1. The Steady Basic State for the case of zero surface tension ($\Gamma = 0$): as $Pr \to \infty$, a uniformly valid asymptotic expansion solution is found under arbitrary undercooling.

$$\begin{aligned} T(\xi,\eta) &= T_*(\eta) + \frac{1}{\ln\frac{2Pr}{\eta_0^2(1+U_\in fty)}+1-\gamma_0}T_0(\eta) + \frac{1}{Pr\ln Pr}T_1(\xi,\eta) \\ &+ \frac{1}{Pr^2\ln Pr}T_2(\xi,\eta)+\cdots \quad (3.1)\\ \eta_s(\xi) &= 1 + \frac{1}{\ln\frac{2Pr}{\eta_0^2(1+U_\infty)}+1-\gamma_0}h_0 + \frac{1}{Pr\ln Pr}h_1(\xi) + \frac{1}{Pr^2\ln Pr}h_2(\xi)+\cdots\,, \quad (3.2)\end{aligned}$$

where γ_0 is the Euler constant. This asymptotic expansion solution is valid in the whole physical region. It satisfies both the interface condition and the far field condition. Thus, when the surface tension equals zero, neglecting all terms of $O\left(\frac{1}{Pr\ln Pr}\right)$, we can consider the steady state solution of dendritic growth with external flow as a nearly similarity solution. It can be approximated by the following similarity solution $\{T_B(\eta), \eta_B\}$:

$$T_B(\eta) = T_*(\eta) + \delta T_0(\eta)\,; \quad \eta_B = 1 + \delta H_0(\eta_0^2) \quad (3.3)$$

where

$$\begin{aligned} T_0(\eta) &= 2U_\infty\eta_0^4 e^{\frac{\eta_0^2}{2}}\left[Q(\eta)+B_0R(\eta)\right] \quad (3.4)\\ Q(\eta) &= \int_\eta^\infty\left[\frac{t}{2}(1-\ln t) - \frac{\ln t}{t}\right]e^{-\eta_0^2t^2/2}dt \quad (3.5)\\ R(\eta) &= \int_\eta^\infty\frac{e^{-\eta_0^2t^2/2}}{t}dt = \frac{1}{2}E_1\left(\frac{\eta_0^2\eta^2}{2}\right) \quad (3.6)\end{aligned}$$

and

$$\delta = \frac{U_\infty}{\left(\ln\frac{2Pr}{\eta_0^2(1+U_\infty)}+1-\gamma_0\right)} \quad (3.7)$$

In the above, B_0 and H_0 are constants depending on η_0^2.

2. The Linear Perturbed State:

(1) the local dispersion relationship

$$\sigma = \Sigma(\xi, k_0) = \frac{k_0}{\tilde{S}^2}\left[\eta_B - \frac{2k_0^2}{\tilde{S}}\right] - i\frac{\xi}{\tilde{S}^2}\,k_0\,, \quad \tilde{S} = \sqrt{\eta_B^2 + \xi^2}. \tag{3.8}$$

(2) the quantum condition for global trapped wave modes:

$$\frac{1}{\varepsilon}\int_0^{\xi_c}\left(k_0^{(1)} - k_0^{(3)}\right)d\xi = (2n + 1 + \frac{2}{3} + \frac{\theta_0}{2})\pi - \frac{i}{2}\log\alpha_0; \quad n = (0, \pm1, \pm2, \pm3, \cdots) \tag{3.9}$$

(3) the global neutrally stable state:

$$\sigma_R^*(\varepsilon_*) = 0, \quad (n = 0).\,; \quad \varepsilon_* = \varepsilon_*(\delta) \tag{3.10}$$

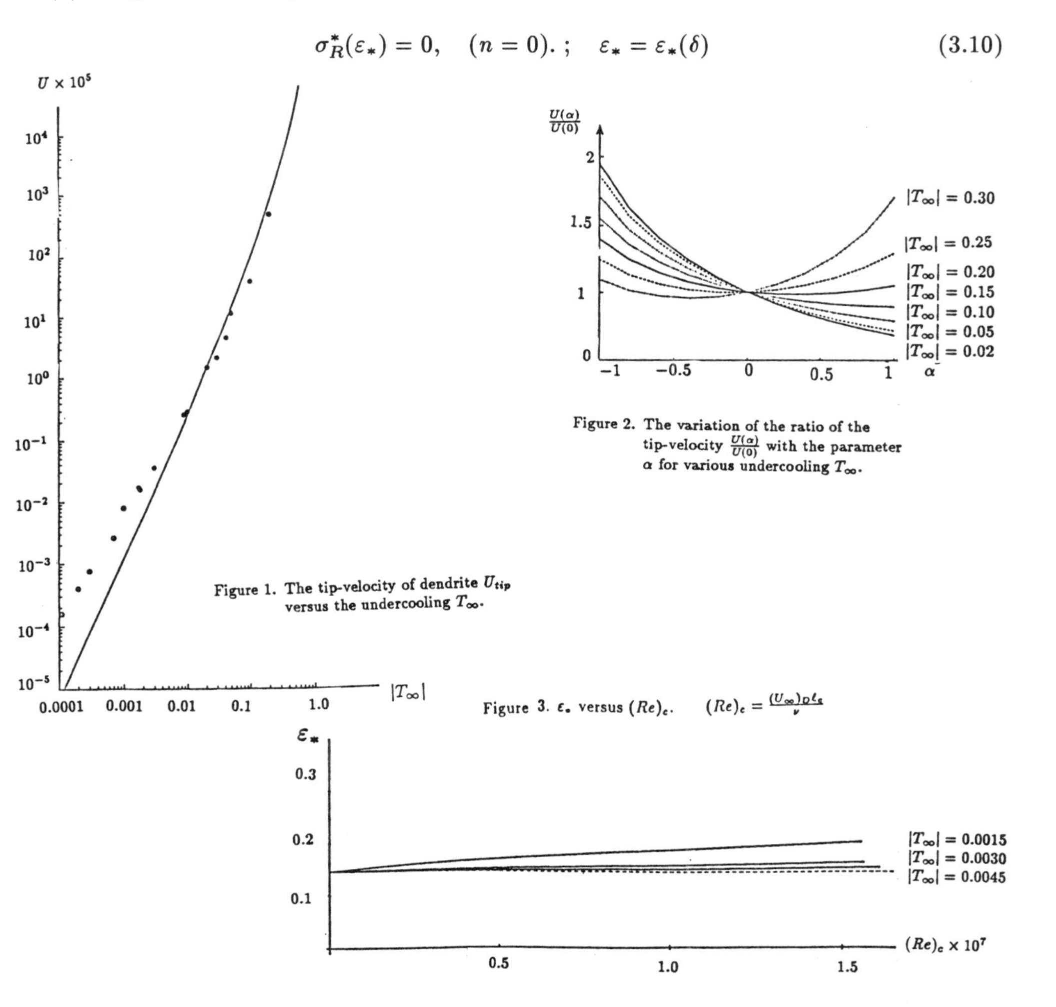

Figure 1. The tip-velocity of dendrite U_{tip} versus the undercooling T_∞.

Figure 2. The variation of the ratio of the tip-velocity $\frac{U(\alpha)}{U(0)}$ with the parameter α for various undercooling T_∞.

Figure 3. ε_* versus $(Re)_c$. $\quad (Re)_c = \frac{(U_\infty)_D \ell_c}{\nu}$

NONLINEAR DYNAMICS IN CELLULAR SOLIDIFICATION IN PRESENCE OF DEFECTS

H. JAMGOTCHIAN* (**), R. TRIVEDI* and B. BILLIA**
* *Ames Laboratory*
USDOE and Department of Materials Science and Engineering,
Iowa State University, Ames IA 50011, USA
** *Laboratoire de Physique Cristalline,*
Faculté des Sciences de St Jérôme, Case 151
Avenue Escadrille Normandie-Niemen, 13397 Marseille Cedex 13, France

ABSTRACT. Interaction of grain boundaries and acetone bubbles with the interface microstructure is analyzed during the directional solidification of thin samples of succinonitrile-acetone mixtures. At subboundaries, depending on the growth parameters, cells are associated in pairs to form doublets which may invade the adjacent grains. When the doublets are stable, the selection of cellular spacing is investigated through the time evolution of both the tip and groove spacing distributions. As acetone is volatile, bubbles are often nucleated which grow with an elongated shape, so that solid envelopes get attached to them to form duplex structures. In the beginning of the cellular range, finite amplitude cells are observed only a distance from the bubbles. At higher acetone concentration, highly correlated sidebranches appear and amplify on the duplex structure, which means that a duplex structure can undergo an oscillatory dendritic transition.

1. INTRODUCTION

The formation of one-dimensional cellular arrays over the entire solid-liquid interface is analyzed during the directional solidification of thin samples of succinonitrile-acetone mixtures in a Hele-Shaw cell. Imperfections are often present in the sample that dynamically interact with the solidification front. Two major defects are considered: grain boundaries in solid and acetone bubbles.

2. CELLULAR DOUBLET GROWTH

As the solid-liquid interface is cusped at a grain boundary (GB), the morphological instability begins at the GB locations (Fig.1) where two small asymmetric cells symmetrically develop on both sides of each GB, thus forming a distribution of doublets. While the amplitude of the doublets increases, the instability propagates laterally into the grains in the form of wave packets. Between two grain boundaries, new wave packets, that are rather regularly spaced, continuously appear until the invasion is complete.

S. H. Davis et al. (eds.), Interactive Dynamics of Convection and Solidification, 105–108.

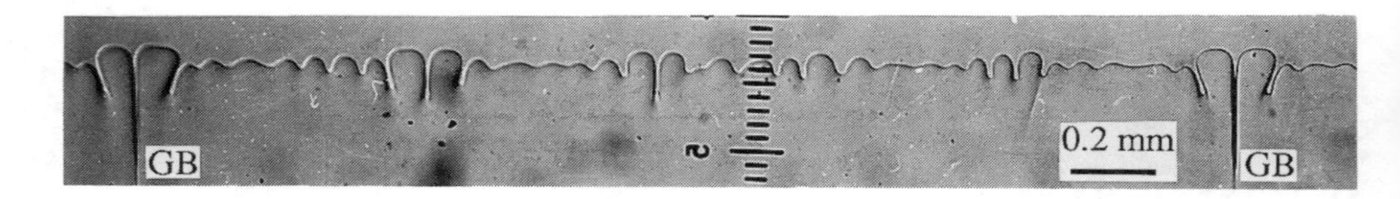

Fig. 1. Birth and lateral spreading of the doublet structure from grain boundaries. Succinonitrile - 0.5 wt % acetone, V = 1 μm / s, G = 30 K / cm.

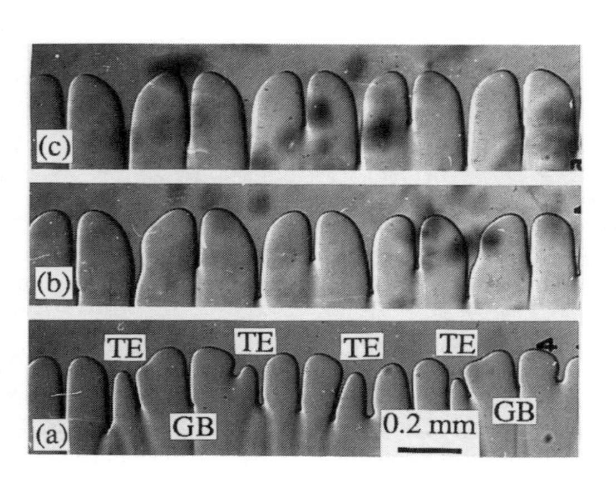

Fig. 2. Formation of an array of cellular doublets a) 40 mn, b) 83 mn and c) 140 mn of growth. Succinonitrile - 0.5 wt % acetone V = 0.75 μm / s, G = 28 K / cm.

The doublet microstructure, which may be an example of pattern selection due to the boundary conditions [1], may be stable or unstable. Figure 2 shows a time evolution of a cellular array in the range where doublets are stable. In the nonlinear enlargement of the doublets, the small cells in between are first eliminated by overgrowth or groove fainting (Fig.2a). Then, the doublets get more and more similar (Fig.2b) and the final microstructure of the solidification front is an array of cellular doublets (Fig.2c). In addition, the tip spacing λ_t at the advancing interface needs to be differentiated from the usual primary spacing λ_g, i.e. the distance between adjacent grooves or cellular spacing in the solid.

The selection of doublet microstructure can be investigated by following the time evolution of the various spacing distributions. Figure 3 shows the final histograms of the tip and groove spacings over the whole solid-liquid interface, for the case corresponding to Fig.2c. Although the pattern looks very regular, it follows from the λ_g distribution that selection of the cellular primary spacing is not sharp. The λ_t distribution, which is at the beginning fairly symmetric, ultimately exhibits two peaks. The highest peak at 110 μm, which is very narrow, corresponds to the intradoublet tip spacing. The other peak at about 200 μm, which comparatively looks flat and very broad, corresponds to the interdoublet tip spacing. It is worth noticing that the intradoublet tip spacing, which is an intrinsic characteristic of the doublets, is precisely defined whereas the interdoublet tip spacing, which characterizes the perfection of the doublet array after the dynamical elimination of cells, is not, which is the very cause of the dispersion of λ_g.

Under certain growth conditions, the cells constituting the doublets become too large to be stable to tip splitting, so that there is an early abortion of the process of formation of a

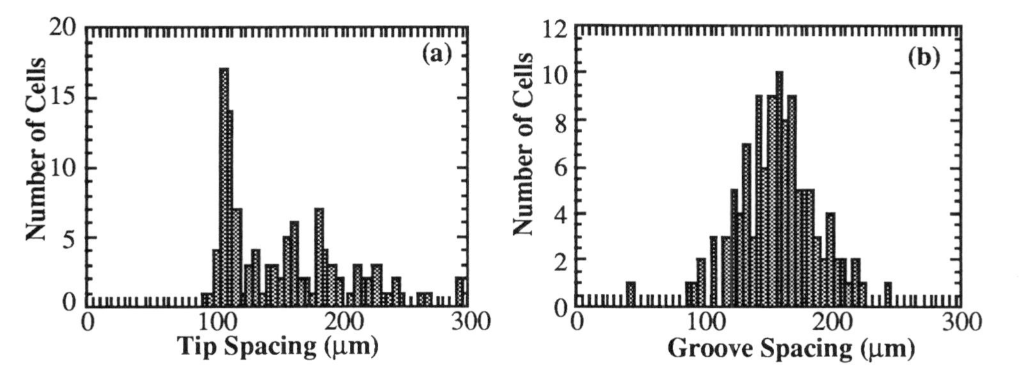

Fig. 3. a) Tip spacing and b) groove spacing distributions after 140 mn of growth. Succinonitrile - 0.5 wt % acetone, V = 0.75 μm / s, G = 28 K / cm.

doublet microstructure. From the observations reported here, and a series of experiment on succinonitrile-0.15 wt % acetone alloys, the stable range of doublet growth can be roughly estimated in between 1.5 and 3 times the critical velocity for morphological instability of a planar solidification front, i.e. about the minimum in the curve giving the variation of primary spacing with velocity. Besides, the formation of other multiplets is rather often observed, which is a function of the number of stable cells in the wave packets, triplets and quadruplets being the most common.

3. BUBBLE GROWTH AND DUPLEX STRUCTURES

As acetone is volatile, the nucleation of acetone bubbles is easy in the intercellular grooves [2]. Then, while growing, bubbles migrate towards the tips due to the disjoining force [3], where they get attached to the solidification front. There exits a range of solidification velocities, in which bubbles finally grow with an elongated body terminated by a spherical cap. Then, the growth velocity of a bubble is on the average equal to the velocity of the solid-liquid interface, but the tip position is function of the growth parameters and bubble size. A coupling occurs with the interface and duplex structures develop, upon the formation of solid sheaths around the elongated bubbles (white line in Fig.4), when there is a liquid groove in the transverse direction.

In the beginning of the cellular regime, at low acetone concentration and growth rate (Fig.4), finite amplitude cells are observed only at a distance from the bubble. Indeed, as the bubble absorbs acetone, it induces a lateral gradient of solute or, in other words, a ramp in the level of morphological instability which may result in a planar-cellular transition along the interface. It should be noticed that the bubble diameter varies almost periodically with a period of the order of 5 mn. Although in the literature, such low frequency variations are preferentially attributed to fluctuations of the temperature regulation, it presently cannot be excluded that they may be concentration driven.

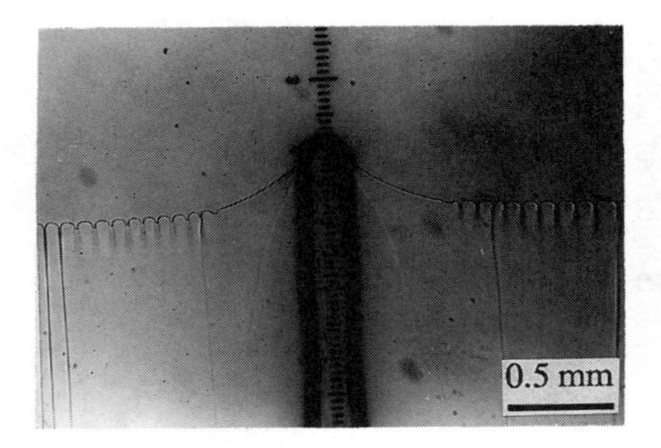

Fig. 4. Elimination of morphological instability in the neighborhood of an acetone bubble. Succinonitrile - 0.1 wt % acetone, $V = 2\ \mu m / s$, $G = 35\ K / cm$.

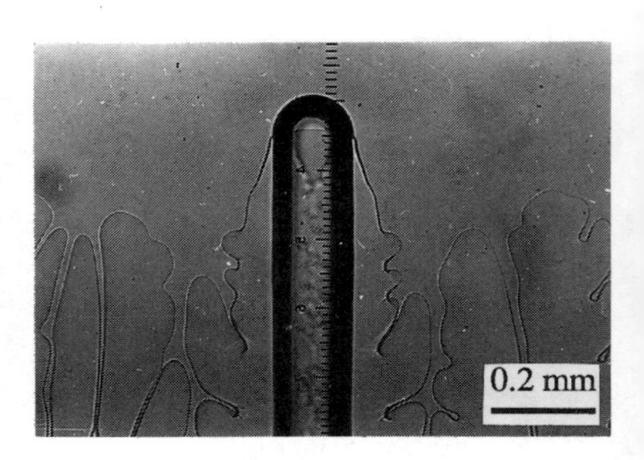

Fig. 5. Coherent sidebranching of a dendritic duplex structure. Succinonitrile - 0.5 wt % acetone, $V = 2\ \mu m / s$, $G = 30\ K / cm$.

When the acetone concentration is increased, the duplex structure gets slightly ahead of the average interface and sidebranches appear and amplify, that are highly correlated even far in the nonlinear range (Fig.5). A common period (about 1.5 mn) between the bubble protrusion and dendritic sidebranches strongly suggests that the solid cap of the duplex structure has localized resonant modes with the acetone bubble, which initiate the symmetric sidebranching. Several phenomena could be at the origin of these modes, among which an internal oscillator. Indeed, as an acetone bubble is a sink for acetone and the neighboring solidification front a source, it is tempting to consider that these antagonist effects will compete, eventually leading to the formation of an internal oscillator, with at least a proper frequency. Then, the resulting oscillations would be damped below the dendritic transition for that frequency, and amplified above. Obviously, complementary experimental work is required to discriminate between the possible mechanisms, if one is dominant, or determine their respective weights, in the reverse situation where external parametric forcing and internal mechanisms would combine.

5. ACKNOWLEDGEMENTS

This work has benefited from a CNRS-NSF post-doctorate Fellowship (H. Jamgotchian) and a NATO Research Grant (B. Billia).

6. REFERENCES

[1] Y. Pomeau and S. Zaleski, J. Phys. (France) 42 (1981) 515
[2] J.M. Laherrère, H. Savary, R. Mellet and J.C. Tolédano, Phys. Rev. 41A (1990) 1142
[3] A. A. Chernov and D. E. Temkin, in Curent Topics in Materials Science, Volume 2. E. Kaldis Ed. (North-Holland, Amsterdam, 1977) p.1

SURFACE KINETICS AND GROWTH MORPHOLOGIES OF NH_4Cl

S. KOSTIANOVSKI and S. G. LIPSON
Physics Department,
Technion- Israel Institute of Technology,
Haifa 32000, Israel.

Abstract. We first present some results of experiments performed on the morphology of dendritic crystals growing from a supersaturated solution during free fall. Then we show an quantitative investigation of the diffusion field around growing dendritic tips as observed in an interference microscope. The results give a picture of some aspects of the kinetic growth effects at the crystal-fluid interface.

In this contribution we shall present experiments on two different aspects of the dendritic growth of NH_4Cl grown from supersaturated solution in the temperature range $0-10°C$. Our previous experiments on this system (1,2) showed that growth at low temperatures, below a surface roughening temperature, is dominated by surface kinetic effects. We first demonstrate the development of growth forms similar to those of a snowflake (also determined by surface kinetics) when the right conditions are created. The we go on to present measurements of the solute field around a crystal growing under an interference microscope which highlight details of the boundary conditions at the interface.

Dendrites grown in free fall

The substrate or initial conditions can have great effect on the development or supression of certain morphologies of the crystals, as we recently demonstrated (2). We have also tried to grow the crystals with the minimum constraints possible. Ideally, this would mean a homogeneously nucleated crystal growing by diffusion in an infinite medium in the absence of gravity. The closest approximation we could achieve in the laboratory was to allow the crystal to grow while falling freely through the solution (snowflake conditions). The field it experiences is then not diffusive but

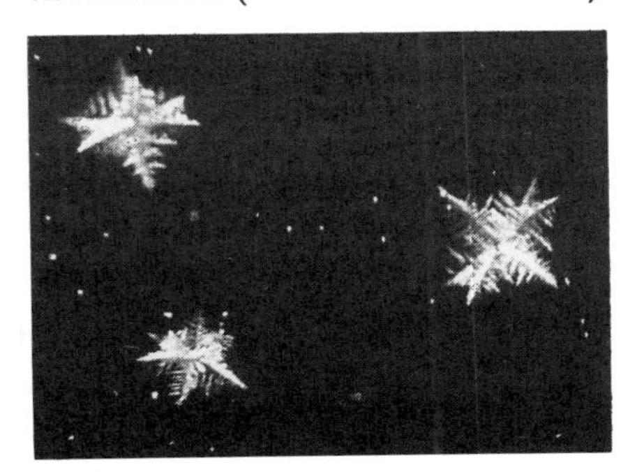

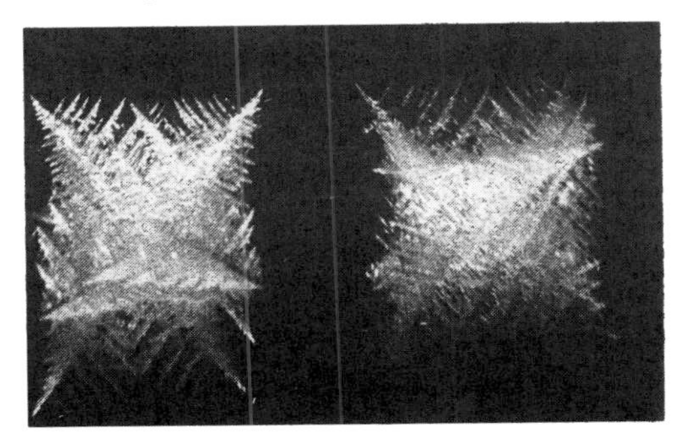

Figure 1. Dendritic crystals grown in free fall: (a) growth axis (100)- octohedral external shape; (b) growth axis (111)- cubic external shape.

S. H. Davis et al. (eds.), Interactive Dynamics of Convection and Solidification, 109–111.

convective. Interaction with the fluid causes the crystal to rotate and thus maintains a field which is spherically symmetrical. Two examples are shown in fig. 1; one has a preferred growth axis along (100) and the other (111). Each gave a dendritic crystal within a well-defined regular polyhedral envelope. The existence of a regular polyhedral envelope suggests that the dendrite tips grow at constant velocity, despite multiple branching, and are little affected by changes in the supersaturation, which must be maximum at the corners. It is easy to see that multiple branching in the forward growth direction along a single preferred orientation at constant velocity gives rise to an envelope of the observed type. This condition arises at large supersaturation if the crystal grows by homogeneous two-dimensional surface nucleation.

Surface kinetic effects.

The second investigation involves the details of growth kinetics of dendritic crystals under well-defined diffusion-limited conditions. The assumption of liquid-solid or crystal- solution local equilibrium at the surface, modified only by the Gibbs- Thomson curvature correction, was questioned by Ben-Jacob et al (3) who proposed that at finite growth velocity there should be an additional supercooling proportional to the local growth velocity. They called this the "kinetic term". Further work showed the anisotropy of both the surface-tension and the kinetic terms to be critical in promoting morphological selection by means of a solvability criterion (4).

We have used an optical interference technique (1) to create a map of the solution concentration around the crystal as it grows. Full details of the optical analysis and tomographic method used today will shortly be published (5). Here we present a sample of the type of results which are obtained. Fig. 2 shows an interferogram and three contour maps of concentration at successive times in the plane of the axis of the growing dendritic tip, assumed for convenience to have axial symmetry. The separation of the contours is 0.1% in concentration. The first thing to notice is the increasing degree of supersaturation at the interface as the tip of the crystal is approached. This is a clear indication of a kinetic term in the growth equation. Then, changes in the solute field can be discerned even though the shape appears to be translationally invariant. A possible interpretation of the effects seen here is an incipient tip-splitting, which does not develop because of anisotropy. The accepted description of tip-splitting is that the high curvature at the tip of the crystal results in a local reduction of the supercooling. This encourages growth at neighbouring points where the curvature is less. High supercooling and anisotropy suppress this mechanism. Fig. 3 shows similar results for crystals grown with small $CuSO_4$ impurity, which show faceted tips and presumably have more complicated surface kinetics.

We have suggested that observed variations in the tip velocity (1,6) could come about in a crystal growing below its roughening temperature by the surface oscillating between the atomically smooth faceted structure and a kinetically roughened one. It is possible that the observed changes in the solute field arise from the same mechanism, and the signs of an instability are visible earlier in the solute field than in the crystal shape.

This work was supported by the Fund for Basic Research administered by the Israel Academy for Arts and Sciences.

REFERENCES.

1. E. Raz, S. G. Lipson and E. Polturak, Phys. Rev., **A40,** 1088 (1989)

2. E. Raz, S. G. Lipson and E. Ben-Jacob, J. Cryst. Growth, **108**, 637 (1990)
3. E. Ben-Jacob et al, Phys. Rev. Lett., **53**, 2210 (1984)
4. D. A. Kessler J. Kopelik and H. Levine, Advan. Phys., **37**, 255 (1989)
5. S. Kostianovski and S. G. Lipson, Applied Optics, to be published.
6. H. Honjo, S. Ohta and Y. Sawada, Phys. Rev. Lett., **55**, 841 (1985).

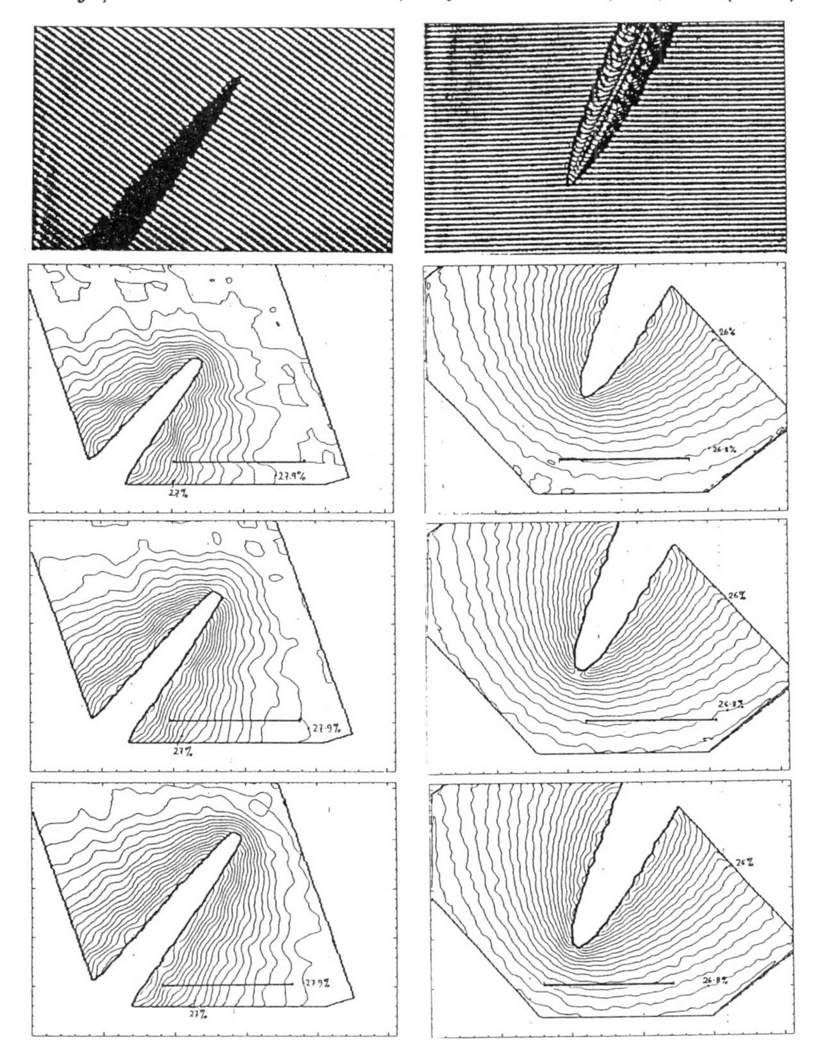

Figure 2 (left). Solute field around growing dendritic tips of pure NH_4Cl, at time intervals of 0.08 sec. The tip velocity is 140 μm/sec. The scale bar is 50μm.

Figure 3 (right). Solute field around growing dendritic tips of NH_4Cl with 0.1% $CuSO_4$, at time intervals of 0.16 sec. The tip velocity is 41 μm/sec.

THE DYNAMICS OF MUSHY LAYERS

M.G. WORSTER
Department of Engineering Sciences & Applied Mathematics
and Department of Chemical Engineering
Northwestern University
Evanston, IL 60208 USA

ABSTRACT. The development of mathematical models describing mushy zones is reviewed. Particular attention is paid to the transport of mass, heat and species in these reacting, two-phase media. Dynamical interactions between solidification in mushy regions and three different types of convection are analyzed: convection due to shrinkage or expansion upon change of phase; and buoyancy driven convection driven either by thermal gradients or by solutal gradients. Directions for future reseach into the dynamics of mushy regions are suggested.

1. Introduction

Regions of intimately coexisting liquid and solid phases, called "mushy regions" are ubiquitous during the solidification of multi-component systems. They can be viewed as the consequence of morphological instabilities of would-be, planar solid–liquid phase boundaries (Mullins & Sekerka, 1964), and serve to reduce or eliminate regions of constitutional supercooling in the system (Worster, 1986; Fowler, 1987) that arise due to the slow diffusion of chemical species relative to heat. The micro-scale morphology of mushy layers varies considerably with the chemical system being solidified (figure 1) but all are characterized by the length scale of internal phase boundaries being very much smaller than the macroscopic dimensions of the layer. This is a key feature upon which mathematical models of mushy regions are based.

It is of great importance to understand the interactions between solidification and flow of the melt, since fluid flow transports heat, which influences the rate of solidification, and transports solute, which causes segregation of the constituents of the melt. A wide range of striking fluid-mechanical effects during solidification are discussed and beautifully illustrated in a review article by Huppert (1990). Here we shall focus specifically on interactions between fluid flow and mushy regions and consider the effects of flow caused by three different physical mechanisms: the flow of interdendritic melt due to the expansion or shrinkage that occurs as one phase changes to another; thermal convection in the region of melt exterior to the mushy layer driven by undercooling at the mush–liqiud interface; and compositional convection driven by the rejection of one component of the alloy during solidification. In each case, we shall see that the flow is an inevitable consequence of

S. H. Davis et al. (eds.), Interactive Dynamics of Convection and Solidification, 113–138.

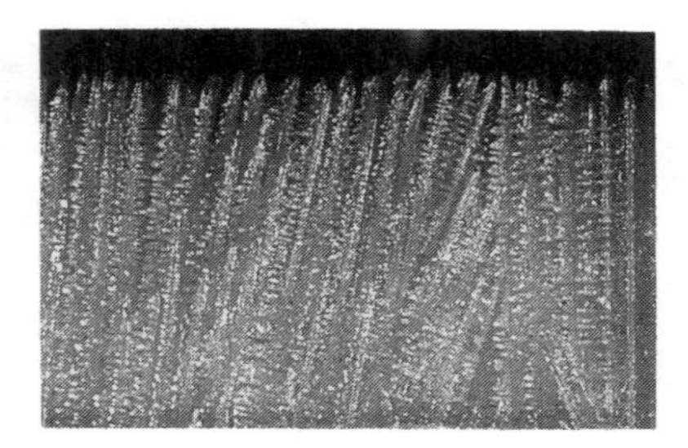

Figure 1. (a) Side view of the dendrites in a mushy layer of ammonium-chloride crystals. (b) End-on view of ice platelets in a mushy layer grown from an aqueous solution. Although different chemical systems have quite different micro-scale morphologies, the crystals in each case are much smaller than the overall dimensions of the mushy layer.

the process of solidification and examine the effect of the flow on macrosegregation of the alloy. Compositional buoyancy typically dominates thermal buoyancy in multi-component systems, though either can be the primary cause of convection depending on the geometry of the mould and the position of its cooled boundaries. On Earth, gravitational convection, whether compositionally or thermally driven, has much larger effects on a casting than does the convection due to solidification shrinkage. However, in a micro-gravitational environment, solidification shrinkage can play the dominant rôle.

Two examples of experimental castings are shown in figure 2. In each case, one of the horizontal boundaries of the mould is cooled to below the eutectic temperature, and a composite solid layer forms, separated from the melt by a mushy layer. The style of convection is quite different in the two cases, and results in the final castings having different textures and compositional variations. It is these sorts of variations that theoretical models aim to explain.

2. Mathematical Modelling of Mushy Regions

Theoretical and numerical models of mushy regions typically seek to provide descriptions of the evolving two-phase media on the macro scale, much larger than the mean spacing between solid particles. Seemingly different models of mushy regions have been formulated independently by metallurgists, engineers and applied mathematicians. On closer examination, one appreciates that the differences are related more to language and terminology than to physical content. For many years, metallurgists have made use of the Scheil equation and the lever rule (Flemings, 1974; Kurz & Fisher, 1986) to deduce microsegregation from experimental measurements of the temperature of a casting. Macrosegregation was similarly estimated from the Local Solute Redistribution Equation (Flemings & Nereo, 1967). These equations allow the evaluation of various properties of castings given measured values of the evolving temperature field. The prediction of the evolution of a casting

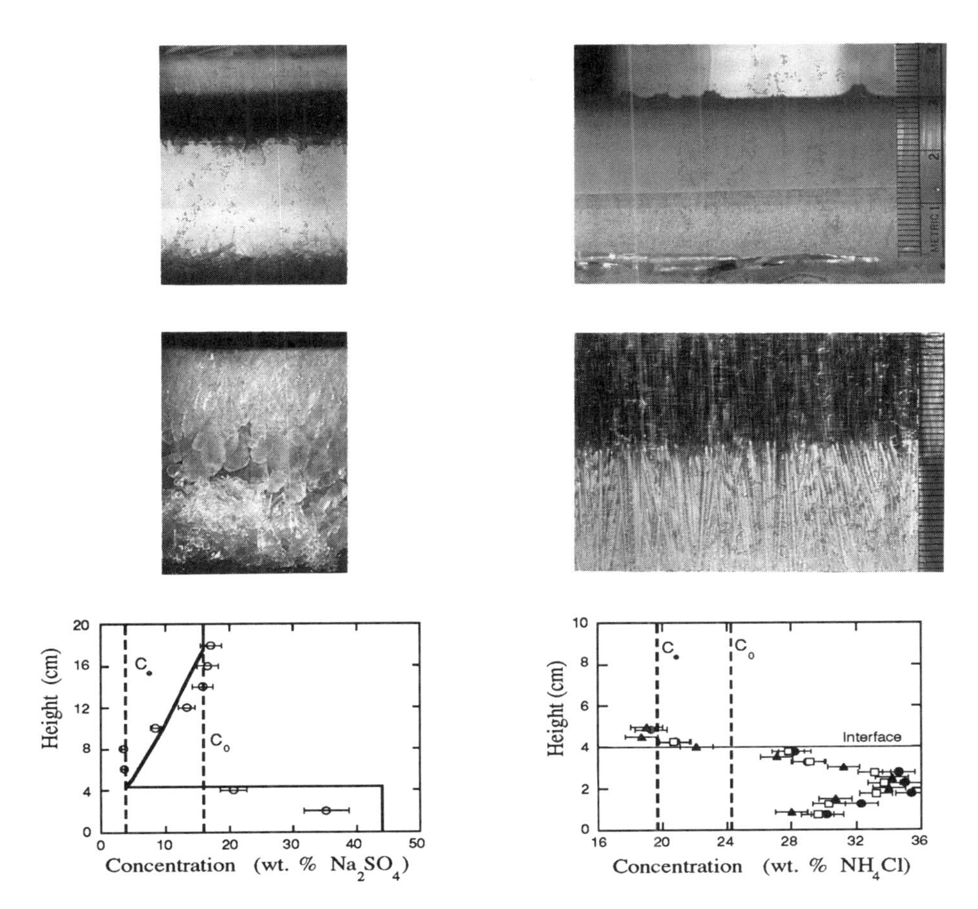

Figure 2. Characteristically different forms of macrosegregation are generated by different types of convection. a) Experiment in which an aqueous solution of sodium sulphate, of initial composition C_0 and eutectic composition C_e, was completely solidified by cooling from above (Kerr *et al.*, 1990c). From the top down, there is a layer of composite, eutectic solid, a mushy layer of sodium-sulphate crystals, a region of melt that is convecting, and a layer of equiaxed crystals on the floor of the tank. b) The final solidified block, showing a distinct change of texture at the columnar-equiaxed transition. c) The symbols show the compositional variation measured in the final solid block. The solid line is the prediction of a mathematical model (Kerr *et al.*, 1990c). d) Experiment in which an aqueous solution of ammonium chloride was completely solidified by cooling from below. A composite eutectic solid layer underlies a mushy layer of ammonium-chloride crystals. The interstitial fluid in the mushy layer is convecting and escapes from the layer through chimneys. e) Close-up of the final solidified block. The structure is all columnar but there is a textural change when ice succeeds ammonium chloride as the primary solidifying phase. f) The compositional variation measured in the final casting. The different symbols correspond to three different corings of the same casting.

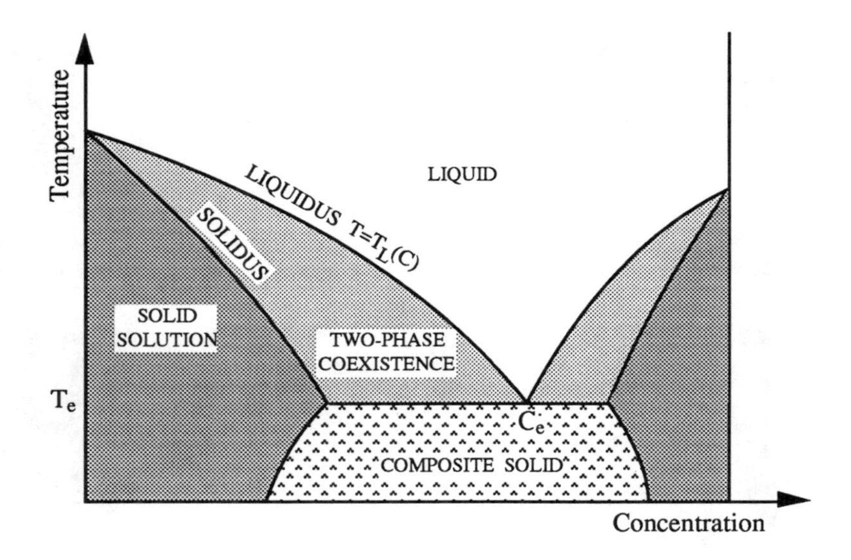

Figure 3. The equilibrium phase diagram of a simple binary alloy. The shaded regions indicate what phases are in thermodynamic equilibrium in a sample of given bulk composition and uniform temperature. If the system remains at equilibrium during cooling then liquid of a particular concentration begins to solidify once the temperature falls below the liquidus curve and is completely solid once the temperature falls below the solidus curve or below the eutectic temperature T_e. When the temperature T and bulk composition are between the liquidus and solidus curves then liquid of the liquidus composition $C_L(T)$ is in equilibrium with solid of the solidus composition given by $kC_L(T)$, which defines the local segregation coefficient $k(C)$.

from the external parameters of the process required additional equations describing fluid flow (Mehrabian *et al.*, 1970) and heat transfer (Fujii *et al.*, 1979). In more recent years, as computers have become more powerful, fully coupled equations describing the transport of mass, heat, momentum and species have been developed and utilized in predictive models of solidifying systems (Szekely & Jassal, 1978; Bennon & Incropera, 1987; Thompson & Szekely, 1988; and see articles by Amberg, Beckermann and Voller in these proceedings, † and references therein). A more philosophical approach was taken by Hills *et al.* (1983) who formulated a very general set of governing equations, based on diffusive mixture theory, that are consistent with fundamental thermodynamical principles. Somewhat different again is the reductionist approach adopted by Huppert & Worster (1985) and Worster (1986), who formulated very simple models that yet contain sufficient information to isolate and explain particular features of the solidification process, and enable quantitative comparisons with laboratory experiments.

† Articles appearing in these proceedings will henceforward be indicated simply by a raised dagger †.

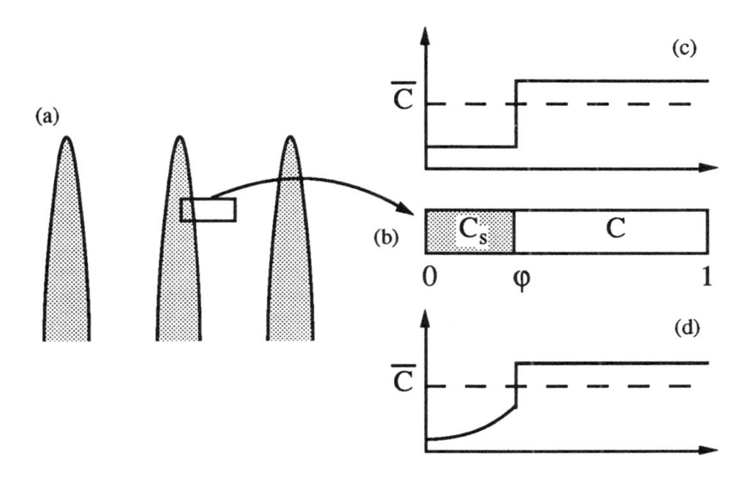

Figure 4. (a) A simplified schematic of a dendritic region. The inset (b) shows the local one-dimensional model used to derive the Scheil equation. (c) The variation of concentration through the solid and liquid regions when there is complete back diffusion. (d) Concentration variation when there is no back diffusion. In each case, the dashed line $\overline{C}$ represents the bulk composition of the element.

2.1 THE SCHEIL EQUATION AND THE LEVER RULE

The mushy layer is a two-phase medium whose properties are determined in large part by the local volume fraction of solid $\phi(\mathbf{x}, t)$, which must be determined as part of the solution of the mathematical model. In most theories to date, the growth or dissolution of solid in the interior of mushy regions is assumed to occur instantaneously, so that the layer is everywhere in local thermodynamic equilibrium. Therefore, the local concentration of the interstitial liquid C and the local temperature T are related to one another through the liquidus

$$T = T_L(C), \tag{2.1}$$

given by the equilibrium phase diagram (figure 3).

The Scheil equation and its close relative, the lever rule, give relationships between ϕ and C. Alternatively, once the equilibrium assumption (2.1) is invoked, these equations can be used to infer the solid fraction once the local temperature is known. These equations can be derived (Kurz & Fisher, 1986) from a local one-dimensional problem (figure 4b) identified from a simplified picture of a dendritic region (figure 4a). The lever rule follows directly from conservation of solute under the assumption of complete local equilibrium in which the solid phase has uniform composition $C_s = kC$, where $k(C)$ is the local value of the segregation coefficient (figure 4c). A simple mass balance, ignoring any change of specific volume on change of phase, gives

$$\phi = \frac{1}{p}\left(1 - \frac{\overline{C}}{C}\right), \qquad \text{where} \qquad p = 1 - k, \tag{2.2}$$

and $\overline{C}$ is the bulk composition of the one-dimensional element (figure 4b).

The lever rule (2.2) is only appropriate if the timescale for diffusion of solute in the solid phase (so-called 'back diffusion') is rapid compared with the timescale for the macroscopic evolution of the mushy region. In fact, a more reasonable approximation in most circumstances is that there is negligible back diffusion, in which case the composition of the solid phase varies across each dendrite; $C_s = C_s(\phi')$, $0 \leq \phi' \leq \phi$. In this case, the composition of the solid phase varies with position, and equilibrium is only imposed at the local solid–liquid interface (figure 4d). Two additional assumptions are made in the derivation of the Scheil equation, namely that the segregation coefficient k is constant, and that the local bulk composition of the element $\overline{C}$ does not change during the evolution of the mushy region. Given these assumptions, it is straightforward to derive the Scheil equation

$$\phi = 1 - \left(\frac{\overline{C}}{C}\right)^{1/p}. \tag{2.3}$$

from conservation of mass of solute.

It has been pointed out (Kurz & Fisher, 1986) that, according to the Scheil equation, $C \to \infty$ as $\phi \to 1$, and that this unphysical singularity is avoided in practice by the process of back diffusion. However, the Scheil equation should rather be interpreted as saying that the solid fraction is bounded away from unity in practical situations, in which the temperature, and hence the local concentration, is always finite, so that the singularity is never encountered.

Neither the lever rule nor the Scheil equation take account of any global redistribution of solute and they give poor approximations whenever flow of the interstitial fluid is significant. The lever rule can be used (provided the assumption of infinite back diffusion is appropriate) if one can determine how the bulk composition is altered by the flow. However, the Scheil equation is never valid once the local bulk composition varies with time. Another way of understanding this is to realise that, if there is no back diffusion or only finite back diffusion, the solid fraction depends not only on the instantaneous value of the bulk composition but on the entire history of its evolution.

2.2 THE MUSHY LAYER AS A CONTINUUM

The need to predict the macroscopic redistribution of solute within the mushy layer requires the development of appropriate transport equations (Flemings & Nereo, 1967; Flemings, 1981; Hills *et al.*, 1983; Worster, 1986; Bennon & Incropera, 1987). Such equations are formulated from fundamental conservation laws applied to 'infinitesimal' control volumes that nevertheless are considered to encompass representative samples of both phases. In this sense, the mushy region is considered as a new continuum phase. Therefore, the resulting description cannot resolve any details on the scale of the spacing between dendrites. Put more positively, the macroscopic predictions of the models that emerge are independent of the structure of the micro scale and can therefore be applied in a wide range of circumstances. For the sake of simplicity, we consider only cases when the solid phase is immobile. The principal assumptions of the model are that, within each infinitesimal control volume, the temperature T is uniform across the solid and liquid phases, the composition of the liquid phase C is uniform and there is no back diffusion of solute within the solid phase.

The properties of the mushy region are local mean properties. For example, the local density is

$$\overline{\rho} = \phi\rho_s + (1-\phi)\rho_l, \tag{2.4}$$

where ρ_s and ρ_l are the densities of the liquid and solid phases, which are assumed to be constant. Mass is only transported by advection of the interstitial fluid, so conservation of mass is expressed by

$$\frac{\partial\overline{\rho}}{\partial t} + \nabla\cdot(\rho_l\mathbf{U}) = 0, \tag{2.5}$$

where $\mathbf{U}$ is the flow rate of the interstitial fluid per unit area. From equation (2.5), it can be determined that the veloctiy field generally has a non-zero divergence given by

$$\nabla\cdot\mathbf{U} = (1-r)\frac{\partial\phi}{\partial t} \tag{2.6}$$

where $r = \rho_s/\rho_l$. Such a velocity can be generated solely in response to the expansion or contraction that occurs due to the difference in density between the liquid and solid phases, and does not require the exertion of any external force such as gravity. The transport caused by this interstitial flow strictly invalidates the Scheil equation and the lever rule, though these often provide adequate approximations in practical situations, especially if the density ratio r is close to unity.

With the approximations stated above, the local mean concentration of solute $\overline{C}$ in the mushy region is given by

$$\overline{\rho C} = \rho_s\int_0^{\phi} C_s(\phi')\,d\phi' + \rho_l(1-\phi)C. \tag{2.7}$$

Conservation of solute requires that

$$\frac{\partial}{\partial t}\overline{\rho C} + \nabla\cdot(\rho_l C\mathbf{U}) = \rho_l\nabla\cdot(\overline{D}\nabla C), \tag{2.8}$$

where $\overline{D}$ is the local mean solutal diffusivity of the mushy layer. Equation (2.8) can be expanded and combined with equation (2.6) to give

$$(1-\phi)\frac{\partial C}{\partial t} + \mathbf{U}\cdot\nabla C = \nabla\cdot(\overline{D}\nabla C) + r(1-k)C\frac{\partial\phi}{\partial t}. \tag{2.9}$$

This is a diffusion-advection equation with a source term related to the rate of expulsion of solvent as the local solid fraction ϕ increases.

We shall see later that it is appropriate to neglect the diffusion of solute within the mushy layer provided that the ratio of the solutal diffusivity D to the thermal diffusivity κ is small. With this approximation, and incorporating the equilibrium condition (2.1), equation (2.9) can be written as

$$\frac{\partial\chi}{\partial C} = -\frac{1}{r(1-k)}\left[1 + \frac{\mathbf{U}\cdot\nabla T}{\dfrac{\partial T}{\partial t}}\right]\frac{\chi}{C}, \tag{2.10}$$

where $\chi = 1-\phi$ is the local volume fraction of liquid. This is the Local Solute Redistribution Equation (LSRE), first derived by Flemings & Nereo (1967). They used the equation to determine the macrosegregation in a casting from estimates of the flow field $\mathbf{U}$ and measurements of the temperature field T.

Note that, in the special case of no flow of the interstitial fluid, the LSRE (2.10) can readily be integrated to recover the Scheil equation (2.3).

An equation governing the temperature field in the mushy region is most readily and systematically derived in terms of the local enthalpy $\overline{H}$, where

$$\overline{\rho}\overline{H} = \phi\rho_s H_s + (1-\phi)\rho_l H_l, \tag{2.11}$$

and H_s and H_l are the local enthalpies per unit mass of the solid and liquid phases respectively. In terms of enthalpy, the equation expressing conservation of heat is identical in form to the equation for conservation of solute (2.8) and is given by

$$\frac{\partial}{\partial t}\overline{\rho}\overline{H} + \nabla \cdot (\rho_l H_l \mathbf{U}) = \nabla \cdot (\overline{k}\nabla T), \tag{2.12}$$

where $\overline{k}$ is the mean thermal conductivity of the mushy region. Expanding equation (2.12) using equations (2.6) and (2.11) yields the equation governing the temperature field,

$$\overline{c}\frac{\partial T}{\partial t} + c_l \mathbf{U} \cdot \nabla T = \nabla \cdot (\overline{k}\nabla T) + \rho_s L \frac{\partial \phi}{\partial t}, \tag{2.13}$$

where

$$\overline{c} = \phi c_s + (1-\phi)c_l,$$

and

$$c_{s,l} = \rho_{s,l}\frac{dH_{s,l}}{dT}$$

are the specific heat capacities per unit volume of the solid and liquid phases. The latent heat of solidification per unit mass is defined by $L = H_l - H_s$. Note that in general, L is a function of both temperature and concentration. However, if the phase change always occurs at the equilibrium temperature, given by equation (2.1), then L can be viewed as a function solely of the concentration C of the interstitial liquid.

The three equations (2.6), (2.9) and (2.13) are all coupled through ϕ_t, the rate of change of solid fraction. To complete the model, an evolution equation for ϕ is required. Alternatively, ϕ can be determined implicitly by invoking an assumption of instantaneous reaction, which leads to the application of the equilibrium liquidus relationship (2.1) throughout the mushy region. Thus the temperature T and concentration C are essentially the same variable, mathematically speaking, and the diffusion term in equation (2.9) can safely be neglected, without causing a singular perturbation, when the solutal diffusivity is much smaller than the thermal diffusivity, as is usually the case.

When flow is driven by external agents, such as a gravitational field, or when shrinkage occurs in more than one dimension, a dynamical equation is required for the velocity field in addition to the kinematic equation (2.6). In the fully liquid region, the appropriate equation is the Navier-Stokes equation for viscous fluid flow. In the mushy region, there is considerable resistance to flow as the fluid passes between the dendrites. It seems appropriate to model the mushy region as a porous medium, the simplest description of which is given by Darcy's equation

$$\mathbf{U} = \Pi \left[(\rho_l - \rho_0)\mathbf{g} - \nabla p \right], \tag{2.14}$$

where ρ_0 is a reference value of the fluid density, p is the hydrodynamic pressure of the interstitial fluid, and Π is the permeability of the medium. Many investigators, particularly those conducting numerical simulations, have replaced equation (2.14) with hybrids of Darcy's equation and the Navier-Stokes equation (Bennon & Incropera, 1987; Voller *et al.*, 1989; Nandapurkar *et al.*, 1989). This has the advantage of allowing numerical solution of the governing equations on a single computational domain. However, it introduces additional physical parameters to the description of the system that must be estimated before realistic computations can be made.

The transport properties of the mushy region $\overline{D}$, $\overline{k}$, and Π are all functions of the local solid fraction ϕ and of the microscopic morphology of the medium (Beran, 1968). For practical purposes, it has been found that simple volume-fraction weighted averages $\overline{D} = (1-\phi)D_l$ and $\overline{k} = (1-\phi)k_l + \phi k_s$ are adequate to describe the mean solutal diffusivity and thermal conductivity. No experimental determination has yet been made of the appropriate form of the permeability of a mushy region, though a variety of expressions have been used in numerical and analytical calculations and in the interpretation of experimental results. The most common have been the Kozeny-Carmen relationship $\Pi = (1-\phi)^3/\phi$ (Bennon & Incropera, 1987; Chen & Chen, 1991; Tait & Jaupart, 1992) and simple power laws such as $\Pi = (1-\phi)^2$ (Roberts & Loper, 1983; Fowler, 1985) and $\Pi = (1-\phi)^3$ (Worster, 1992).

2.3 INTERFACIAL CONDITIONS

When solving a problem involving solidification, there are prescribed boundary conditions applied at the surfaces of the mould. In addition, there are free internal interfaces between solid and liquid regions or between mushy and liquid regions, for example. The motion of these interfaces is partly determined by conservation laws that can be derived directly from the governing equations for mushy regions. Equations (2.6), (2.9) and (2.13) can be integrated over a small volume spanning the interface, in a frame of reference moving with the normal velocity of the interface V, to yield

$$[\mathbf{n}\cdot\mathbf{U}] = -(1-r)V[\phi], \tag{2.15}$$

$$r(1-k)C[\phi]V = \left[\overline{D}\,\mathbf{n}\cdot\nabla C\right] \tag{2.16}$$

and

$$\rho_s L[\phi]V = \left[\overline{k}\,\mathbf{n}\cdot\nabla T\right], \tag{2.17}$$

where [ ] denotes the change in the enclosed quantity across the interface. Conditions (2.16) and (2.17) are derived under the assumption that the temperature and the composition of the liquid phase are continuous across a mush–liquid interface, and can be used at a solid–liquid interface with C interpreted as the concentration in the liquid region at the interface. However, there may be discontinuities of the solid fraction across the interface.

If the jump in ϕ is prescribed, for example $[\phi] = 1$ across a solid–liquid interface, then conditions (2.15)–(2.17) are sufficient to determine the evolution of the interface, in the absence of any free energy associated with the curvature of the interface. In the case of an interface between a mushy layer and either a completely solid region or a completely liquid region, another equation is required in order to determine $[\phi]$. Since the governing equations are only first-order for ϕ, only one jump condition can be applied in a given problem. Some authors have imposed continuity of ϕ simultaneously at the mush–liquid interface and at the solid–mush interface, at the expense of one of the conservation laws (2.15)–(2.17). Another common practice is to impose continuity of ϕ at mush–liquid interfaces, though there seems no good reason for doing this *a priori*. An alternative condition, suggested by Worster (1986), is the "marginal equilbrium condition" that the normal temperature derivative at the mush–liquid interface be equal to the normal derivative of the local liquidus temperature,

$$[\mathbf{n} \cdot \nabla T] = [\mathbf{n} \cdot \nabla T_L(C)] \equiv T_L'(C)\,[\mathbf{n} \cdot \nabla C]\,. \tag{2.18}$$

Worster (1986) shows that this condition sometimes leads to $[\phi]$ being non-zero and that in such cases setting $[\phi] = 0$ as the interfacial condition instead of (2.18) renders the governing equations insoluble. However, under most common operating conditions, equation (2.18) coupled with equations (2.16) and (2.17) implies that, in fact, $[\phi] = 0$ at advancing mush–liquid interfaces. Indeed, equation (2.18) shows that the temperature and concentration gradients have equal orders of magnitude in the limit as $D/\kappa \to 0$, where κ is the thermal diffusivity, so the right-hand side of equation (2.16) is negligible in this limit. This same equation thus shows that $[\phi] = 0$ when the diffusivity ratio is small, while equation (2.17) then implies that the temperature gradient is continuous across the interface.

An alternative approach to tracking the interfaces between the phases is to solve the governing equations on a single computational domain. This method, which is used mainly for numerical calculations, obviates the need for interfacial conditions. In such calculations, it is usual for the enthalpy to be used as a dependent variable in place of the temperature and for the position of the interfaces to be found *a posteriori* from calculations of the local solid fraction.

3. Phase-Change Convection (Solidification Shrinkage)

One of the purposes of the mathematical model presented in section 2 is to allow quantitative predictions to be made of macrosegregation within a casting. Such undesirable separation of the constituents of an initially uniform melt during casting is effected by various convective processes within the melt. It is sometimes possible to reduce or even to eliminate convection due to buoyancy forces by cooling from below, provided that the primary solidifying phase leaves behind a dense residual (Huppert & Worster, 1985). Alternatively, the alloy could be cast in a micro-gravitational environment such as inside an orbiting space station. However, none of these contingencies will eliminate the flow of melt due to the expansion or contraction that occurs as a result of the solid and liquid phases having different specific volumes.

It has long been realized that solidification shrinkage is an important mechanism causing macrosegregation in both metallurgical systems (Flemings & Nereo, 1967, 1968) and in geological systems such as magma chambers (Petersen, 1987). In addition there have been quantitative estimates made of the extent of macrosegregation using modifications of the Scheil equation coupled with estimates of the local temperature field during solidification (Flemings & Nereo, 1968; Mehrabian *et al.*, 1970). More accurate calculations can be made by solving the full set of coupled transport equations given in the previous section. In complicated mould geometries, numerical solution of the equations must be sought (Beckermann†), but if the mould is cooled from a single planar boundary then one-dimensional, similarity solutions can be found (Vas'kin, 1986; Chiareli & Worster, 1992). These solutions, which serve to illustrate the dependence of the degree of macrosegregation on the external control parameters, will be described briefly here.

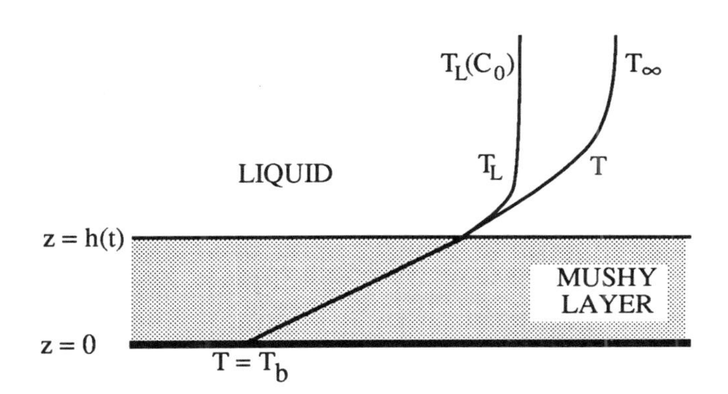

Figure 5. A schematic diagram of a mushy layer growing from a cooled plane boundary when the temperature of the boundary T_b is higher than the eutectic temperature. The temperature T and the local liquidus temperature $T_L(C)$ are illustrated. Far from the cooled boundary, the temperature is T_∞ and the concentration is C_0.

The one-dimensional geometry to be analyzed is illustrated in figure 5. A mushy layer grows in the positive z-direction and has depth $h(t)$ after a time t has elapsed. The boundary $z = 0$ is maintained at the fixed temperature T_b, while the melt far from the interface has temperature T_∞ and concentration C_0. The governing equations and interfacial conditions presented in section 2 admit a similarity solution, in which the dependent variables are functions of the single variable

$$\eta = \frac{z}{2\sqrt{\kappa_l t}}, \tag{3.1}$$

where κ_l is the thermal diffusivity of the liquid phase, and the height of the mush–liquid interface is given by

$$h(t) = 2\lambda\sqrt{\kappa_l t}, \tag{3.2}$$

where λ is a constant to be found as part of the solution. With this transformation of variables, the governing equations reduce to a set of ordinary-differential equations for $T(\eta)$, $C(\eta)$, $\mathbf{U}(\eta)$ and $\phi(\eta)$. In figure 6, the bulk composition $\overline{C}$ at the base of the mushy layer ($\eta = 0$) is plotted as a function of the density ratio $r = \rho_s/\rho_l$. The difference between this bulk composition and the initial concentration of the melt gives a measure of the degree of macrosegregation that has taken place during solidification. It can be seen that the bulk concentration of the casting increases with the density ratio, as shrinkage causes advection of solute into the mushy layer.

In addition to causing macrosegregation, the flow of interstitial fluid alters the solid fraction of the mushy layer and hence its permeability to flow generated by external forces. Because of the importance of the permeability in determing the strength of convective flow in mushy layers (see section 5) there has been interest recently in measuring the porosity of mushy layers (Chen & Chen, 1991; Shirtcliffe *et al.*, 1991). The latter authors measured the porosity in a system in which ice was solidified from an aqueous solution of sodium nitrate, and compared their measurements with analytical predictions made by Worster (1986). These predictions, in which the expansion during the solidification of ice is ignored, correspond approximately to what would be determined from the Scheil equation. If one takes into account the redistribution of solute caused by expansion then one obtains much better theoretical agreement with the experimental results (figure 7, Chiareli & Worster, 1992). We see from figure 7 that ignoring expansion causes a 10% error in the prediction of the solid fraction. More importantly, the error in the prediction of the porosity (liquid fraction) near the base of the layer is about 50% in this case, which would make a significant difference to the mobility of the interstitial fluid in the presence of external forces.

4. Thermal Convection of the Melt.

Buoyancy forces generated by gradients of temperature and concentration within the melt are the major cause of convection during solidification on Earth. Within the mushy layer, where local equilibrium (equation 2.1) is imposed, the net buoyancy of the interstitial fluid is

$$\Delta\rho = \beta\Delta C - \alpha\Delta T = (\beta - \alpha\Gamma)\Delta C, \tag{4.1}$$

where α and β are expansion coefficients for temperature and solute, Γ is the slope of the liquidus curve, and ΔT and ΔC are the temperature and concentration variations across the mushy layer. Under the assumption of local equilibrium, the temperature and solute fields are not free to diffuse independently. Therefore, since $\beta/\Gamma\alpha$ is usually larger than unity, gravitational convection within the mushy layer is determined principally by the solute field.

If a casting is cooled from a horizontal, upper boundary and the primary solidifying phase rejects a less-dense residual then there is no tendency for the interstitial fluid in the mushy layer to convect gravitationally. However, thermal convection of the melt can ensue, since it is being cooled from above. The mushy layer itself can be modelled using the equations of section 2, with r set equal to unity if attention is to be focused on the rôle of thermal convection (Kerr *et al.*, 1990a). The effect of thermal convection in the melt is felt only through the interfacial condition (2.17), in which account must be taken of the convective heat flux from the melt to the mushy layer. Such convection influences the rate of growth of the mushy layer (Turner *et al.*, 1986; Kerr *et al.*, 1990a) but does not by itself cause any macrosegregation of the casting.

However, in laboratory experiments in which aqueous solutions were cooled from above, with all other boundaries of the system insulated, significant macrosegregation of the solidified product has been observed (Turner *et al.*, 1986, Kerr *et al.*, 1990c; figure 2c). The interface between the upper region, where the bulk composition decreases with depth, and the lower region, where the composition increases with depth, was also marked by a

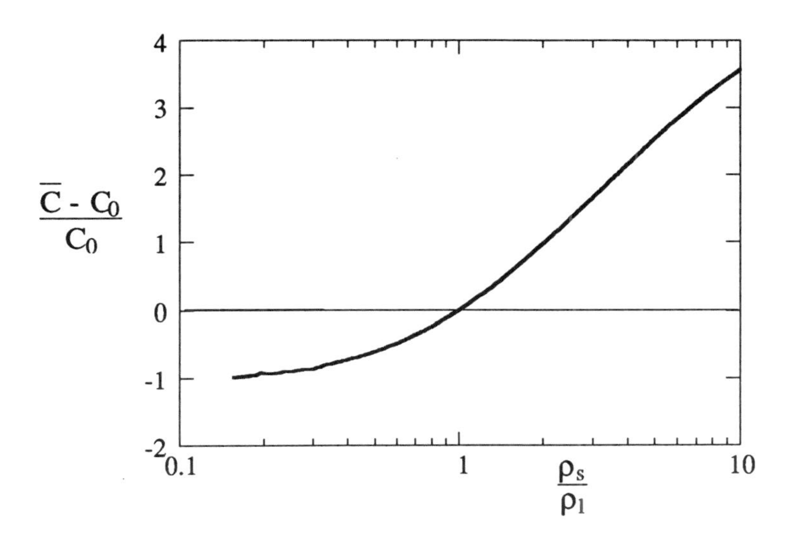

Figure 6. The theoretical macrosegregation caused by solidification shrinkage as a function of the density ratio between solid and liquid. The bulk composition at the base of the mushy layer is $\overline{C}$, while the initial composition of the melt is C_0. The calculations for this graph were made by A.O.P. Chiareli (Ph.D. Northwestern University).

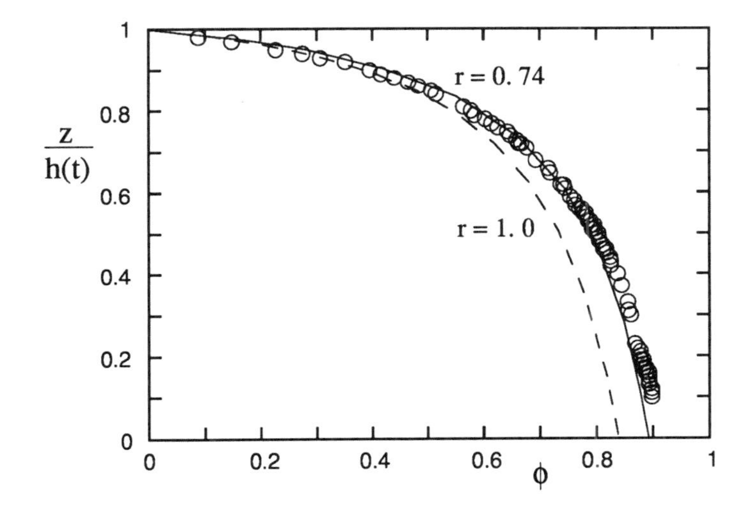

Figure 7. The volume fraction of solid ϕ in a mushy layer as a function of the relative depth in the layer, $z/h(t)$, where $h(t)$ is the height of the layer. The circles are data from experiments of Shirtcliffe *et al.* (1991), using aqueous solutions of sodium nitrate, reinterpreted by Chiareli & Worster (1992). The dashed line is the prediction of a model that negletcts solidification shrinkage ($r \equiv \rho_s/\rho_l = 1$), while the solid line takes full account of the interactions between solidification and the velocity field induced by shrinkage ($r = 0.74$).

so-called "columnar–equiaxed transition". At this interface, vertically-aligned, dendritic crystals attached to the roof of the container met randomly-oriented crystals that had grown in the interior of the melt and had settled and continued to grow at the base of the container.

The columnar-equiaxed transition is very important metallurgically since the different crystal morphologies influence the structural properties of the casting significantly. The growth of equiaxed grains in the interior of the melt (away from the cooled boundaries) cannot be explained by theoretical models that employ all the assumptions of equilibrium thermodynamics since, according to such models, the melt cannot be cooled below its liquidus temperature (Brandeis & Marsh, 1989; Kerr *et al.*, 1990a).

There are two places where equilibrium is imposed in the theoretical model described in section 2: in the interior of the mushy layer; and at the advancing interface between the mushy layer and the liquid region. In reality, there must exist some degree of undercooling (disequilibrium) at all evolving solid–liquid phase boundaries in order to drive solidification, so there must be some disequilibrium associated with the surfaces of individual dendrites within the mushy layer. However, any disequilibrium will also promote morphological instabilities of the dendrite leading to secondary and tertiary side branches. The side-branching activity within the mushy layer will tend to reduce the level of disequilibrium as the consequent increase in specific surface area of micro-scale phase boundaries promotes the release of latent heat and solute into the interstices. Thus the level of disequilibrium can be kept very small in the interior of the mushy layer, and the assumption of local equilibrium there is a good one. Note that the morphological instabilities that lead to side branching are inhibited by the surface energy associated with the curvature of the solid–liquid interface. Therefore, the assumption of internal equilibrium will be less good in systems that have large interfacial energies, especially those systems that display faceted rather than dendritic crystals.

The "interface" between the mushy region and the liquid region does not have a well-defined position on the micro scale. Rather it is a region of some small but finite thickness inhabited by the tips of dendrites. The specific surface area of solid–liquid phase boundaries in this interfacial region may be quite small since significant side-branching only occurs at some distance from the tips of dendrites. Thus the level of disequilibrium may be significant at the mush–liquid interface. It is possible to explain the growth of equiaxed grains in the interior of the melt and the macrosegregation that can additionally occur by relaxing the condition of interfacial equilibrium while maintaining the assumption of internal equilibrium.

A model incorporating a prescribed constant level of undercooling at the interface was proposed by Clyne (1981) and was subsequently extended to include a dynamically variable undercooling by Flood & Hunt (1987). These authors modelled the mushy region using the Scheil equation and an equation equivalent to (2.13) to determine the local temperature. We can describe the way in which disequilibrium was accounted for by these authors, using the terminology of the present paper, as follows. In the limit of zero diffusivity ratio ($D/\kappa \to 0$), the condition of marginal equilibrium at the mush–liquid interface (2.18) implies that the temperature of the interface is given by

$$T_i = T_L(C_\infty), \tag{4.2}$$

where C_∞ is the composition of the liquid far from the interface (strictly, the composition just outside the compositional boundary layer). Clyne (1981) replaced this condition by

$$T_i = T_L(C_\infty) - T_U, \tag{4.3}$$

where T_U is an assumed-constant value of the interfacial undercooling, while Flood & Hunt (1987) used the condition

$$V = F\left[T_L(C_\infty) - T_i\right]. \tag{4.4}$$

In the kinetic condition (4.4), V is the normal velocity of the interface and $F(\Delta T)$ is some function of the undercooling $\Delta T = T_L(C_\infty) - T_i$. In fact, Flood & Hunt (1987) used the function $F(\Delta T) = \mathcal{G}(\Delta T)^2$, which is appropriate for crystal interfaces growing by screw dislocations (Kirkpatrick, 1975). A similar analysis was carried out by Kerr *et al.* (1990b) using the function $F(\Delta T) = \mathcal{G}\Delta T$, which is more appropriate for molecularly rough crystals. In addition, Kerr *et al.* (1990b) conducted experiments in which a mushy layer of ice crystals was grown from a mixture of water and isopropanol, and were able both to confirm the linearity of the kinetic growth law for that system and to determine the value of the coefficient $\mathcal{G}$.

Although disequilibrium is only imposed locally at the mush–liquid interface, thermal convection of the melt can sweep undercooled liquid from the neighbourhood of the interface into the bulk of the region of melt. Thus the whole melt region becomes supercooled, and any nuclei within the melt can grow to produce large crystals. The heat transfer from these crystals is not directed (as it is in the mushy layer), so the crystals are randomly oriented (equiaxed).

In the model of Flood & Hunt (1987), the equiaxed crystals in the melt were assumed to grow from pre-inserted nucleation sites and remained fixed in space. Therefore, although, the authors were able to draw conclusions regarding the columnar-equiaxed transition, their model did not produce any macrosegregation. By contrast, Kerr *et al.* (1990b) made the assumption that all crystals grown in the interior of the melt settled instantaneously to the bottom of the container to form a uniform layer of solid there. This settling, and the consequent enrichment of the melt of the secondary component causes macrosegregation of the form illustrated in figure 2c. The assumption of instantaneous settling results in an over prediction of the degree of macrosegregation, as does the assumption that the settled crystals form a solid layer rather than a porous pile. As to the cause of macrosegregation in this case; while it is true that there would be none without the settling of crystals or, equivalently, the convection of depleted melt from the neighbourhood of the interior crystals, the root mechanism is the interactive coupling of thermal convection of the melt with the kinetic undercooling of the crystals at the mush–liquid interface.

5. Compositional Convection in the Mushy Region

Buoyancy-driven convection within mushy layers can occur when the composition of the interstitial fluid is unstably stratified. Such is the case for example when an alloy is solidified from below and the primary solidifying phase leaves a less dense residual. Convection of this type was first observed by Copley *et al.* (1970) in experiments in which an aqueous solution of ammonium chloride was solidified from below to form a mushy layer of ammonium-chloride crystals. Similar experiments have since been reported by a number of authors (Roberts & Loper, 1983; Sample & Hellawell, 1984; Chen & Chen, 1991; Tait & Jaupart,

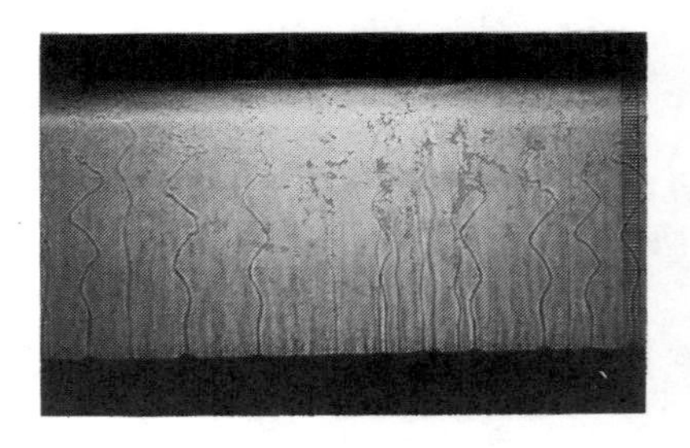

Figure 8. Photographs of the two different types of convection seen in experiments in which solutions of ammonium chloride are cooled and solidified from below. a) Early in the experiment double-diffusive fingers emanate from the vicinity of the mush–liquid interface. b) Later, isolated plumes rise through vents (chimneys) in the mushy layer into the overlying solution.

1992). The evolution of such experiments follows three distinct phases. Initially, a uniform mushy layer grows with a planar interface, and double-diffusive, finger convection rises from the interface a few centimetres into the solution (figure 8a). Once the mushy layer has reached a certain depth, the top of the layer becomes hummocked and chimneys begin to form, through which emanate plumes of buoyant fluid (figure 8b). As the strength of convection through chimneys becomes stronger, the finger convection wanes and eventually disappears. The later stages of the experiments are marked by a gradual decrease in the number of chimneys as the mushy layer deepens.

A full theoretical understanding of these experiments has still not been acheived, though aspects of the observed phenomena have been at least partially explained. One of the principal questions that has intrigued researchers in this field is the mechanism of chimney formation. Perhaps guided by observations of similar structures in fluidized beds (for example in a pot of boiling rice), it was suggested that chimneys might be produced mechanically by convective upflow breaking off the side branches of dendrites (Roberts & Loper, 1983). Hellawell (1987) noticed that he could induce the formation of a chimney by sucking fluid vertically through a pipette near the upper surface of the mushy layer. This observation led to the conjecture that upflow in the double-diffusive fingers above the mushy layer might be the root cause of chimney formation. Mehrabian *et al.* (1970) used the LSRE to deduce the incipiant formation of chimneys in the interior of a mushy region whenever the vertical velocity exceeds the local velocity of the isotherms. The mechanism of formation in this case is the local dissolution of solid (dendrites) within the layer. This idea was echoed by Fowler (1985) who argued additionally that this condition on the velocity field would first be met at the mush–liquid interface and that, therefore, chimneys should be observed growing downwards from the interface into the interior. Dissolution has been shown to be a possible mechanism for chimney formation in large-scale numerical

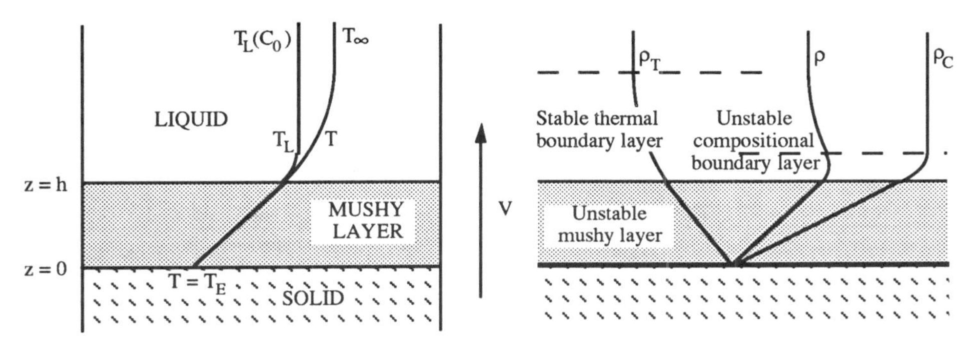

Figure 9. a) A schematic diagram of a binary melt being solidified at a constant rate from below. The temperature is fixed at the eutectic temperature at the horizontal position $z = 0$ in a frame moving with the constant solidification speed V. The temperature field T and the local liquidus temperature $T_L(C)$ are illustrated. b) The density distributions caused by the temperature field ρ_T, the solute field ρ_C and the total density of the liquid ρ. The density field in the mushy layer is statically unstable in a gravitational field directed vertically downwards, as is the density in the compositional boundary layer just above the mush–liquid interface.

calculations by Bennon & Incropera (1988) and in a linear-stability analysis by Worster (1992). All this does not rule out the possibility of a substantial rôle being played by mechanical shearing of side branches, however. Recent work by Glicksman† shows how local dissolution and ripening effects cause a necking of the side branches, possibly to a point where they can be easily torn off by a weak flow. In addition, observations by Sarazin & Hellawell† clearly show crystal fragments being carried upwards out of chimneys by the convecting plumes.

5.1 STABILITY THEORY

The prediction of when, rather than how, chimneys form can begin to be answered by an analysis of the stability of a mushy layer to the onset of buoyancy-driven convection. The governing equations presented in section 2 have a steady solution in a frame of reference moving with a prescribed vertical speed V, (see figure 9a). In the absence of any flow, the steady temperature and concentration fields give rise to the density field illustrated in figure 9b. The mushy layer is unstably stratified and, in addition, there is a narrow unstable boundary layer just above the mush–liquid interface.

The first analysis of the convective stability of a mushy layer was conducted by Fowler (1985), who analyzed a special limit of the governing equations, in which the compositional boundary layer has negligible thickness and the solid fraction in the mushy layer is zero. The stability problem reduces to that of the stability of a fixed porous medium with a solid lower boundary and a condition of zero pressure applied to the upper boundary. Chen & Chen (1988) presented an analysis of the stability of a fluid layer above a fixed porous medium, with a uniform, unstable temperature gradient applied across them. This arrangement mimics the compositional boundary layer above the mushy layer illustrated in figure 9b. They found a bi-modal marginal-stability curve similar in form to that shown in figure 10. A stability analysis using the full transport equations for the mushy layer was conducted by Nandapurkar *et al.* (1989). However, they did not allow perturbations of the solid fraction, and thereby suppressed any interaction between the convective flow and the growth of solid. They considered a single set of physical parameters and found just one mode of convection that did not penetrate the mushy layer.

Fully coupled interactions between flow and solidification were also considered in a normal-mode analysis by Worster (1992). He found that, in general, the marginal-stability curve is bimodal (figure 10), with the structure of the flow in the two modes as shown in figure 11. One mode, the "mushy-layer mode", has a horizontal wavelength comparable to the depth of the mushy layer, while the other, the "boundary-layer mode" has a wavelength comparable to the thickness of the compositional boundary layer above the mush–liquid interface. In effect, Fowler (1985) had analyzed a special case of the mushy-layer mode, while Nandapurkar *et al.* (1989) had found the boundary-layer mode. Either mode can be the more unstable depending on the parameters of the system (Worster, 1992), though for the parameter values of a typical laboratory experiment, the boundary-layer mode is found to be unstable long before the mushy-layer mode.

Since the flow in the boundary-layer mode does not penetrate the mushy layer, this mode has little influence on the solid fraction within the layer. By contrast, the mushy-layer mode causes the perturbations to solid fraction shown in figure 12. The interface is elevated in regions of upflow and the solid fraction through most of the depth of the layer is decreased there. These features are consistent with the form of chimneys observed in laboratory experiments, and suggest that it is the mushy-layer mode of convection that is primarily responsible for the formation of chimneys.

The sequence of events in a typical laboratory experiment might be explained as follows. Very soon after the start of an experiment, the boundary-layer mode becomes unstable and gives rise to double-diffusive plumes above the mushy layer. Sometime later, when the mushy layer is deeper, the Rayleigh number associated with the layer exceeds the critical value for the mushy-layer mode, and convection throughout the layer is initiated. This convection causes the interface to be elevated in regions of upflow, and depressed elsewhere, which gives rise to the hummocky appearance of the mushy layer. If the mushy-layer mode is sub-critically unstable then large perturbations, such as sucking, might trigger the early onset of the mode, which would explain the observations of Hellawell (1987).

If it is true that the mushy-layer mode gives rise to chimneys then it is of some importance to determine the conditions under which it is unstable. The Rayleigh number determining the convective stabilty of the mushy layer is

$$R_m = \frac{\beta \Delta C g \Pi^* H}{\kappa \nu}, \tag{5.1}$$

where g is the acceleration due to gravity, ν is the kinematic viscosity of the fluid, H is a length scale, such as the unperturbed depth of the mushy layer, and Π^* is a representative value of the permeability of the layer. Quantitative prediction of the onset of convection in a physical situation is complicated by the difficulty of determining the permeability of a mushy layer. Tait & Jaupart (1992) made estimates of the permeability in their experiments with ammonium chloride by measuring the mean separation of primary dendrite arms and applying an equation derived for the permeability of a vertical array of cylinders. This allowed them to estimate the critical Rayleigh number from observations of when the mushy layer first appeared hummocky. Some of their results are reproduced in figure 13,

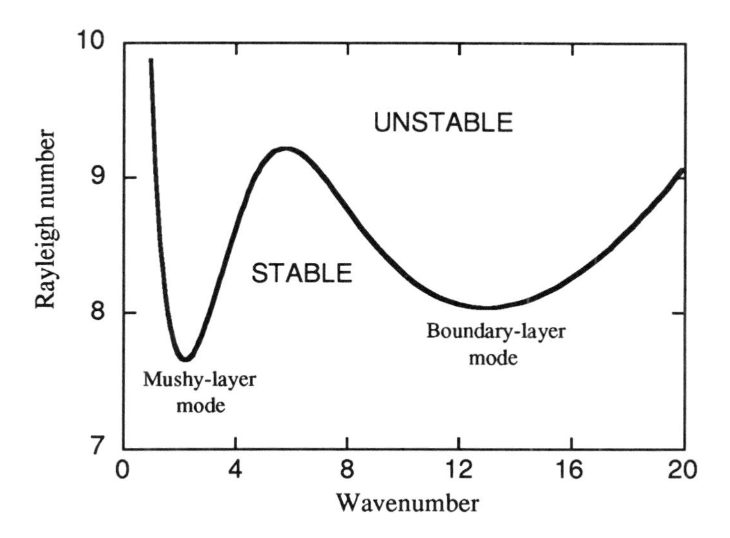

Figure 10. The marginal-stability curve for the onset of compositional convection in the system illustrated in figure 9. The two minima correspond to distinct modes of convection, as illustrated below. This graph was calculated for the diffusivity ratio $D/\kappa = 0.025$ and the ratio $H^2/\Pi^* = 10^5$, where $H = \kappa/V$ is the length scale for thermal diffusion and Π^* is a reference value of the permeability of the mushy layer. Either mode can be the more unstable depending on the values of these parameters (Worster, 1992).

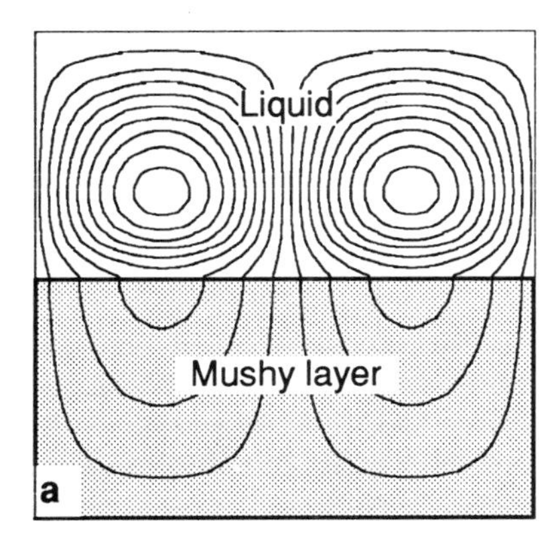

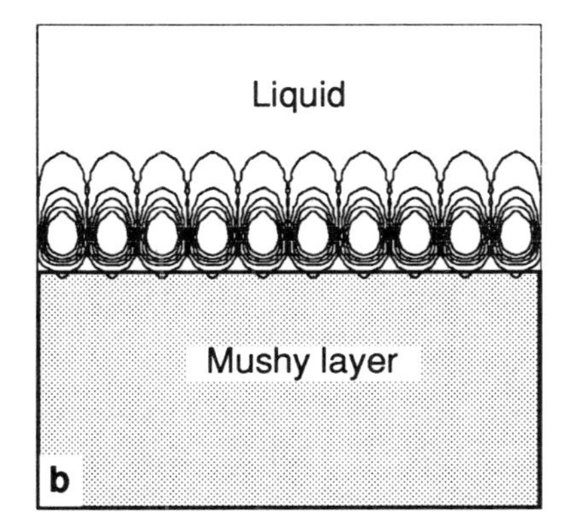

Figure 11. Streamlines for the marginally-stable modes corresponding to the two minima in figure 10. a) The mushy-layer mode has a wavelength comparable to the depth of the mushy layer and the flow penetrates the layer. b) The boundary-layer mode has a wavelength comparable to the thickness of the compositional boundary layer above the mush-liquid interface. The flow in this mode barely influences the underlying mushy layer.

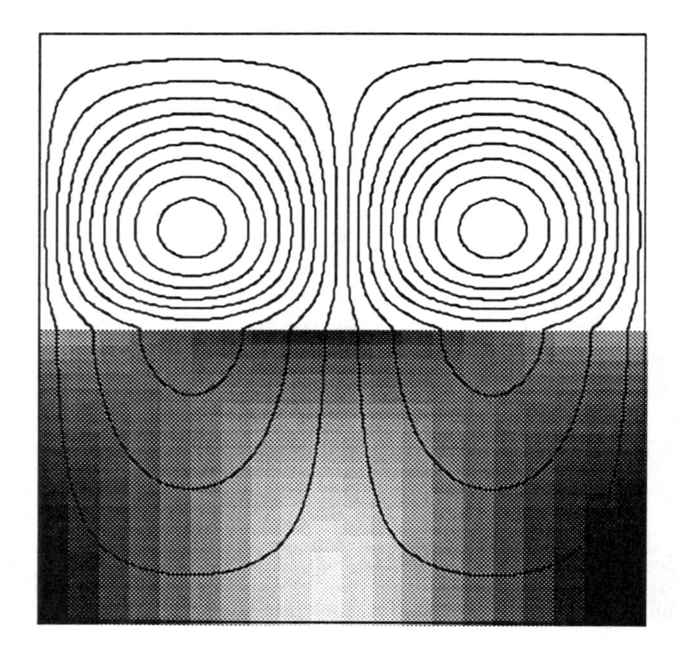

Figure 12. The perturbations to the solid fraction caused by the mushy-layer mode of convective instability. Dark regions show where the solid fraction has increased while lighter regions indicate where the crystals have dissolved. In regions of upflow, the solid fraction is generally decreased while the mush–liquid interface is elevated (indicated by the dark patch near the top of the mushy layer).

where they are compared with the predictions of the linear-stability analysis of Worster (1992). There is some scatter in the data but the general trend that the critical Rayleigh number decreases as the far-field temperature increases is found both in the theory and the experiments.

5.2 FULLY-DEVELOPED CHIMNEYS

As important as knowing when chimneys will occur is to know what their effect is once their occurence is unavoidable. For example, the convective flow through chimneys causes an exchange of solute between the mushy layer and the overlying region of melt that results in macrosegregation of the casting (figure 2f). In order to make predictions of the extent of macrosegregation it is necessary to be able to calculate the flux of solute through the chimneys. The flow through the mushy layer feeding the upflow through chimneys was first addressed by Roberts & Loper (1983). They showed that the flow through each chimney is similar to that through a thermosyphon (Lighthill, 1953) and calculated the flow through the surrounding mushy layer, whose solid fraction was taken as a fixed function of depth. These calculations were extended by Worster (1991), who analyzed the flow in the asymptotic limit of large Rayleigh number. He found that, in this limit, there is a

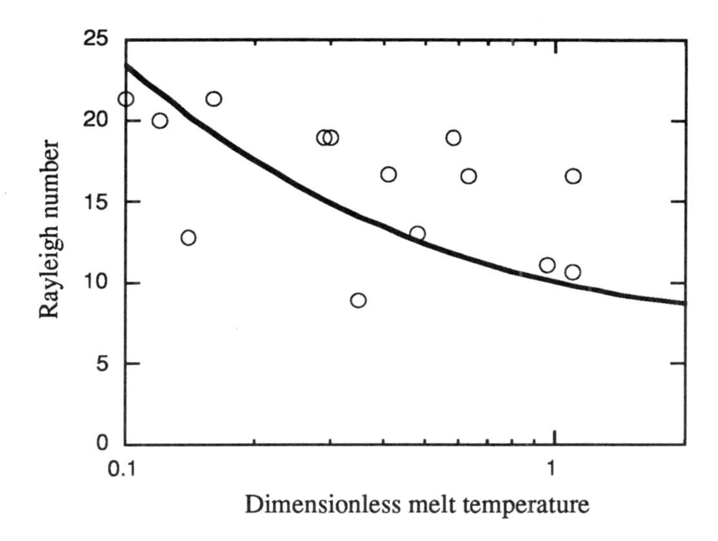

Figure 13. The critical Rayleigh number as a function of the far-field temperature of the melt given parameter values appropriate for an aqueous solution of ammonium chloride. The solid line shows the prediction of a stability analysis. The data are from experiments by Tait & Jaupart (1992).

thermal boundary layer around each chimney, in which the solid fraction is larger than in the rest of the mushy layer as a result of additional cooling by the fluid rising through the chimney. This increased crystal growth can be seen in experiments, and is the likely cause of the 'volcanic vent' around the top of each chimney (figure 8b). The asymptotic analysis allows calculation of the fluxes of mass, heat and solute through each chimney. The overall strength of the convective exchange between the mushy layer and the overlying liquid depends, in addition, on the areal number density of chimneys.

From his asymptotic analysis, Worster (1991) determined a family of solutions for the temperature, composition and solid fraction in the mushy layer, parameterized by the number density of chimneys. The fluxes of heat F_T and solute F_C from the mushy layer to the overlying fluid are given by

$$F_T = -\rho C_p \mathcal{F} V (T_\infty - T_L(C_0)) \tag{5.2}$$

and

$$F_C = -\tfrac{1}{2}\mathcal{F} V (C_0 - C_b), \tag{5.3}$$

where T_∞ and C_0 are the temperature and composition of the melt, C_b is the composition of the interstitial fluid at the base of the mushy layer, V is the rate of solidification, and $\mathcal{F}$ is a parameter proportional to the number density of chimneys. These expressions can be used to determine a relationship between the temperature and the composition of the liquid region

$$T = T_L(C_0) + (T_\infty - T_L(C_0))\left[1 - \frac{C - C_0}{C_0 - C_b}\right]^2 - (T_L(C_0) - T_b)\left(\frac{C - C_0}{C_0 - C_b}\right)^2. \tag{5.4}$$

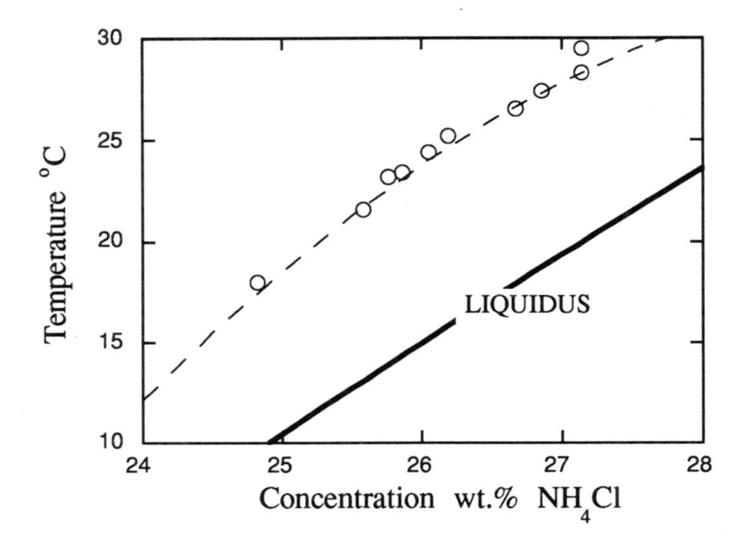

Figure 14. The evolution of the temperature and composition of the melt above a mushy layer caused by convection through chimneys in the layer. The data are from an experiment using an aqueous solution of ammoinium chloride. The dashed line is the predition of an asymptotic analysis in the limit of large Rayleigh number (Worster, 1990, 1991).

This expression is plotted in figure 14, where it is compared with measurements from an experiment using an aqueous solution of ammonium chloride. There is good agreement between the theoretical results and the experimental measurements, but the result is of limited use. In order to make individual predictions of the evolution of the temperature and solute fields, and hence the rate of solidification and the extent of macrosegregation, it is necessary to find some way to predict the number density of chimneys, or alternatively to determine the parameter $\mathcal{F}$ directly.

However, expression (5.4) and figure 14 do serve to illustrate a particular feature of chimney convection, namely that the temperature of the melt remains above the liquidus temperature. Huppert (1990) pointed out that this behaviour is distinct from what is found in experiments using some other aqueous solutions, in which the melt became supercooled. In these experiments, convection is only seen emanating from the vicinity of the mush–liquid interface, so it might be associated with the boundary-layer mode. As discussed in section 4, significant disequilibrium (undercooling) can exist at the mush-liquid interface, which could be carried into the interior of the melt by the boundary-layer mode of convection. By contrast, when chimneys are present, the flow is primarily downwards into the mushy layer, where the fluid can be restored to near-equilibrium conditions before being ejected through chimneys. During ascent through a chimney, the fluid reaches thermal equilibrium with its surroundings but its composition remains unchanged. It therefore emerges from the chimney undersaturated (superheated). Whether the melt becomes supercooled or not will have a profound influence on the form of macrosegregation and the possibility of columnar-equiaxed transitions.

6. Future Directions

In this review, I have tried to highlight with a few specific examples key interactions that can occur between soldification in mushy layers and the flow of the melt. It is apparent that significant advances have been made in our understanding of mushy layers through a combination of mathematical models and quantitative laboratory experiments. In particular, it now seems certain, from the excellent agreement obtainable between theoretical predictions and experimental data, that mushy layers in which the interstitial fluid is stagnant, or which only moves in response to shrinkage, are well characterized and quantified by the continuum theories. However, while many qualitative aspects of buoyancy-driven convection through the interstices of mushy layers are understood, considerable work still needs to be done before accurate quantitative predictions can be made.

Perhaps the most important subject that needs to be addressed is the appropriate form of the momentum equation in mushy layers and the determination of the associated transport parameters. Even in the simplest cases when Darcy's equation is appropriate, there is a need to determine how the permeability varies as the mushy layer evolves. In general, the permeability Π of a porous medium depends on its porosity χ (liquid fraction) and on the specific surface area M of its internal, wetted surfaces (phase boundaries).

Experimental determinations of the porosity of mushy layers are already being undertaken (Chen & Chen, 1991; Shirtcliffe *et al.*, 1991) and the experimental techniques for doing so will undoubtedly improve in the next few years. Measuring the specific surface area of internal phase boundaries will pose greater difficulties but may not be beyond modern computerized imaging techniques, for example. Even once these two properties of the medium are determined, obtaining the permeability is not straightforward. Many different relationships for the permeability in terms of the porosity and the specific surface area have been proposed, a popular choice being the Kozeny equation (Bear, 1988) which gives $\Pi = c_0\chi^3/M^2$, where c_0 is a constant.

From a theoretical point of view, the variation of porosity has already been taken into account in theories of mushy layers but little attention has yet been given to the evolution of the specific surface area of internal phase boundaries. There are at least two competing influences on the surface area: morphological instabilities that lead to side branches; and surface-energy effects that lead to coarsening. The instabilities of individual, isolated dendrites have been addressed in the past (Langer, 1980) though one suspects that the characteristics of these instabilities will be changed considerably once the dendrite forms part of an array within a mushy region. There have also been experimental (Marsh & Glicksman, 1987) and theoretical (Voorhees, 1985) studies of coarsening in single-component systems. The theoretical studies have a statistical nature that is suitable for application to mushy layers but there is a need to extend them to binary systems and to systems that are evolving due to imposed macroscopic temperature gradients.

Evolution of the internal phase boundaries will only take place in the presence of micro-scale gradients of temperature and/or solute concentration. This signals the need to relax two of the principal assumptions upon which current theories of mushy layers are based, namely the uniformity of the interstities and local equilibrium of the mushy region. We have already seen that local disequilibrium at a mush–liquid interface, coupled with convection, has important consequences for macrosegregation of a casting. Depending on the geometry of the casting and the nature of the crystals, flow of the melt will penetrate different depths of a mushy region and it is therefore important to know the extent of significant disequilibrium. Loper† has formulated a non-equilibrium theory of a mushy region which might form the basis for future investigations.

Disequilibrium is also a significant issue in the study of melting within mushy regions, which has yet received little attention. Some aspects of the melting of binary solids have been analyzed by Woods (1992), and the melting of side braches of dendrites has been investigated experimentally by Glicksman†. Localized melting of the solid phase within mushy regions is important for the formation of chimneys, for example, and similar phenomena occur during the slow percolation of groundwater thorugh porous rocks (Phillips, 1991). Whereas morphological instabilities act to keep the specific surface area of internal phase boundaries large during solidification, no such effect occurs during melting, so the micro-scale gradients of temperature and solute concentration that cause internal disequilibrium can be sustained.

One of the major strengths of the theoretical models of mushy regions that have been developed and utilized during the past several years is their independence of the morphology of internal phase boundaries. This has rendered them universally applicable regardless of the chemical species being solidified. To address the challenges of the future (disequilibrium, permeability to fluid flow, coarsening and melting) much more specific information about the chemical system will be required. In addition, there will be an increasing need to include micro-scale phenomena into models of macro-scale processes. Interactions between scientists from the various disciplines dealing with solidification and transport processes will be greatly advantageous in this endeavour.

Acknowledgements

I am grateful to S.H. Davis and H.E. Huppert for their helpful review of early drafts of this review. My research is partially supported by the Thermal Systems Program of the National Science Foundation.

References

Bear, J. (1988) *Dynamics of Fluids in Porous Media.* Dover.

Bennon, W.D. & Incropera, F.P. (1987) A continuum model for momentum, heat and species transport in binary solid–liquid phase change systems — I. model formulation. *Int. J. Heat Mass Transfer* **30**, 2161–2170.

Bennon, W.D. & Incropera, F.P. (1988) Numerical analysis of binary solid-liquid phase change using a continuum model. *Numerical Heat Transfer* **13**, 277–296.

Beran, M.J. (1968) *Statistical Continuum Theories.* Interscience Publishers, New York.

Brandeis, G. & Marsh, B.D. (1989) The convective liquidus in a solidifying magma chamber: a fluid dynamic investigation. *Nature* **339**, 613–616.

Chen, F. & Chen, C.F. (1988) Onset of finger convection in a horizontal porous layer underlying a fluid layer. *Trans. ASMEC: J. Heat Transfer* **110** 403–407.

Chen, F. & Chen, C.F. (1991) Experimental study of directional solidification of aqueous ammonium chloride solution. *J. Fluid Mech.* **227** 567–586.

Chiareli, A.O.P. & Worster, M.G. (1992) On the measurement and prediction of the solid fraction within mushy layers. (preprint, Northwestern University)

Clyne, T.W. (1981) In:Proc. 2nd Intern. Conf. on Numerical Methods in Thermal Problems, Venice, Italy 240–256.

Copley, S.M., Giamei, A.F., Johnson, S.M. & Hornbecker, M.F. (1970) The origin of freckles in binary alloys. *IMA J. Appl. Maths* **35**, 159–174.

Flemings, M.C. (1974) *Solidification Processing.* Mc.Graw Hill.

Flemings, M.C. (1981) in *Modeling of Casting and Welding Processes* (eds. Brody, H.D. & Arpelian, D.), pp. 533–548. The Metallurgical Society of the American Institute of Mining, Metallurgical and Petroleum Engineers, Warrendale.

Flemings, M.C. & Nereo, G.E. (1967) Macrosegregation, part I. *Trans. Met. Soc. AIME* **239**, 1449–1461.

Flemings, M.C. & Nereo, G.E. (1968) Macrosegregation, part III. *Trans. Met. Soc. AIME* **242**, 50–55.

Flood, S.C. & Hunt, J.D. (1987) A model of a casting. *Appl. Sci. Res.* **44**, 27–42.

Fowler, A.C. (1985) The formation of freckles in binary alloys. *IMA J. Appl. Maths* **35**, 159–174.

Fowler, A.C. (1987) Theories of mushy zones: applications to alloy solidification, magma transport, frost heave and igneous intrusions. In *Structure and Dynamics of Partially Solidified Systems* (ed. D.E. Loper), pp. 161–199. NATO ASI Series. Martinus Nijhoff.

Fujii, T., Poirier, D.R. & Flemings, M.C. (1979) *Metall. Trans. B* **10B**, 331–339

Hellawell, A. (1987) Local convective flows in partially solidified alloys. In *Structure and Dynamics of Partially Solidified Systems* (ed. D.E. Loper), pp. 3–22. NATO ASI Series. Martinus Nijhoff.

Hills, R.N., Loper, D.E. & Roberts, P.H. (1983) A thermodynamically consistent model of a mushy zone. *Q. J. Mech. Appl. Maths* **36**, 505–539.

Huppert, H.E. (1990) The fluid dynamics of solidification. *J. Fluid Mech.* **212**, 209–240.

Huppert, H.E. & Worster, M.G. (1985) Dynamic solidification of a binary melt. *Nature* **314**, 703–707.

Kerr, R.C., Woods, A.W., Worster, M.G. & Huppert, H.E. (1990a) Solidification of an alloy cooled from above. Part 1. Equilibrium growth. *J. Fluid Mech.* **216** 323–342.

Kerr, R.C., Woods, A.W., Worster, M.G. & Huppert, H.E. (1990b) Solidification of an alloy cooled from above. Part 2. Non-equilibrium interfacial kinetics. *J. Fluid Mech.* **217** 331–348.

Kerr, R.C., Woods, A.W., Worster, M.G. & Huppert, H.E. (1990c) Solidification of an alloy cooled from above. Part 3. Compositional stratification within the solid. *J. Fluid Mech.* **218** 337–354.

Kirkpatrick, R.J. (1975) Crystal growth from the melt: a review. *American Mineralogist* **60**, 798–814.

Kurz, W. & Fisher, D.J. (1986) *Fundamentals of Solidification.* Tans. Tech. Publications.

Lighthill, M.J. (1953) Theoretical considerations on free convection in tubes. *Quart. J. Mech. Appl. Math.* **6**, 398–439.

Langer, J.S. (1980) Instabilities and pattern formation in crystal growth. *Rev. Mod. Phys.* **52**, 1–28.

Marsh, S.P. & Glicksman, M.E. (1987) Evolution of lengthscales in partially solidified systems. In *Structure and Dynamics of Partially Solidified Systems* (ed. D.E. Loper), pp. 25–35. NATO ASI Series. Martinus Nijhoff.

Mehrabian, R., Keane, M. & Flemings, M.C. (1970) Interdendritic fluid flow and macrosegregation: influence of gravity. *Metall. Trans.* **1**, 1209–1220.

Mullins, W.W. & Sekerka, R.F. (1964) Stability of a planar interface during solidification of a dilute binary alloy. *J. Appl. Phys.* **35**, 444–451.

Nandapurkar, P., Poirier, D.R., Heinrich, J.C. & Felicelli, S. (1989) Thermosolutal convection during dendritic solidification of alloys: Part 1. Linear stability analysis. *Metallurgical Transactions B* **20B**, 711–721.

Petersen, J.S. (1987) Crystallization shrinkage in the region of partial solidification: implications for silicate melts. In *Structure and Dynamics of Partially Solidified Systems* (ed. D.E. Loper), pp. 417–435. NATO ASI Series. Martinus Nijhoff.

Phillips, O.M. (1991) *Flow and Reactions in Permeable Rocks.* Cambridge University Press.

Roberts, P.H. & Loper, D.E. (1983) Towards a theory of the structure and evolution of a dendrite layer. *Stellar and Planetary Magnetism*, 329–349.

Sample, A.K. & Hellawell, A. (1984) The mechanisms of formation and prevention of channel segregation during alloy solidification. *Metallurgical Transactions A* **15A**, 2163–2173.

Shirtcliffe, T.G.L., Huppert, H.E. & Worster, M.G. (1991) Measurement of solid fraction in the crystallization of a binary melt. *J. Crystal Growth* **113**, 566–574.

Szekely, J. & Jassal, A.S. (1978) An experimental and analytical study of the solidification of a binary dendritic system. *Metall. Trans. B* **9B**, 389–398.

Tait, S. & Jaupart, C. (1992) Compositional convection in a reactive crystalline mush and the evolution of porosity. *J. Geophys. Res.* (in press).

Turner, J.S., Huppert, H.E. & Sparks, R.S.J. (1986) Komatiites II: Experimental and theoretical investigations of post-emplacement cooling and crystallization. *J. Petrol.* **27**, 397–437.

Thompson, M.E. & Szekely, J. (1988) Mathematical and physical modelling of double-diffusive convection in aqueous solutions crystallizing at a vertical wall. *J. Fluid Mech.* **187**, 409–433.

Vas'kin, V.V. (1986) Solving solidification equations for a binary alloy by similarity. *Izvestiya Akademii Nauk SSSR, Metally* **1** 83–87. Published by Allerton Press.

Voller, V.R., Brent, A.D. & Prakash, C. (1989) The modelling of heat, mass and solute transport in solidification systems. *Int. J. Heat Mass Transfer* **32**, 1719–1731.

Voorhees, P.W. (1985) The theory of Ostwald ripening. *J. Stat. Phys.* **38**, 231–252.

Woods, A.W. (1992) Melting and dissolving. *J. Fluid Mech.* **239**

Worster, M.G. (1986) Solidification of an alloy from a cooled boundary. *J. Fluid Mech.* **167**, 481–501.

Worster, M.G. (1990) Structure of a convecting mushy layer. *Appl. Mech. Rev* **43, 5** S59–S62.

Worster, M.G. (1991) Natural convection in a mushy layer. *J. Fluid Mech.* **224**, 335–359.

Worster, M.G. (1992) Instabilities of the liquid and mushy regions during solidification of alloys. *J. Fluid Mech.* **237**, 649–669.

CONVECTION IN THE MUSHY ZONE DURING DIRECTIONAL SOLIDIFICATION

C. F. CHEN
Department of Aerospace and Mechanical Engineering
The University of Arizona
Tucson, AZ 85721 USA

ABSTRACT. Directional solidification experiments with 26% NH_4Cl-H_2O solution were carried out to study convection in the mushy layer. Results show that prior to the onset of plume convection, there is no convective motion in the mush, even though there is vigorous finger convection emanating from the mush-liquid interface. After plume convection is fully developed, the flow in the mush is mainly downward, with some lateral feeding into the chimneys, where the upward fluid motion is confined. Further experiments in a Hele Shaw cell revealed interesting behavior of the flow in the chimney.

1. Introduction

In a series of experiments conducted earlier (Chen and Chen 1991), we have shown that when an aqueous solution of NH_4Cl is directionally solidified from below, salt finger convection occurred in the fluid region just above the mushy layer. When the mushy layer attains a certain thickness, chimneys start to appear, with upward-flowing plumes of lighter fluid. Recently, Worster (1992) analyzed the onset of compositional convection in a solidifying mushy layer underneath a layer of melt liquid. He showed that there exist two modes of instability: the boundary layer mode and the mushy layer mode. In the former, convection is confined in a thin boundary layer in the liquid above the interface, with a critical wavelength comparable to the boundary layer thickness. In the latter, convection is due to the instabilities in the mushy layer, with a critical wavelength comparable to the thickness of the mushy layer. This result is analogous to that for the onset of thermal convection in superposed fluid and nonreacting porous layers discussed by Chen and Chen (1988). Worster's results imply that during the early stages of a directional solidification experiment, convection is confined within the liquid region. Only when the mushy layer reaches a certain thickness will there be convection within the layer. These experiments were carried out to test this theory and to gain some knowledge of the convection.

2. Experiments in a Three-Dimensional Tank

This series of experiments was carried out in a tank of dimensions 15 × 19.5 × 30 cm tall. All sides of the tank were made of plexiglass and the bottom of copper, which was kept at -9°C by a constant-temperature circulator. The motion of the fluid in the mush was observed through flow visualization. A dye-injection system was used to introduce a layer of dye solution at the interface between the mush and the bulk liquid region near a 19.5-cm-wide transparent wall. The dye solution was made by dissolving a small

S. H. Davis et al. (eds.), Interactive Dynamics of Convection and Solidification, 139–141.

amount of $KMnO_4$ in the NH_4Cl solution withdrawn at the interface region to minimize the density difference. Measurements made with a Paar DMA 60 Density Meter showed that the dye solution was 0.078% heavier than the undyed solution.

First, an experiment was conducted to determine the sinking rate of the dyed solution in an inactive mush. The results are shown in Fig. 1a, in which the average vertical position of the dyed layer is shown over a period of 70 min. The straight line obtained by linear regression in shown on the graph. The slope is -0.156 mm/min.

The next set of flow visualization experiments was performed in a mushy layer during the solidification process. Dye solution was introduced at 35 min, 1 hr 25 min, and 2 hr 22 min into the experiment. A time-lapse movie at 1 frame/sec was made to record the motion of the dye trace. The average positions of the dye layer are shown in Figs. 1b and 1c. In Fig. 1b, the data from the first two time segments are shown. Superposed on the data points are straight lines with slope -0.156 mm/min. It is seen that during the earliest period, when the mush is free of chimneys, the dye sinking rate is the same as that in the inactive mush (Fig. 1a), signifying that there is no motion in the mush. It is to be noted that, at this time, there was vigorous finger convection just above the interface. A significant portion of the injected dye was carried away by the upward motion of the finger convection. The data from the second period (1 hr 25 min to 1 hr 50 min), in which the chimneys are developing, are also shown in Fig. 2b. The average net downward convection rate is 0.21 mm/min. During the fully developed chimney convection stage (2 hr 20 min to 3 hr 20 min), the downward convection is much faster, as shown in Fig. 1c. The average net sinking rate is 0.53 mm/min at 2:26, decreasing to 0.14 mm/min at 2:55. Throughout the entire experiment, there was no indication of upward flow in the mush. All upward flow was confined within the chimneys.

3. Experiments in a Hele Shaw Cell

In order to gain a clear view of the onset of chimneys and the flow within each chimney subsequently, a series of experiments was conducted in a Hele Shaw cell 150 × 220 × 1 mm. Similar to the three-dimensional tank, all sides of the cell were made of plexiglass, while the bottom was made of copper kept at a low temperature by a constant-temperature circulator. The mean fluid velocity in a Hele Shaw cell $\overline{V}$ and the filtration velocity V in a porous medium are, respectively,

$$\overline{V} = -\frac{d^2}{12\mu}\nabla p \quad \text{and} \quad \mathbf{V} = -\frac{K}{\mu}\nabla p$$

where d is the cell thickness, μ the viscosity, p the pressure, and K the permeability. For the present cell, $d^2/12 \sim 10^{-3}cm^2$ and $K \sim 10^{-5}cm^2$ in the mushy zone of a solidifying NH_4Cl system (Chen and Chen 1991). It is therefore reasonable to expect that the narrowness of the slot does not unduly affect the flow in the mushy layer. However, motion in the bulk fluid will be restricted, which in turn will affect the solidification process.

Three preliminary experiments were run with the bottom temperature maintained at -12°C, -21°C, and -25°C. With the highest temperature, some chimneys were initiated, but all became blocked before being fully developed. With the cooler temperatures, fully developed chimneys were generated. With the coldest temperature, the start of the crystallization process was a bit chaotic due to the rapid cooling effect. Subsequent experiments were carried out at -21°C.

The decrease of the bottom temperature and the growth of the mushy layer with time are shown in Fig. 2. Compared to the data of Chen and Chen (1991), the growth in the Hele Shaw cell is slower. Initial stages of chimney formation started between 40 and 50 min, with some channels formed near the top of the mush. These channels deepened as time went on. Some of these became blocked, while others became fully developed chimneys with an extensive root system near the bottom. Generally, there were four chimneys in the tank at this stage.

At the early stages of the mushy growth, the upward-convecting fluid from the mush contained tiny crystals of NH_4Cl. These crystals served as nuclei for further crystallization, and eventually they became heavy enough to drop back to the mush. Tiny crystals were also present in the upward-flowing plumes in the chimneys. Because of remelting, large pieces of crystals broke off the chimney wall and descended in the plume. Some of these became lodged in the crevices in the wall, thus making the periphery of the chimney more densely packed, as was observed in the CT scans (Chen and Chen 1991).

Acknowledgments

The financial support of NASA through Grant NAG3-1268 is gratefully acknowledged. These experiments were performed in the GFD Laboratory at the Research School of Earth Sciences, The Australian National University, during June and July of 1991. I wish to thank Professor Stewart Turner for his hospitality and Tony Beasley, Ross Wilde-Browne, and Derek Corrigan for their able assistance in the lab.

References

Chen, C. F. and Chen, F. (1991) 'Experimental study of directional solidification of aqueous ammonium chloride solution', J. Fluid Mech. 227, 567-586.

Chen, F. and Chen, C. F. (1988) 'Onset of finger convection in a horizontal porous layer underlying a fluid layer', J. Heat Trans. 110, 403-407.

Worster, M. G. (1992) 'Onset of compositional convection during solidification of a mushy layer from a binary alloy cooled from below', J. Fluid Mech. 237, 649-669.

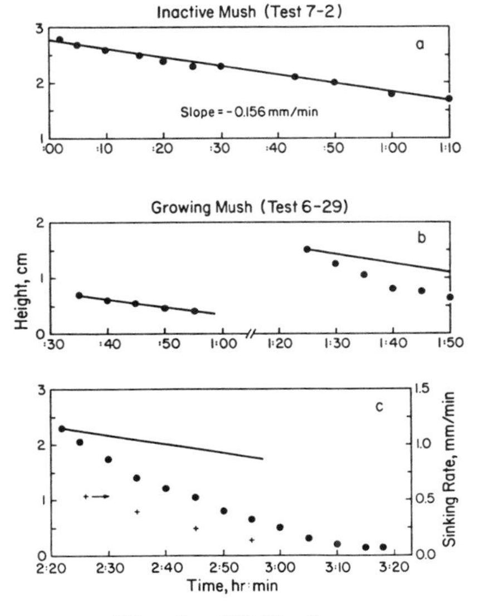

Fig. 1. 3D Tank

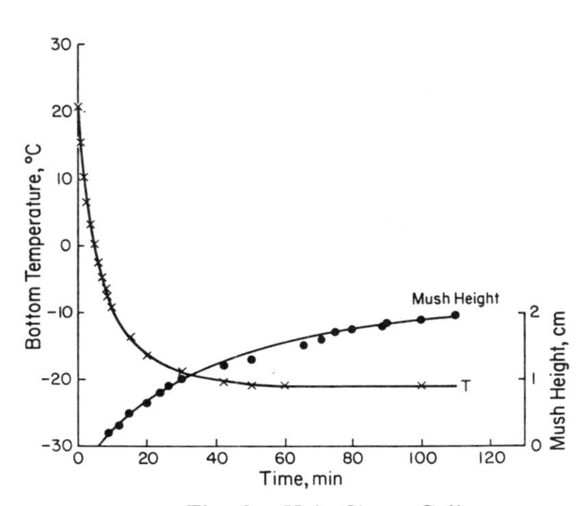

Fig. 2. Hele Shaw Cell

STUDIES OF CHANNEL- PLUME CONVECTION DURING SOLIDIFICATION

J. R. SARAZIN * and A. HELLAWELL
Department of Metallurgical and Materials Engineering
Michigan Technological University
Houghton, Michigan 49931 USA
**Ames Laboratory*
Iowa State University
Ames, Iowa 50010 USA

ABSTRACT. Convective plume flows and associated channels were examined in opaque metallic, and in transparent aqueous and organic systems, covering a range of Prandtl numbers from $\sim 10^{-2}$ to ~ 30. Measurements have been made of plume-channel compositions and those of their surroundings, by direct sampling of transparent liquids and by electron microprobe analysis of fully solidified metallic alloys, after the events. Plume liquid temperatures were also measured in the transparent materials and flow rates were measured from video records of particle movements. The plume-channel widths and spacings are remarkably similar in all these materials despite their physical differences. Analysis shows that the flow rates for the transparent systems fit a Poiseuille relation, modified for slip at the plume boundaries. Extrapolation of this analysis, using measured channel compositions and estimated buoyancy terms, predicts plume flow velocities in metals to lie between $1 \cdot 10^{-1}$ and $2 \cdot 10^{-1}$ m$\cdot$ s^{-1}. The factors that determine the pattern selection of the plume-channel dimensions, compositions and flow rates are discussed with reference to thermal equilibrium, the onset of turbulence and the restrictions to the liquid re-entrainment needed to sustain such flow.

Introduction.

Natural convection occurs when density differences arise in different parts of a system from both temperature and/or compositional variations. Of particular interest is the geometrical configuration where the vertical density gradients due to temperature and composition $[d\rho/dz]$ are of opposite sign with the compositional contribution dominating ie. $[d\rho/dz]_T < -[d\rho/dz]_C$, resulting in a density inversion. Although the density inversion is compositionally unstable, thermal effects may result in metastability to convection.

Eventually, by double-diffusive convection [1] , stability breaks down and in systems that freeze over a temperature range forming a two phase, solid/liquid "mushy zone" , liquid perturbations result in convection in the form of descrete plumes of solute rich liquid (See Figure 1) which rise vertically upward, leaving a channel or hole in the mushy zone [2]. The solute rich liquid in the channel is the last to freeze leaving a macroscopic defect commonly known as a channel, freckle or A-segregate [3,4].

Channel-plumes have been examined in both metallic and transparent analog systems. The physical characteristics of channels in these systems are remarkably similar, despite large differences in their properties. Measured plume flow velocity and liquid temperatures from transparent materials have been used to extrapolate to the flow velocity in opaque liquid metals.

Materials and Methods

Channels were examined in the metallic lead-tin (Pb-Sn) and lead-Antimony (Pb-Sb) eutectic systems by metallographic sectioning of ingots after solidification. The convective plumes resulting in channels could be examined insitu in the transparent analog ammonium chloride-water (NH_4Cl-H_2O) and succinonitrile-ethanol (SCN-EtOH) systems. In all systems the solute (Sn,Sb,H_2O or EtOH) rejected from the primary dendritic solid (Pb,NH_4Cl or SCN) decreases the

S. H. Davis et al. (eds.), Interactive Dynamics of Convection and Solidification, 143–145.

Figure 1. Shadowgraph of solute rich plumes in the aqueous NH_4Cl-70wt.% H_2O Plume Diameter ~1 mm

liquid density. Alloys from all systems were solidified with the growth direction vertically upwards utilizing previously employed methods.[2,3,5-7]

Channel-plume compositions were measured in the metallic alloys by electron microprobe analysis across the channels, postmortem and in transparent materials by refractive index measurements of liquid samples obtained from the plumes during solidification using a small motor-driven syringe.

Temperature measurements in transparent materials were obtained by passing a fine wire K-type thermocouple through the plumes at various heights above the interface during solidification.

Plume flow velocities were also determined in the transparent systems by timing the translation of particles, either indigenous or induced, over a fixed distance on a video monitor.

Results

The results of the measurements of the channel-plume scale and compositional difference for the three systems as well as velocity and temperature data for the transparent systems are summarized in Table 1. Physical dimensions are remarkably similar despite differences in the material properties.

Discussion

Analysis of plume flow data for the transparent analogs shows the maximum velocity to fit a Poiseuille type relation, modified to eliminate the no-slip boundary condition at the plume wall [6,7]. In all these systems the observed or estimated flow rates are from $1 \cdot 10^{-1}$ to $2 \cdot 10^{-1}$ $m \cdot s^{1}$ significantly above the rate necessary for the attainment of full thermal equilibrium, but well below that for the onset of turbulence, determined by a balance of buoyancy and viscosity and limited by the necessary re-entrainment of liquid through the semi-permeable mushy zone.

TABLE 1 Channel/Plume Data: Summary of Results

	Metallic (10 wt.%Sn)	Aqueous (70 wt.% H_2O)	Organic (15 wt.% EtOH)
Height above the growth front, H m	?	>.1	~.1
Depth below the growth front, D m	$>5 \cdot 10^{-3}$	$>5 \cdot 10^{-3}$	$>5 \cdot 10^{-3}$
Channel/Plume spacing, L m	5 to $10 \cdot 10^{-3}$	5 to $10 \cdot 10^{-3}$	5 to $10 \cdot 10^{-3}$
Channel/Plume radius, r m	5 to $7 \cdot 10^{-4}$	5 to $8 \cdot 10^{-4}$	$\sim 5 \cdot 10^{-4}$
Composition difference between channel/plume and bulk ΔC, wt.%	1 to 3	1-2	$\lesssim 1$
Temperature difference between plume and bulk, ΔT, K	?	1-4	<1
Mean plume flow velocity $\bar{v}$, $m \cdot s^{-1}$	?	6 to $9 \cdot 10^{-3}$	$\lesssim 1 \cdot 10^{-3}$
Primary dendrite spacing under prevailing conditions, λ m	$3 \cdot 10^{-4}$	$5\text{-}6 \cdot 10^{-4}$	5 to $6 \cdot 10^{-4}$

Acknowledgments

The authors would like to acknowledge the financial support of the National Science Foundation, Grant No. DMR-88-15049 and the National Aeronautics and Space Administration, Grant No. NAG-3-560.

References

1. J.S. Turner, (1973) 'Buoyancy Effects in Fluids', Cambridge University Press, London
2. A.K. Sample and A. Hellawell, (1984) ' Mechanisms of Formation and Prevention of Channel Segregation During Alloy Solidification', Metall. Trans , 15A, pp. 2163-2173.
3. S.M. Copley, A.F. Giamei, S. M. Johnson and M.F. Hornbecker, (1970), 'The Origin of Freckles in Unidirectionally Solidified Castings', Metall. Trans. ,1, pp. 2193-2204.
4. R.J. McDonald and J.D. Hunt, (1969) 'Fluid Motion Through the Partly Solidified Zone of a Casting and its Importance in Understanding A-Type Segregation' , Trans. TMS-AIME, 245, pp.1993-1997.
5. J.R. Sarazin and A. Hellawell, (1988),'Channel Formation in Pb-Sn, Pb-Sb and Pb-Sn-Sb Alloys and Comparison with the System NH_4Cl-H_2O', Metall. Trans., 19A, pp.1861-1871.
6. J.R. Sarazin , (1990) PhD Dissertation, 'Channel formation in Metallic, Aqueous and Organic Systems', Michigan Technological University.
7. A Hellawell, J.R. Sarazin and R.S. Steube,'Channel Formation in Partly Solidified Systems', . (in review)

THE ONSET OF FRECKLING IN THE SOLIDIFICATION OF BINARY ALLOYS

A.C. FOWLER AND P. EMMS
Mathematical Institute,
Oxford University,
24-29 St Giles',
Oxford OX1 3LB,
England.

ABSTRACT. A mathematical model is presented for the solidification of a binary alloy when cooled from below. When light fluid is released, compositional convection can occur in the dendritic, mushy zone as well as in the fluid above. Here we summarise efforts to study the onset of convection in the mush, by parameterising the vigorous convection in the liquid above, with a view to prediction of the circumstances under which flow channels (freckling) can occur.

1. Introduction

Since the early work on the formation of freckles in alloy castings (McDonald and Hunt 1969, 1970, Copley et al. 1970) there has been much interest in the dynamics of convective fluid motion within and above a 'mushy' region of partially solidified alloy material. Descriptions of the process are of interest in metallurgy (e.g., Sarazin and Hellawell 1988), in studies of convective motions in the earth's core (Loper and Roberts 1981), and in magma chambers (Tait and Jaupart 1992). Theoretical studies of convective motion have been undertaken by Worster (1991, 1992): in the second paper, he studies the onset of convective motion in a combined mush/liquid system, and distinguishes between two types of motion, one in the liquid, and one in the mush, with very different length scales. At high values of the Lewis number $Le = \kappa/D$, it is found that convection is initiated first in the liquid, consistent with observations of Sample and Hellawell (1984). Double-diffusive, 'finger' convection develops, and then convection in the mush is observed through the formation of channelised flow, which is initiated when the fluid flow in the mush exceeds the solidification rate. Worster (1991) studies this fully developed flow.

In this note, we summarise recent efforts to develop a pseudo-linear theory for the onset of convection in a mushy region, where we suppose the amplitude of motion in the mush is 'small', but that convection in the liquid has already been initiated, and is in fact vigorous. Thus we attempt a boundary layer analysis of the convection in the liquid, which is then parameterised to allow a description of the motion in the mush. We offer no apologies for using every approximation available, in the hope of producing a tractable theory. In particular, we assume the melt composition is close to eutectic, which enables the model for the mush to be reduced to that of convection in a porous medium.

S. H. Davis et al. (eds.), Interactive Dynamics of Convection and Solidification, 147–150.

2. Mathematical Model

The model presented here is similar to that of Fowler (1985), Worster (1986) or Hills et al. (1983). We consider solidification of a binary melt, with a liquid region $z > z_l$, a dendritic, or 'mushy' region $z_s < z < z_l$, and a fully solid region $z < z_s$. In the mush, we neglect density variations (specifically, $(\rho_s - \rho_l) \ll \rho_s$): the equations are then

$$\begin{aligned}
\nabla.\mathbf{u} &= 0, \\
\rho c_p \frac{dT}{dt} - L\frac{\partial}{\partial t}[\rho(1-\chi)] &= k\nabla^2 T, \\
\rho\frac{d}{dt}[\chi c + (1-\chi)s] + \nabla.[\rho(1-\chi)(c-s)\mathbf{u}] &= \nabla.[\rho\chi D\nabla c], \\
\frac{\partial}{\partial t}[\rho(1-\chi)s] &= \lambda c\frac{\partial}{\partial t}[\rho(1-\chi)], \\
\mathbf{u} &= -(k_p/\mu_l)[\nabla p + \rho g\mathbf{k}], \\
\rho &= \rho_0[1 - \alpha(T - T_0) - \beta c],
\end{aligned} \tag{1}$$

where c and s are liquid and solid compositions, χ is liquid mass fraction, T is temperature, Γ is the liquidus slope, $\mathbf{u}$ is the liquid flux, d/dt is the material derivative, $\mathbf{k}$ is a unit vertical vector and other symbols have their usual meanings (k_p is the permeability).

Equations of motion in the liquid are

$$\begin{aligned}
\nabla.\mathbf{u} &= 0, \\
\frac{dT}{dt} &= \kappa\nabla^2 T, \\
\frac{dc}{dt} &= D\nabla^2 c, \\
\rho\frac{d\mathbf{u}}{dt} &= -\nabla p + \mu_l\nabla^2\mathbf{u} - \rho g\mathbf{k},
\end{aligned} \tag{2}$$

and these are coupled across the mush by the usual jump-derived boundary conditions, which we omit. A point of note is the necessity to pose an 'extra' condition for χ on z_l, and we take this to be $\chi = 0$. The justification for this choice is not clear, however, and Worster (1986) prefers a condition of marginal supercooling.

3. Nondimensionalisation

We scale the variables as

$$\mathbf{x} \sim \kappa/V, \quad \mathbf{u} \sim V, \quad t \sim \kappa/V^2, \quad p - p_{hydro} \sim \kappa/k^*, \quad T - T_\infty^L \sim \nu\Gamma, \tag{3}$$

where V is a 'typical' value of $\dot{z}_l$, p_{hydro} is the hydrostatic pressure, T_∞^L is the far-field liquidus temperature, $\nu = c_E - c_\infty$ is the difference between eutectic and far field concentrations, and we have defined the permeability as $k_p = k^*\chi^2$. Now also rescale by putting

$$\chi = 1 - \nu\phi, \quad c - c_\infty \sim \nu, \tag{4}$$

and taking $\nu \ll 1$. At leading order, we then have in the mush,

$$\nabla.\mathbf{u} = 0,$$

$$\mathbf{u} = -\nabla p + Rc\mathbf{k},$$

$$(c_\infty + St)\phi_t = [1 + St/c_\infty]\frac{dc}{dt} = \nabla^2 c, \tag{5}$$

just the equations of convection in a porous medium. ϕ uncouples, St is the Stefan number $St = L/c_p\Gamma$, and R is the Rayleigh number

$$R = \Delta\rho g k^*/\mu_l V, \tag{6}$$

where $\Delta\rho$ is the density difference between eutectic and far field liquid.

The main simplification in deriving (5) is the limit $\nu \to 0$, and also we have supposed $Le \gg 1$, so that compositional diffusion can be ignored (an apparently regular approximation).

4. Reduction

The idea to parameterise the correspondingly scaled liquid equations (2) is in parallel with Roberts' (1979) boundary layer theory, generalised for double-diffusive convection. In terms of the scaled variables, we find that the compositional boundary layer above the mush is of thickness $\delta \sim Le^{-1/3}$, while the velocity is of order $\mathbf{u} \sim (R/Le\gamma)^{1/4}$, where $\gamma = k^*V^2/\kappa^2$ ($1/\mathcal{H}$ of Worster 1992). The method of solution is then as follows. In the mush $0 < z < z_l$ (as it is found that $\dot{z}_s \sim \nu/St$),

$$[1 + St/c_\infty]\frac{dc}{dt} = \nabla^2 c,$$

$$c = 1 \quad \text{on} \quad z = 0, \quad c = 0 \quad \text{on} \quad z = z_l, \tag{7}$$

so that if z_l is known, one can compute the compositional flux

$$-\frac{\partial c}{\partial n} = g. \tag{8}$$

In the liquid, we hypothesise that the convective finger width is such as to balance thermal advection with conduction. This width is then of order $\epsilon = (\gamma Le/R)^{1/4}$, typically ~ 0.1. We then find that the compositional convection problem uncouples from that of the heat flow, and the temperature satisfies

$$T_t + \frac{1}{\epsilon}\mathbf{u}.\nabla T = \nabla^2 T, \tag{9}$$

where $\mathbf{u} = \mathbf{u}(\mathbf{x}/\epsilon)$ is the (rescaled) convective flow field, and t and $\mathbf{x}$ in (9) are scaled as for the mush. This is a multiple scale problem, but at leading order, T satisfies

$$T_t = T_{xx} + (1 + D_T)T_{zz}, \quad T = 0 \text{ on } z_l, \quad T \to \Delta_\infty \sim O(1) \text{ at } \infty. \tag{10}$$

The enhanced diffusivity D_T arises through Taylor dispersion, and $D_T \propto \left\langle |\mathbf{u}|^2 \right\rangle$. The extra condition $\overline{\partial T/\partial n} = g$ (the overbar denotes a spatial average) allows determination of z_l. Thus, compositional convection does not affect the diffusional nature of the temperature field, and this is our major conclusion. Before the onset of convection in the mush, a similarity solution is appropriate, and we find that $z_l \sim At^{1/2}$, where A depends on Δ_∞, D_T and St/c_∞. A stability analysis for the mush then predicts that convection is initiated at a critical value of the dimensional mush thickness of

$$z_l > \frac{c_\infty}{(c_\infty + St)} \frac{\mu_l \kappa R_c}{\Delta\rho g k^*}, \tag{11}$$

where for a linear compositional profile and a quasi-static analysis, $R_c = 27.1$, though this is modified in the present situation. This analysis is consistent with results of Tait and Jaupart (1992, fig. 16).

References

Copley, S.M., A.F. Giamei, S.M. Johnson and M.F. Hornbecker 1970 The origins of freckles in unidirectionally solidified castings. Metall. Trans. **1**, 2193-2204.

Fowler, A.C. 1985 The formation of freckles in binary alloys. IMA J. Appl. Math. **35**, 159-174.

Hills, R.N., D.E. Loper and P.H. Roberts 1983 A thermodynamically consistent model of a mushy zone. Quart. J. Mech. Appl. Math. **36**, 505-539.

Loper, D.E. and P.H. Roberts 1981 Compositional convection and the gravitationally powered dynamo. In: Stellar and Planetary Magnetism, ed. A.M. Soward, pp. 297-327; Gordon and Breach, New York.

McDonald, R.J. and J.D. Hunt 1969 Fluid motion through the partially solid regions of a casting and its importance in understanding *A*-type segregation. Trans. Metall. Soc. AIME **245**, 1993-1997.

McDonald, R.J. and J.D. Hunt 1970 Convective fluid motion within the interdendritic liquid of a casting. Metall. Trans. **1**, 1787-1788.

Roberts, G.O. 1979 Fast viscous Bénard convection. Geophys. Astrophys. Fluid Dynamics **12**, 235-272.

Sample, A.K. and A. Hellawell 1984 The mechanisms of formation and prevention of channel segregation during alloy solidification, Metall. Trans. **15A**, 2163-2173.

Sarazin, J.R. and A. Hellawell 1988 Channel formation in $Pb - Sn$ and $Pb - Sn - Sb$ alloy ingots and comparison with the system $NH_4Cl - H_2O$. Metall. Trans. **19A**, 1861-1871.

Tait, S. and C. Jaupart 1992 Compositional convection in a reactive crystalline mush and the evolution of porosity. J. Geophys. Res., n press.

Worster, M.G. 1986 Solidification of an alloy from a cooled boundary. J. Fluid Mech. **167**, 481-501.

Worster, M.G. 1991 Natural convection in a mushy layer. J. Fluid Mech. **224**, 335-359.

Worster, M.G. 1992 Instabilities of the liquid and mushy regions during solidification of alloys. J. Fluid Mech. **237**, 649-669.

NONEQUILIBRIUM EFFECTS IN A SLURRY

D. E. LOPER
Geophysical Fluid Dynamics Institute
Florida State University
Tallahassee, FL 32306 USA

ABSTRACT. This is a synopsis of a dynamical theory of a nonequilibrium slurry recently developed by Loper (1992).

1. Introduction

A slurry is a liquid alloy containing a suspension of solid particles which may melt or freeze as thermodynamic conditions dictate. In equilibrium the thermodynamic state of a slurry containing two constituents is determined by three variables. Commonly these are chosen to be pressure, p, temperature, T, and overall composition, ξ (i.e., mass fraction of the less dense of the two constituents). In a general nonequilibrium theory, such as that developed by Loper and Roberts (1978), there are two additional independent thermodynamic variables measuring the departure from equilibrium, one for melting/freezing processes and one for change of composition of the phases. If we assume that the composition of the solid phase is constant, this removing one of the two additional variables, and the state is determined by four independent thermodynamic variables. A dynamical theory of a nonequilibrium slurry has recently been developed (Loper, 1992); this paper is a synopsis of that theory.

2. The nonequilibrium variable and its evolution equation

The independent thermodynamic fourth variable, Γ, measuring the departure from phase equilibrium is defined by the differential relation

$$d\Gamma = d\xi^{L} + [(L/T)\, d - \delta\, dp\,]/\bar{\mu}\,\xi^{L} \tag{1}$$

where δ is the change of specific volume upon melting at constant composition, L is the latent heat of fusion and $\bar{\mu} = \partial\mu^{L}/\partial\xi^{L}$ is the change of chemical potential of the liquid phase with change of liquid composition, ξ^{L}. In equilibrium ($\Gamma = 0$), (1) reduces to the differential form of the liquidus relation for a binary alloy. The sign of Γ has been chosen such that when $\Gamma > 0$ the thermodynamic state of the liquid phase lies above the liquidus surface in p, T, ξ^{L} space (i.e., on the hot side of the liquidus surface) and melting promotes relaxation to equilibrium. Thus Γ may be thought of as a dimensionless generalized temperature, and will be seen to govern the rate of melting or freezing of the slurry.

The rate-limiting process of relaxation to phase equilibrium is assumed to be compositional diffusion to and from the small slurry particles. The process of microscale

S. H. Davis et al. (eds.), Interactive Dynamics of Convection and Solidification, 151–154.

diffusion of material is parameterized in the macroscale theory, leading to a Landau-type relaxation term in the equation of evolution of Γ. Departures from phase equilibrium are driven by changes in temperature due to thermal diffusion and chances in pressure due to vertical motion. The Boussinesq form of the equation of evolution may be expressed as

$$\frac{D\Gamma}{Dt} + \frac{\Gamma}{t_\Gamma} = \kappa_o \nabla^2 \tau + \frac{\mathbf{u} \bullet \hat{\mathbf{z}}}{H_\Gamma} \tag{2}$$

where a subscript o denotes a constant value. Here $\tau = T(L/T\overline{\mu}\xi)_o$ is a dimensionless measure of the temperature, $\mathbf{u}$ is the barycentric velocity, $\hat{\mathbf{z}}$ is a unit vector anti-parallel to gravity, κ is the thermal diffusivity, t_Γ is the freezing relaxation time and H_Γ is the pressure-freezing scale height.

In the limit that the mass fraction of solid phase, ϕ, is much smaller than unity, the freezing relaxation time is given by

$$\frac{1}{t_\Gamma} = 2\overline{D}\left[1 + \left(\frac{L^2}{C_p T \overline{\mu} \xi^2}\right)_o\right]\left(\frac{6\pi^2 N^2 \rho \phi}{\rho^s}\right)^{1/3}\left[1 + \hat{\lambda}\phi^{1/3}\right] \tag{3}$$

where $\overline{D}$ is the material diffusivity in the liquid phase, ρ^s is the density of the solid phase, C_p is the specific heat, N is the number density of solid particles,

$$\hat{\lambda} = 0.624\left(\frac{(\rho^s - \rho)g}{6\pi N \nu \overline{D} \rho^s}\right)^{1/3} \tag{4}$$

is a parameter measuring the importance of particle sedimentation on the diffusion process, ν is the kinematic viscosity of the liquid and g is the magnitude of the local acceleration of gravity. Although it has been assumed that $\phi << 1$, $\hat{\lambda}$ may be large so that the last term of (3) is not necessarily small.

The pressure-freezing scale height is given by

$$\frac{1}{H_\Gamma} = \left[\frac{g}{\overline{\mu}\xi}\left(\rho\delta - \frac{\alpha L}{C_p}\right)\right]_o \tag{5}$$

In writing this term, the nearly hydrostatic assumption has been made:

$$p = p_H + p_1 \tag{6}$$

with

$$\nabla p_H = -\rho_o [1 + \overset{*}{\beta}_o (p - p_o)] g_o \hat{\mathbf{z}} . \tag{7}$$

and $|p_1| << |p_H|$. Here $\overset{*}{\beta}_o$ is the isothermal compressibility of the slurry in equilibrium.

3. The remaining governing equations

Variations in the dimensionless temperature, τ, are governed by normal thermal diffusion, by changes in pressure due to vertical motion (inducing an adiabatic temperature gradient) and by release or absorption of latent heat as solid particles melt or freeze:

$$\frac{D\tau}{Dt} = \kappa_o \nabla^2 \tau - \frac{\mathbf{u} \bullet \hat{\mathbf{z}}}{H_\tau} - \frac{\Gamma}{t_\tau}. \tag{8}$$

Here

$$H_\tau = \left(\frac{C_p \bar{\mu} \xi}{\alpha g L}\right)_o \tag{9}$$

is the adiabatic scale height and the thermal relaxation time is

$$\frac{1}{t_\tau} = 2\bar{D}\left(\frac{L^2}{C_p T \bar{\mu} \xi^2}\right)_o \left(\frac{6\pi^2 N^2 \rho \phi}{\rho^s}\right)_o^{1/3} \left[1 + \hat{\lambda} \phi^{1/3}\right]. \tag{10}$$

Due to the constraint of phase equilibrium, the equation governing conservation of the light constituent of the alloy lacks the usual Fickian diffusive term. Variations of composition of the slurry are induced by diffusion of either temperature or the nonequilibrium variable and by sedimentation of solid particles:

$$\frac{D\xi}{Dt} = \bar{D}_o \nabla^2 (\Gamma - \tau) - V \frac{\partial \phi}{\partial z} \tag{11}$$

where V is a sedimentation speed given by

$$V = \left(\frac{5 g \xi^2 \rho \bar{\delta}^*}{9 \nu}\right)_o \left(\frac{\rho^s}{6\pi^2 \rho N^2}\right)_o^{1/3} \phi^{2/3}. \tag{12}$$

The mass fraction of solid is related to the other variables by the relation

$$\xi_o \partial\phi/\partial z = -\partial\xi/\partial z + \partial(\Gamma - \tau)/\partial z - (g \rho \delta / \bar{\mu} \xi)_o. \tag{13}$$

The set of governing equations is completed by the equations of conservation of mass and momentum. The first is the familiar equation of incompressible flow

$$\nabla \bullet \mathbf{u} = 0. \tag{14}$$

The buoyancy term in the momentum equation includes the effect of composition, temperature and the lack of phase equilibrium on the density:

$$D\mathbf{u}/Dt = -\nabla p_1/\rho_o + \hat{g}\left[(\xi - \xi_o) + \tilde{\alpha}(\tau - \tau_o) + \eta \Gamma\right] \hat{\mathbf{z}} + \nu_o \nabla^2 \mathbf{u}. \tag{15}$$

where $\hat{g} = \rho_o \bar{\delta}_o^* g_o$, $\tilde{\alpha} = (\alpha^* T \bar{\mu} \xi^L / \rho \delta^* L)_o$ and $\eta = [\delta / \bar{\delta}^* \xi^2]_o$. The buoyancy term has been scaled with the compositional buoyancy; $\hat{g}$ is a reduced gravity and $\bar{\delta}^*$ is the change of specific volume upon melting at constant pressure and temperature in phase equilibrium. The parameter $\tilde{\alpha}$ measures the relative importance of thermal buoyancy compared with compositional buoyancy; α^* is the thermal expansion coefficient when phase equilibrium prevails. The dimensionless parameter η is new, measuring the change in density that accompanies a change in the nonequilibrium variable Γ. That is, in the absence of phase equilibrium, the mass fraction of solid is different from the equilibrium value. Since the solid has a different density than the liquid, this causes a change in the density.

4. Stability

When viewed in a frame of reference moving vertically with speed $\mathbf{u} = w_a\hat{\mathbf{z}}$ where

$$w_a = \frac{\left(C_p^*\overline{\mu}\xi^2\right)_o V_o \phi_b}{g_o(\rho\delta C_p - \alpha L)_o + \left(C_p^*\overline{\mu}\xi^2\right)_o \phi_b},$$

the governing equations admit a steady solution representing sedimentation in the presence of prescribed vertical gradients of both liquid composition, ζ, and solid mass fraction, ϕ_b:

$$\Gamma = \Gamma_a, \qquad p_1 = 0, \qquad \phi = \phi_o + \phi_b z,$$

$$\tau - \mathring{\tau}_o = \Gamma_a + (z - w_a t)\zeta - (g\rho\delta/\overline{\mu}\xi)_o z + (V_o\phi_b t - \phi_b z)\xi_o,$$

$$\xi - \xi_o = -(z - w_a t)\zeta - \xi_o V_o \phi_b t$$

where Γ_a is a known constant.

The linear perturbation equations yield, in the absence of buoyancy forces, four modes representing viscous decay, thermal decay, local relaxation to equilibrium and neutral sedimentation. It was previously found (Loper and Roberts, 1987) that in equilibrium ($t_\Gamma = 0$) the stability character of the sedimentation mode, when buoyancy forces are included, is anomalous for long-wavelength. That is, the normal situation of increasing ζ leading to destabilization is reversed for long-wavelength disturbances, with increasing ζ tending to *stabilize* the system. In the non-equilibrium case, Loper (1992) found that increasing the relaxation time t_Γ from zero first increases the range of wavelength over which the anomalous behavior occurs, but as t_Γ tends to infinity, the effect of ϕ_b dominates that of ζ and the normal situation prevails in which an increase of ϕ_b tends to destabilize the system. For more details see Loper (1992).

References

Loper, D. E. & P. H. Roberts (1978) On the motion of an iron-alloy core containing a slurry. I. General theory. *Geophys. Astrophys. Fluid Dynam.*, **9**, 289-321.

Loper, D. E. & P. H. Roberts (1987) A Boussinesq model of a slurry, in D. E. Loper (ed.) *Structure and Dynamics of Partially Solidified Systems*, Martinus-Nijhoff, Dordrecht, pp. 506.

Loper, D. E. (1992) A nonequilibrium theory of a slurry, *Continuum Mechanics and Thermodynamics*, to appear.

NEW EXPERIMENTS ON COMPOSITIONAL CONVECTION

S. R. TAIT and C. JAUPART
Institut de Physique du Globe, 4, Place Jussieu,
75252 Paris Cedex 05, France

Compositional convection in a reactive crystalline mush is a process of importance in numerous fields. Theoretical foundations for governing equations consistent with thermodynamic principles have been laid out by Hills et al. (1983). Constitutive laws must be postulated to describe the interactions between solid and liquid. The equations have been reduced to a porous medium, Darcy-flow model, but experimental verification of this approach is not yet available. In the mush, the coupling between fluid flow and liquid/solid equilibrium constraints imply that permeability varies both in space and time. Convective upwellings can become focussed into narrow vertical channels devoid of crystals, called "chimneys". The majority of observations on how chimneys develop in a mush come from laboratory experiments with aqueous ammonium chloride solutions (Copley et al., 1970; Sarazin & Hellawell, 1988), however the small solution viscosity leads to several drawbacks. The horizontal scale of motions is close to the spacing of crystals in the mush, which violates the assumptions behind Darcy's law. Compositional convection starts before initial irregularities of crystal nucleation have been smoothed out and before the crystal/liquid interface has reached thermodynamic equilibrium. Another difficulty is the lack of knowledge of the permeability.

We designed a set of experiments to overcome these difficulties. Adding small quantities of a polymer to ammonium chloride solutions increases viscosity without perturbing the phase diagram. This allows us to affect the rate of development and wavelength of convection. Solutions were cooled from below causing the growth of a mush. Convective instabilities both of the chemical boundary layer above the mush and that of the interstitial fluid within the mush were observed. Details of the flow patterns are described elsewhere (Tait and Jaupart, 1989, 1992), but fig.1 indicates how chimney convection differed for experiments with similar thermal conditions but different viscosities.

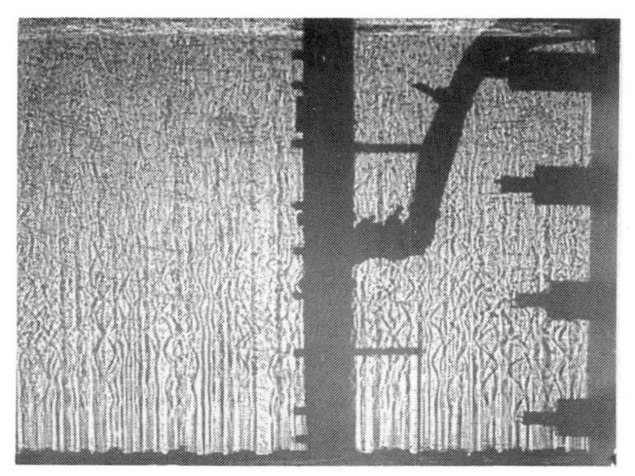
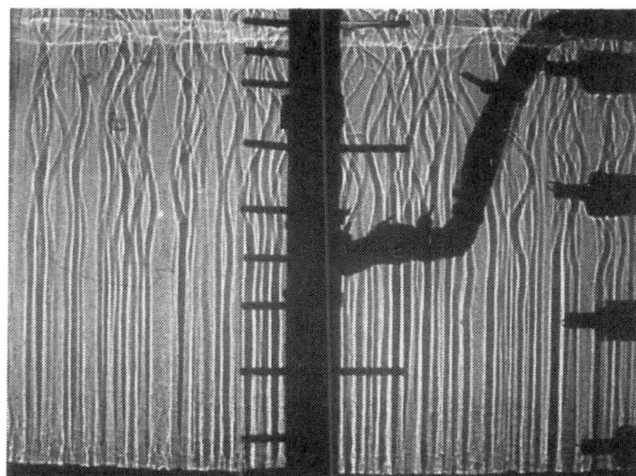

Fig.1 Showing the effect of viscosity on chimney convection. Solution viscosities are a) 1.5 mPa.s. b) 5.5 mPa.s

S. H. Davis et al. (eds.), Interactive Dynamics of Convection and Solidification, 155–158.

At relatively high viscosities, only the boundary layer above the mush became unstable. This occurred when a local Rayleigh number reached a critical value (Ra_c) (Tait and Jaupart, 1989). In a porous medium of constant thickness and permeability, convection occurs when Ra_p exceeds a critical value of 27 (Lapwood, 1948; Nield, 1968). This criterion is hard to test experimentally because the permeability function is unknown. In this study, by increasing the viscosity, we are changing the Rayleigh number at constant permeability and hence are able to check the criterion without making any assumption on the structure of the mush. With sufficiently low viscosities, after a phase of small-scale convection from the boundary layer at the top of the mush, stronger upwellings appeared at the same time as porosity fluctuations in the mush. We took the presence of porosity fluctuations as a first sure sign of porous medium convection. This is admittedly a visual criterion subject to error, however errors are expected to be small compared to the range of mush thicknesses and Rayleigh numbers investigated. At constant superheat the data satisfy the following criterion (fig.2):

$$Ra_p = \frac{g\ \gamma\ \Delta T\ h\ \Pi_o}{\kappa\ \mu} = R_c \tag{1}$$

The critical value is 25 for low superheats (ST), and decreases with increasing superheat.

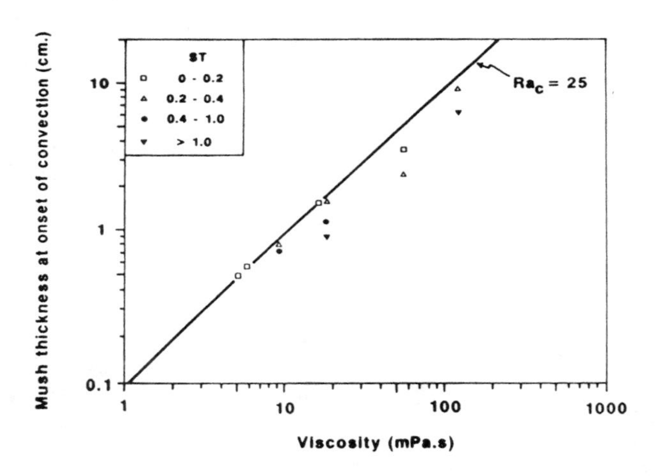

Fig.2 The mush thickness at which convection of the interstitial fluid began. The line has a slope of 1 which is consistent with equation (1).

The mechanism by which porosity fluctuations can eventually focus into chimneys can be understood by considering the equation for composition, transformed into an equation for temperature (T) using the liquidus relation (Tait and Jaupart, 1992):

$$\frac{\partial}{\partial t}[\chi(T-Ts)] = -\mathbf{U}.\mathbf{v}(T-T_S) \tag{2}$$

where **U** =(U,V,W) is Darcy velocity, χ is porosity and T_S the melting point of pure solid. Convective motions were predominantly vertical and the vertical temperature gradient was substantially larger than horizontal gradients. Hence, we can simplify (2) to:

$$\frac{\partial \chi}{\partial t} = - \frac{\chi}{(T-T_S)} \frac{\partial}{\partial t}(T-T_S) - \frac{W}{(T-T_S)} \frac{\partial}{\partial z}(T-T_S) \qquad (3)$$

This shows two contributions to the evolution of porosity: one due to temperature changes and the second due to the advection of solute. Within the mush, the constraint of local thermodynamic equilibrium implies that the system must attempt to smooth out compositional fluctuations associated with vertical motions such that the interstitial fluid remains on the liquidus. Chemical diffusion cannot act over the required scale, and the compositional budget is closed through dissolution and precipitation. In the early stages of convection, upwellings and downwellings are expected to have similar shapes of velocity profile (fig.3a). In downwellings both cooling and advection cause porosity and hence permeability to decrease. Exactly between adjacent upwellings and downwellings, the vertical velocity is zero (fig.3a), only the cooling term acts, and porosity decreases. In upwellings, cooling tends to reduce porosity and advection tends to increase it. In the centre of an upwelling, the velocity has a maximum value and advection may dominate cooling and cause porosity to increase. The dependence of permeability on porosity implies a "feed back" effect, which may ultimately produce chimneys (fig.3c).

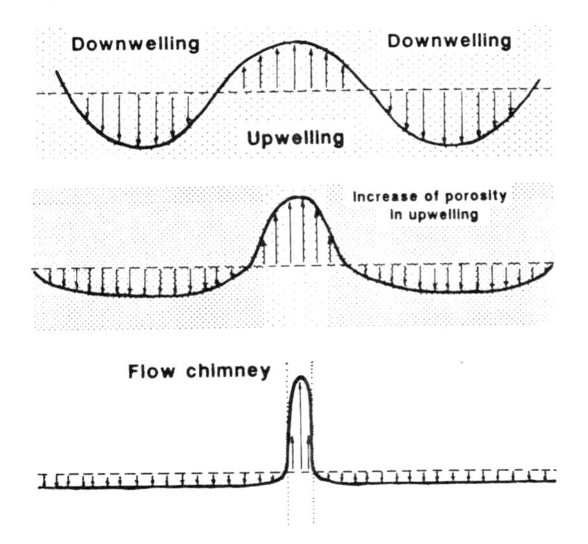

Fig.3 a),b),c) Showing profiles of vertical velocity (W) in the mush. Fluctuations of porosity ultimately focus into chimneys.

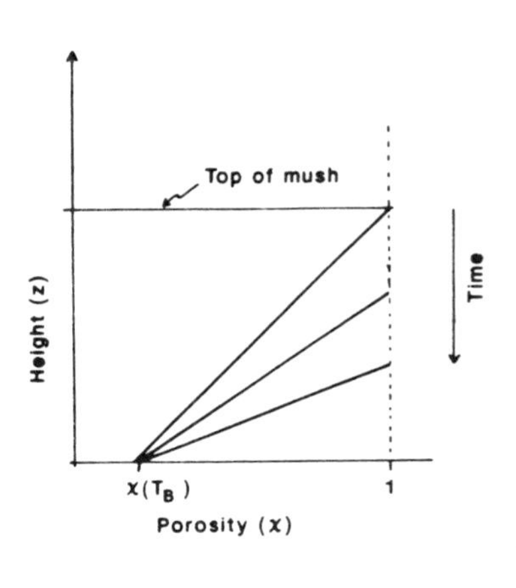

Fig.4 Showing the evolution of porosity in a developing chimney.

Sarazin and Hellawell (1988) have suggested that chimneys nucleate on convective instabilities in the upper boundary layer and propagate downward into the mush. Our observations suggest another interpretation. Convection developed on two distinct scales:

initially on a small scale associated with the compositional boundary layer at the top of the mush, and subsequently on a larger scale, associated with porosity fluctuations in the mush. In solutions with higher viscosities the initial regime of small-scale convection was vigorous, but chimneys did not form because the critical porous medium Rayleigh number was not attained. We observed the formation of chimneys to start with the onset of porous medium convection, independently of the upper boundary layer instability. In short, the formation of chimneys is part and parcel of convection in a reactive porous medium at high Lewis number. The observation that chimneys apparently propagate downwards can be rationalised by noting that a gradient of porosity is present in the mush and that velocity increases with height. Porosity is simultaneously increasing over the whole mush height in a developing chimney, and will first reach 1 at the top of the mush and subsequently at progressively lower levels (fig.4).

The diameter of chimneys was insensitive to superheat and was found to increase with solution viscosity (fig.5). Although convection, chimney spacing and thermal conditions steadily changed with time, in a given experiment, chimney diameter remained constant. When two chimneys coalesced, the resulting chimney had the same diameter as those of the individuals from which it developed. This shows that chimney diameter is determined by a local balance. The fact that chimney diameter is unaffected by variations of chimney spacing emphasizes that these two characteristic length scales are determined by different phenomena.

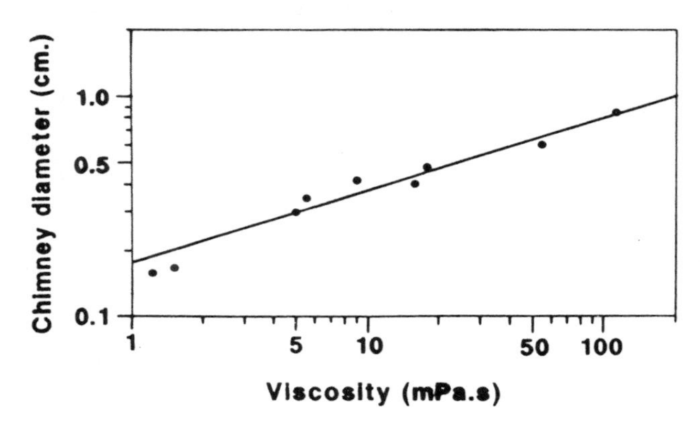

Fig.5 Chimney diameters shown as a function of viscosity.

REFERENCES

Copley S.M., Giamei A.F., Johnson S.M. and Hornbecker M.F. (1970). Metall. Trans. 1, 2193-2204

Hills R.N., Loper D.E., Roberts P.H. (1983). Q.J. Mech. Appl. Math. 36, 505-539

Lapwood E.R. (1948). Proc. Camb. Phil. Soc. 44, 508-521

Nield D.A., (1968). Water Resources Research 4, 553-560

Sarazin J.R. and Hellawell A. (1988). Metall. Trans. 19A, 1861-1871

Tait S. and Jaupart C. (1989). Nature 338, 571-574

Tait S. and Jaupart C. (1992). J.Geophys. Res. (in press)

CONVECTIVE ASPECTS OF SOLIDIFICATION EXPERIMENTS UNDER LOW GRAVITY

S. REX
ACCESS e. V.
Intzestr. 5
5100 Aachen
Germany

ABSTRACT. The application of a low-gravity environment is generally accepted for investigations of convection affected processes. A variety of existing and planned means of access to low gravity is offered to investigators. Some of the main results of the past concerning metal solidification are reported in this article. Different approaches have been made to analyse the experiments sensitivity to residual accelerations. However, the impact of so-called g-jitter is not well understood. It is concluded that the mathematical modeling of the effect of residual and time-dependent accelerations should be supported more intensively by experimental results dedicated directly to this issue.

1. Introduction

The low gravity environment offered by an orbiting spacecraft to perform new classes of solidification experiments has been vehementely in discussion since the days of SKYLAB in 1975. Only a few years ago it was very popular to take the view that "no physicist is being in doubt that the effect of gravity can exactly be calculated in any kind of experiment and that therefore the result of experiments under zero-gravity conditions can be predicted", Krupp and Ruppel (1986). The opposite party some ten years ago used to state that by simply switching of the earth´s gravity field by the use of a spacecraft will immediately enable production e.g. of ultimate crystals and remove all the cumbersome problems of convection.

However, in the course of time both parties become more realistic. The solidification experiments performed in the space program so far mainly proved that there still remains a large lack of understanding concerning the basic physical processes and interdependence of e.g. the dynamic of solidification and convection. It has become obvious that the new possibility to vary the earth´s fundamental parameter gravity has opened up the curtain to an up to now not conceivable field of new and fascinating investigations.

First this paper will give a short review of how to achieve low gravity and what kind of flight opportunities are available today and in the near future. It will then be outlined which area of solidification science research has been tackled in the past with low gravity experiments and what are the main results and consequences. The following chapter deals with the so-called g-jitter. The main question here is how to qualify the reduced gravity environment, up to what levels and frequencies of residual gravity and vibrations may be tollerable and, on the other hand, what would be the influence of very small and/or a modulated driving force for convection. In the final conclusions there will be given some personal recommendations concerning future solidification experiments.

S. H. Davis et al. (eds.), Interactive Dynamics of Convection and Solidification, 159–178.

2. How to lower gravity for experimental conditions

The gravitational field of earth is usually known as to be cumbersome with respect to experimental requirements. Electrical fields may be shielded by use of a Faraday cage, the earth's magnetic field locally by a Helmholtz coil. But the gravitational field, on earth can only be compensated by applying an inertia field of opposite direction. However, for a macroscopic sample this compensation equals only to the center of mass.

Only in case of a body free falling in the earth's gravitational field the compensation by the inertia force works completely. Therefore, as long as there had been no choice to performe experiments under zero-gravity conditions for a longer period, the scientists have restricted themselves to only short duration experiments. These experiments lasting at best some seconds could never usually reach any steady state of the investigated process. This method of free falling experiments recently has become popular again due to the drop tower Bremen providing 4.7 sec. free-fall time regardless of the various opportunities that meanwhile exists to carry out space experiments. Solidification experiments dedicated essentialy to short time containerless processing prefer the use of a classical highly evacuated drop tube. Such a drop tube depending on the special requirements of the experiment can be either tall like the drop tube Grenoble with a total height of 47.1 m or short like e. g. in the ACCESS laboratory providing still 0.6 sec. free-fall time which is sufficient to melt and solidify samples up to 0.5 mm in diameter.

Any rigid body free falling in the earth's gravitational field follows an eliptical orbit even if this body consists of a small laboratory. In the vicinity of the earth's surface this elliptical orbit is transformed to a parabolic flight path. A german rocket campaign called TEXUS for technical experiments under weightlessness (German: Schwerelosigkeit) is continuously going on since 1976 using sounding rockets being launched from northern Sweden one to three times a year. Such a rocket flight following a parabolic flight path up to 250 km in hight offers a reduced gravity environment over some 360 s. before the reentry into the earth's atmosphere. An upgraded version called MAXUS to provide 15 min. low gravity time unfortunately fails on his maiden flight in spring '91, but will be certainly succeed and available on a regular basis in the near future.

The borderline case of an elliptical orbit is represented by a circle, which means in this context the circular low earth orbit for the manned SPACELAB on board the shuttle or for automatic experiment platforms like EURECA or other retrievable carriers. This is the place to provide a low gravity environment for a sufficient period of time to perform all classes of solidification experiments.

In Table 1 the variety of existing and planned means of access to low gravity is listed including some of the carriers of Japan, China and the former USSR which become available recently as well.

The different flight opportunities differ remarkably in the space of time usable for low gravity experiments but also with respect to the quality of this particular environment. As depicted out in Fig. 1 in a first time the attention was solely drawn to the residual g-level on board the different carriers. This also can be seen on the linguistic evolution of the term 'weightlessness' or 'zero-gravity' which has been changed totally just to 'reduced' or 'low gravity environment'. However, soon it became clear that the dominant parameter to characterize the quality of the g-level has to be seen in the frequency spectrum and the amplitude of the residual micro- accelerations, as called by Tiby and Langbein (1984). This will be outlined in more detail in the next but one chapter.

Means	Microgravity Level(g)	Duration	Programmes
TERRESTRIAL			
Drop tubes and towers	10^{-5}	3 to 10 s	MSFC (USA), BREMEN (FRG) Grenoble (France), Kamisunagawa (Japan)
BALLISTIC FLIGHTS			
Aircrafts	10^{-2}	15 to 30 s	KC 135 (USA) Mitsubishi MU300 (Japan), Caravelle (France)
Sounding rockets	10^{-4}	5 to 15 min.	Texus, Maxus (Europe) — Starfire, Connatec (USA) — TR1 (Japan)
ORBITAL FLIGHTS			
Shuttles	10^{-2} to 10^{-4}	7 to 10 days	Spacehab (USA) — Spacelab (Europe) Buran (USSR)
Retrievable capsules	10^{-5}	up to weeks	Photon (USSR) — Service (USA) FSW (P.R. China) — Raumkurier (FRG)
Space stations	10^{-2} to 10^{-4}	up to months	International Space Station (USA, Europe, Japan, Canada) — MIR (USSR)
Automatic satellites	10^{-5} to 10^{-7}	up to months	ISF (USA) — Eureca (Europe) Columbus MTFF (Europe) — SFU (Japan)

MSFC : Marshall Space Flight Centre

Table 1: Existing and planned means of access to microgravity (non exhaustive list), OFTA (1990)

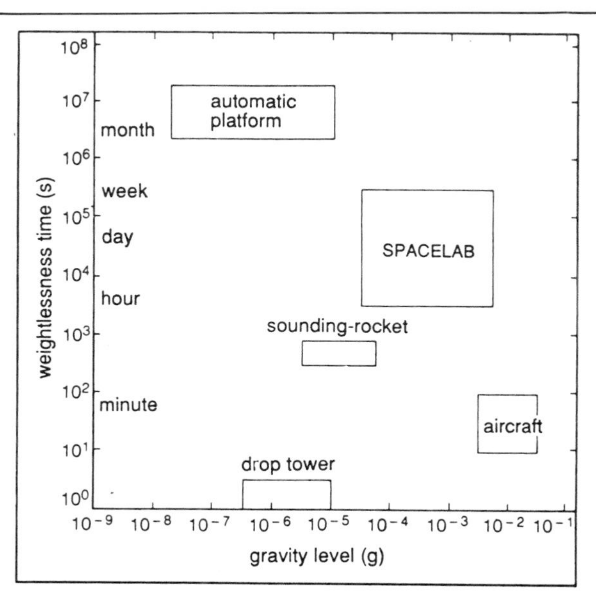

Figure 1: Nominal acceleration levels and effective free-fall times for different low-g flight opportunities

3. Solidification experiments using low gravity

The majority of the results obtained up till now with regard to solidification experiments under low gravity are from the spacelab missions, mainly of the "first spacelab payload missions", FSLP in 1983 and the German mission D1 in 1985. This is due to the fact that sounding rocket missions do only meet the requirements of very few experiments in solidification science. However, rocket flights can be very important for first attempts and technological problems of the experiments performance.

The results are meanwhile are well published and gathered among others in the books of Walter (1987) and Feuerbacher et al. (1986) and recently in a review article of Sahm and Keller (1991). The latter also describe in more detail the different experiment facilities which today are available and being developed for spacelab missions as well as for automatic operations on EURECA. However, all these publications mainly cover the activities of western-Europe and the United States. The results and experiences of the eastern hemispher are much less known and appear just recently in the international journals.

But what can be gained by reducing gravity in solidification experiments? The imagination to switch of gravity in a normal terrestrial laboratory is usually described as follows:

- the metallostatic pressure is cancelled
- no buoyancy nor sedimentation takes place
- density driven convection fails.

In our context the suppression of density driven convection is the most important factor since the central purpose of all solidification-linked low-gravity research is to increase our understanding of interactions between fluid mechanics and the liquid-solid phase transition. This may be realized by separating gravity-induced transport mechanisms from gravity independent processes such as

- diffusion
- Marangoni convection caused by surface tension differences
- convection induced by the volume change at the transformation from liquid to solid

to study the contribution of individual mechanisms to the overall heat and mass transport.

If we consider an alloyed melt which is directionally solidified with a planar interface there is a pile-up of concentration in front of the solidification interface due to the in general less solubility of the growing crystal. This pile-up is produced at the advancing interface and transported into the bulk liquid by diffusion, causing a concentration boundary layer in the melt as first describd by Tiller et al. (1953). In the presence of gravity this purely diffusiv controlled boundary layer in practice can not be observed in metallic melts due to their very low Prandtl- numbers. Even in thermal stable configurations when solidification is performed vertically upwards easily minor lateral temperature gradients occure e.g. due to different thermal conductivities of melt and crucible causing convective motion in the melt.

Thus, Tensi et al. (1989) measured the thickness of the concentration boundary layer of an Al-0.3wt%Cu sample under 1g and reduced gravity conditions. The alloy was solidified in the "Gradient furnace with quenching" (GFQ) during the D1 spacelab mission (Fig. 2). The concentration profile was in close agreement with calculated profiles using the pure diffusion theory.

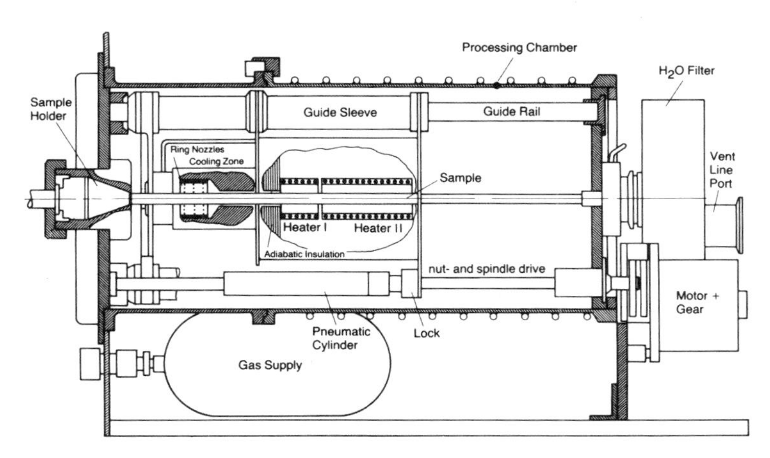

Figure 2: The Gradient Furnace with Quenching (GFQ) is a Bridgman-type furnace with two individually controlled heating zones and a water cooled zone. The solid-liquid interface is located in the adiabatic zone to minimize radial temperature gradients. For quenching, the entire furnace is displaced rapidly, spraying the liquid section of the sample to be quenched with water (courtesy of Dornier Deutsche Aerospace).

The characteristic thickness of the solute boundary layer was 2.35 times larger in space than in the terrestrial reference experiment.

In the case of a solute accumulated at the interface and being specificly lighter than the mean density of the alloy additionally solutal induced convection occurs in the earth's gravitational field. Depending on the particular orientation of the solutal and/or thermal gradients relativly to each other and to the vector of gravitational force different figures of convection in the melt may take place as described e.g. in the book of Turner (1973).

The effect of such thermosolutal convection on macrosegregation was demonstrated by Rex and Sahm (1987). A sample of Al-0.3wt % Mg was solidified in a Bridgman configuration on earth and in the GFQ furnace during the D1 mission. For the low Mg concentration a convectively unstable boundary layer was predicted only for 1 g conditions, as calculated by Coriell (Fig. 3). Heat and mass transport in the melt were purely diffusion controlled under low-g conditions. Fig. 4 shows the axial concentration profiles of the 1 g and low-g specimens. In the ground experiment convective melt motion dimished the Mg pile-up at the solidification front. The concentration profile of the flight sample fulfilled the classical case of segregation without mixing in the melt.

Besides the stability calculations of Coriell et al. (1980 u. 1981) determining the onset of convective mixing in an thermosolutal configuration another approach to the problem was made by Favier and Camel (1983). They investigated in an 'order of magnitude (OM) analysis' the different regimes of convective transport and the related makrosegregation profiles (Fig. 5). This scaling analysis includes in particular the variation of gravity as one of the experiment parameters.

The steady state diffusion boundary layer for a given solidification rate is directly coupled with a drop of the liquidus temperature determined by the equilibrium phase diagram. If the temperature gradient in the melt imposed by the heat flow is too low to compensate the concentration pile-up ahead of the solidification interface a zone of constitutional supercooling in the melt occurs (Fig. 6). The planar solidification front becomes unstable and the morphology changes to cells or dendrites. It

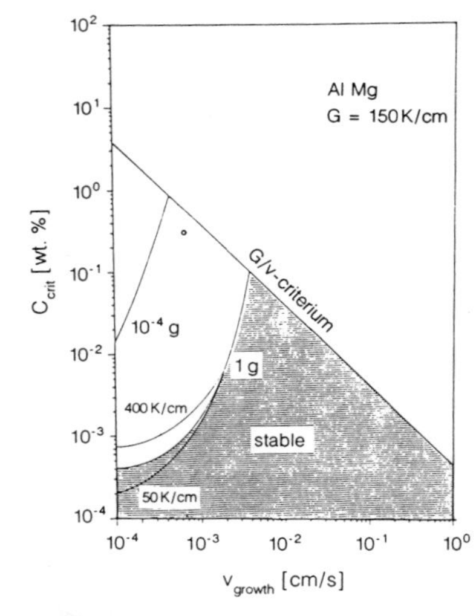

Figure 3: Stability computations were made by S. R. Coriell for AlMg to determine the stability boundaries. The diagram shows the critical concentrations as a function of growth velocity, for the onset of morphological instability (G/v-criterion) and convective mixing. The circle indicates the experimental parameters chosen for the D1 experiment (Rex and Sahm, 1987).

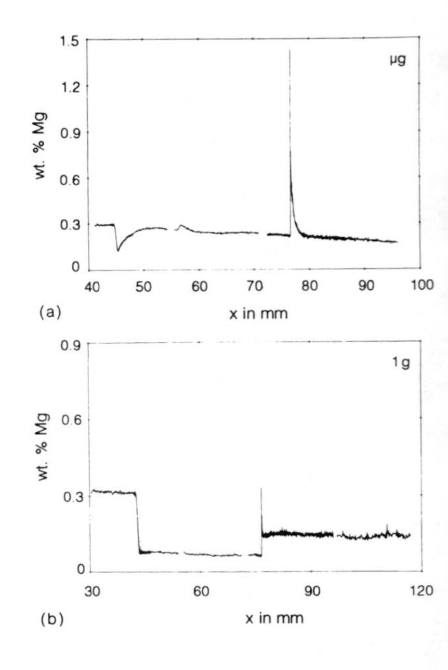

Figure 4: The macrosegregation profile of the μg solidified sample (a) differs substantially from that of a 1g-reference sample. Without convection, only a 4 mm initial transient is needed here to reach nominal composition. The ground sample (b) remains at the low level given by the distribution coefficient. Furthermore, regarding the total loss of Mg, this seems to be clearly enforced by convective mixing in the ground experiment (Rex and Sahm, 1987).

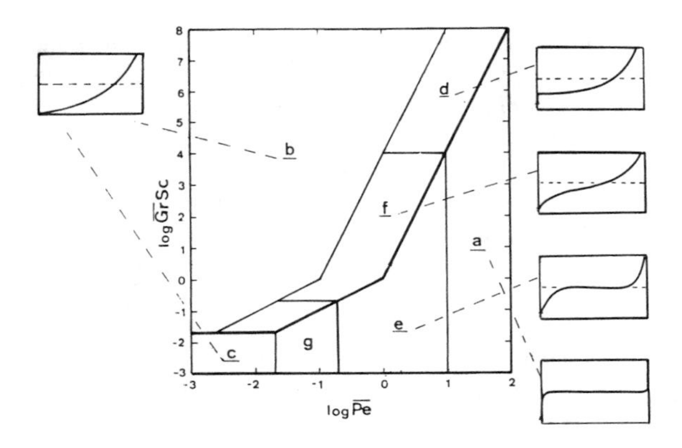

Figure 5: Domains in the log Pe versus log GrSc plane, corresponding to the different longitudinal macrosegregation regimes. The segregation profiles are shown schematically.

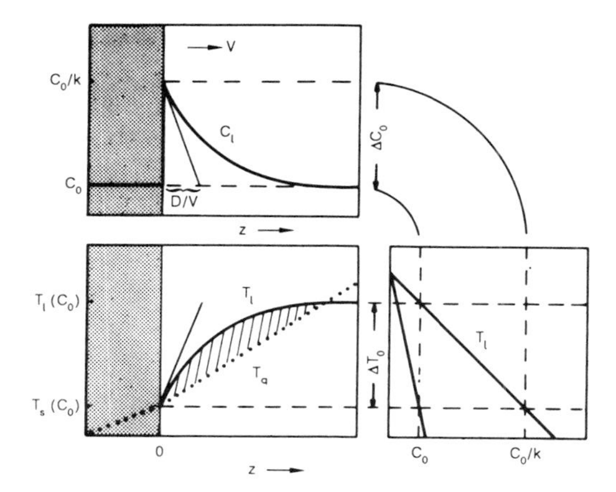

Figure 6: Constitutional supercooling in alloys. The steady-state diffusion boundary layer for a given growth rate is shown in the upper diagram. It corresponds to the concentration pile-up ahead of the solidification front. The temperature profile T1 taken from the phase diagram (lower right), has been transferred to the diagram in the lower left. The actual temperature Tq is lower then T1. The cross-hatched region is called *zone of constitutional supercooling* (Kurz and Fisher, 1984).

is obvious that via the concentration boundary layer the morphological stability boundaries are directly coupled to the convective motion in the melt. The same is true for the wavelength of the instabilities and of the developing microstructure.

A lot of work has already been done to investigate the interaction of thermal or solutal buoyancy driven convection with the solidification morphology using the advantage of a low-g environment. Billia et al. (1987) conducted experiments on the cellular microstructure of directional solidified PbTl samples in the GHF during the D1 mission (Fig. 7). Due to convection in the melt at the interface in concentrated alloys (25, 30 and 40 wt% Tl) the morphology of the cellular pattern was strongly affected. For instance, a comparison of earth and space samples of the Pb-25wt%Tl alloy shows several convection induced rolls that have developed in the ground experiment, destroying the regularity of the cells (Fig. 8). In addition the average cell size is larger in space than on earth. The morphological stability models have been developed considerably, in particular stimulated by the space experiments. The agreement between theory and experiments concerning the cell size under given solidification parameters is, however, still not yet in an acceptable shape. It looks as if fluid flow interaction of the growth mechanisms is not understood completely.

Dendritic growth was investigated during the D1 mission by Tensi and by Camel. The Al-7wt%Si samples of Tensi et al. (1989) were directionally crystallized to study the influence of gravity on the spacing of dendrite arms. The experiments were made at growth velocities of 5 and 8 mm/min. A comparison of the experimental results for 1 g and low-g conditions shows the dendrite arm spacing to be nearly the same for the higher solidification rate; however, for the 5 mm/min growth rate the dendrite arm spacings under low-g conditions are much lower in comparison with the 1 g condition.

Dendritic growth on concentrated Al-26wt%Cu alloys under 1 g and low-g conditions were studied by Camel et al (1987). The results of the flight experiments were in excellent agreement with theoretical predictions: no radial nor longitudinal segregation, except in the final transient, is prevalent, i. e. purely diffusion-controlled conditions.

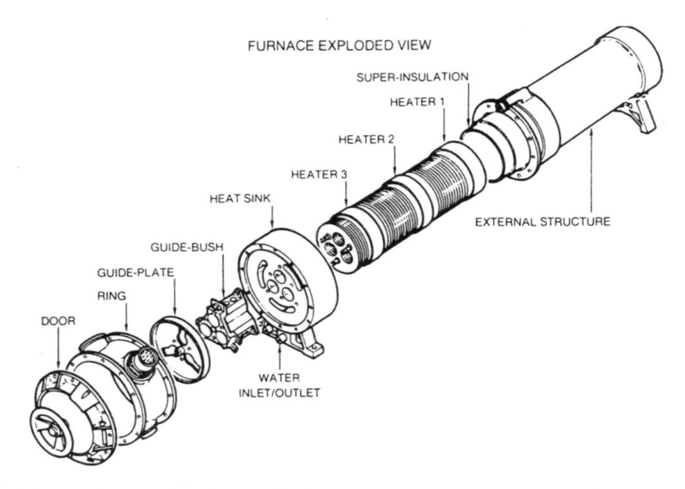

Figure 7: The gradient heating facility (GHF), consists of three independently controlled heaters allowing directional processing while sustainine a defined temperature gradient at the solid-liquid interface. Three such heating systems are integrated in the GHF, allowing three samples to be simultaneously processed (ESA, 1989).

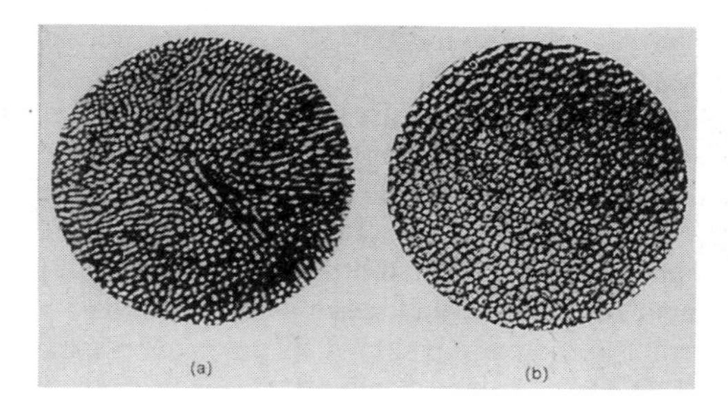

Figure 8: Comparison between corresponding cross sections of Pb-30wt%Tl samples grown vertically (a) on earth (b) in space. On earth the shape of the cells is affected by four convective rolls. In space the cells are more regular and are coarser (Billia et al., 1987).

The primary-arm spacing is found to be five times larger than in the ground experiments. In Figure 9 the ideal regularity and large size of the dendritic array of the flight sample is clearly to be seen. Figure 10 shows the variation of the primary dendrite spacing with growth velocity for this alloy. The low-g result corresponds with the classical diffusion law, whereas the ground sample confirms the scaling law prediction of the growth regime affected by convection. This space experiment validated the modeling of solutal convection in dendritic growth.

Concerning the eutectic microstructure several eutectic alloys were directionally solidified utilizing a variety of growth velocities in 1 g and low-g conditions. To study the influence of convection on solidification morphology interfibre spacing were compared between 1 g and flight experiments. As yet no simple correlations have been possible.

Larger rod separations of a space-grown Al-Al_3Ni eutectic were detected by Favier and de Goer (1984) in samples solidified in sounding rocket flights (TEXUS VI) and spacelab FSLP. Investiga-

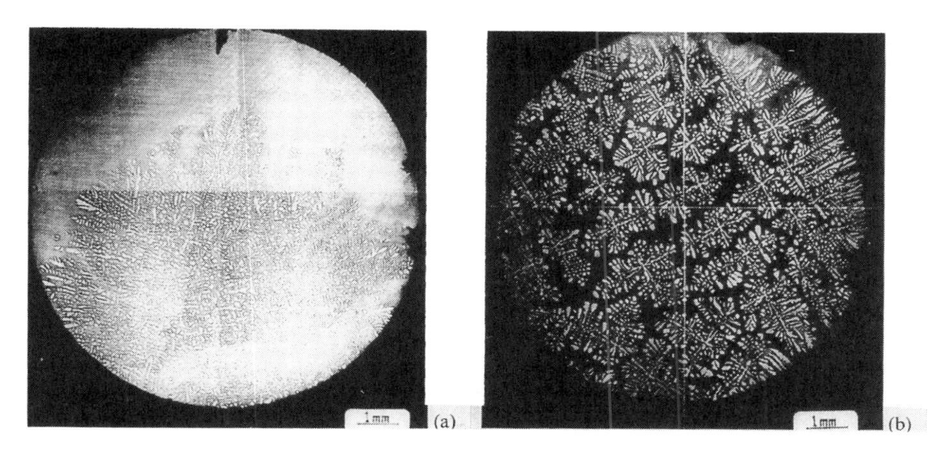

Figure 9: Comparison between corresponding cross-sections of Al-26wt%Cu samples, solidified (a) on the ground and (b) in space, shows the presence of radial segregation in the first case and its absence in the second; the latter also displays coarser and more reglar dendrites (Camel et al., 1987)

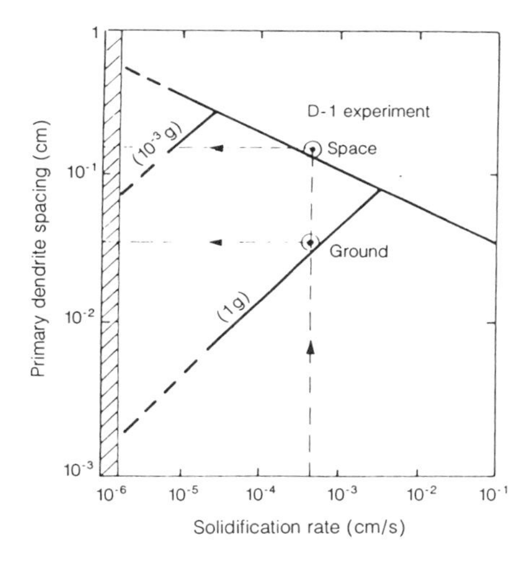

Figure 10: Primary interdendritic spacing versus growth rate: (solid line) theoretical, (•) ground and D1 experiment (Al-26wt%Cu, G = 25 Kcm-1; Camel et al., 1987).

tions with CuAl2-Al on TEXUS IV and VI and confirmed on spacelab FSLP were reported by the same researchers. The interlamellar distance of the eutectic was not affected by the change of the g-level.

An investigation with InSb-NiSb in the same furnace on spacelab FSLP showed the opposite result. The mean inter-fibre distance, measured in 3 equivalent samples solidified at growth rates ranging from 0.15 to 0.5 mm/min, was about 16 % lower in the low-g samples. These results were confirmed first by a sounding rocket experiment (TEXUS X) in which pulse marking was used to

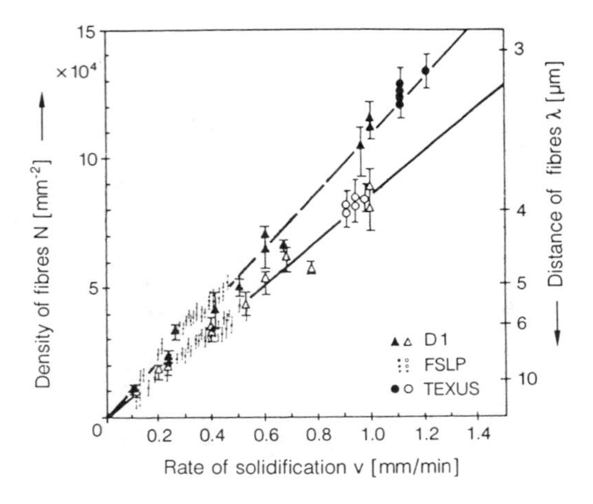

Figure 11: Density of fibers N versus growth rate v for microgravity (filled symbols) against 1 g reference experiments (open symbols) with bottom seeding (Müller and Kyr, 1987).

precisely determine the microscopic solidification rate. A second confirmation and extension of the data was obtained with a D1-experiment. A summary of all measurements is shown in Figure 11, Müller and Kyr (1987).

Smaller fibre separations in space had also been observed in MnBi-Bi eutectics during earlier American rocket flights (SPAR I,VI and IX) by Pirich et al. (1980) and by Tensi et al. (1989) during spacelab D1 using the Al-Si eutectic system. At low crystallization rates (5 mm/min) he found smaller interfibre distances in the low-g than in the 1 g sample. At higher velocities, the microstructure appears to become independent of the g-level (Fig. 12).

Table 2 summarizes the three different apparently contradictory observations which have been gathered in the low-g experiments on eutectic solidification. Consequently these results have provoked a deeper engagement in the models of eutectic growth. Eisa et al. (1986) modified the diffusive model of Jackson and Hunt (1966) and studied the influence of thermal or thermosolutal convection on eutectic solidification. They decided that the absence of convection in space is not sufficient to understand the experimental results.

However Favier and de Goer (1984) extended this idea by allowing slightly off-eutectic solidification. This assumption makes for a drastic increase in the thickness of the diffusive concentration layer, comparable to the planar front solidification. However, the reason for the shift from the initially eutectic composition is not quite clear.

According to their theory, for fibrous systems the fiber separation should increase for hypo-eutectic compositions (e. g. the Al side of Al-Al_3Ni) whereas in hypereutectic alloys it should decrease when buoyancy driven convection is suppressed. Calculations of Müller and Kyr (1987) on the strength of this model in fact agree quite well with the results concerning the InSb-NiSb eutectic. In lamellar eutectics, when the volume percentage of the two phase are approximately equal, according to the model the lamellar distance should be independent of convection conformable with the results of Favier and de Goer (1984) on Al_2Cu-Al eutectic alloy.

Tensi et al. (1989) pointed to the difference of latent heat released by the two phase during eutectic solidification and the differences of the specific density change during phase transition as well (Fig. 13). Both effects are not taken into consideration by the models of eutectic solidification so far.

The low-g experiment results reported so far being far away from being complete, are related nearly exclusively to directional solidification of metals. Nothing was stated about solidification of monotectic and peritectic systems nor of metal-ceramic composites. Nevertheless these subjects of solidification research, as well as the experiments on crystal growth processes of semiconductors,

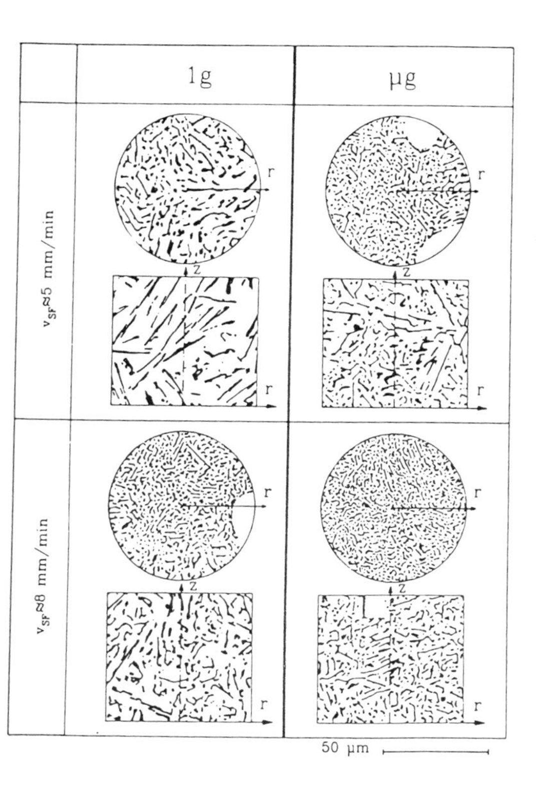

Figure 12: Comparison of Al-Si-eutectic microstructure solidified under μg and 1 g conditions for different growth rates at a constant temperature gradient (G ~ 15 K/mm), carried out on a spacelab-D1 experiment using the GFQ (Tensi et al., 1989)

Investigator		Flight	System	+)	Results
Hasemeyer	[16]	Skylab	Al-Al_2Cu	*l*	lower fault density
Yue	[15]	Skylab	NaCl-NaF	*f*	longer fibres
Yue	[15]	ASTP	NaCl-LiF	*f*	longer fibres
Larson	[18]	ASTP	MnBi-Bi	*f*	smaller fibre distance
Larson	[18]	SPAR I	MnBi-Bi	*f*	smaller fibre distance
Larson	[18]	SPAR VI	MnBi-Bi	*f*	smaller fibre distance
Larson	[18]	SPAR IX	MnBi-Bi	*f*	smaller fibre distance
Favier	[19]	Texus IV, VI	Al-Al_2Cu	*l*	no difference
Favier	[19]	Texus VI	Al-Al_3Ni	*f*	larger fibre distance
Müller	[21]	Spacelab I	InSb-NiSb	*l*	smaller fibre distance
Favier	[20]	Spacelab I	Al-Al_2Cu	*l*	no difference
Favier	[20]	Spacelab I	Al-Al_3Ni	*f*	larger fibre distance
Müller	[21]	Texus X	InSb-NiSb	*f*	smaller fibre distance
Müller	[22]	S/L D-1	InSb-NiSb	*f*	smaller fibre distance
Sprenger	[24]	S/L D-1	Ni/Al-Mo	*f*	larger $(G/R)_{crit}$

+) microstructure: *l* = lamellar, *f* = fibrous

Table 2: Low-gravity experiments on directional solidification of eutectic composites

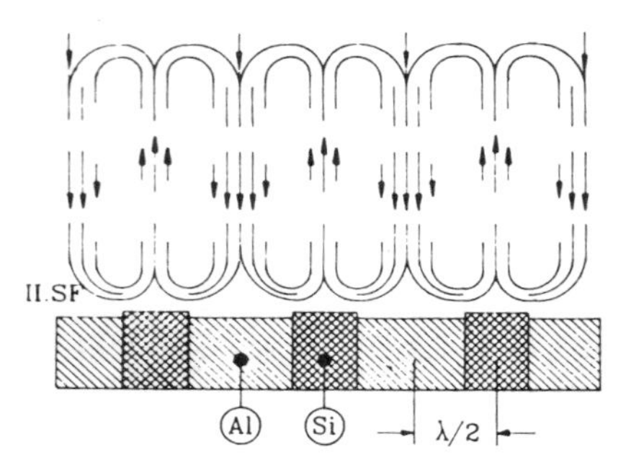

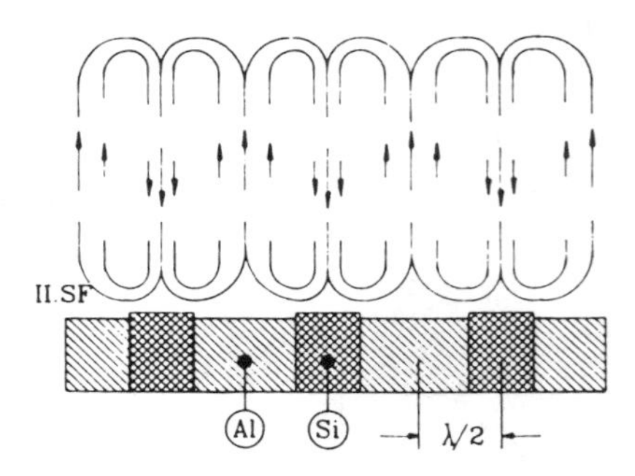

Figure 13 a and b: Convection at the eutectic solidification front (schematically)
a: by the difference of the local volume jumps of Al and Si,
b: by the difference of the latent heat of Al and Si (Tensi et al, 1989)

dealing additionally with the problems of Marangoni convection on free surfaces, have produced excellent results which stimulated the development of improved models.

To finalize this chapter some of the experiments and experimental methods shall be reported and since, being now in preparation, are intended to answer probably the remaining open questions.

Obviously, local information on the solidification process by in situ monitoring of the crystallisation front is of prime importance in fundamental studies. Current pulses inducing periodical striations, thus delineating the growth front, have been applied, mainly to semiconductor samples. A Peltier demarcation technique has been successfully used in a sounding rocket furnace (TEM 02-3) and in the gradient heating facility GHF (Fig. 7). In addition a further in situ diagnostic technique, which makes use of the Seebeck effect, is being offered by the MEPHISTO-facility. The determination of the electro-magnetic force at the two solid-liquid interfaces of the same sample makes it possible to obtain continuous feedback about the interface temperatur (Fig. 14). At present a further diagnostic system using ultra-sound are under investigation either to monitor in situ the convective motion in the melt or the actual position of the solid-liquid interface in time, ESA (1992). Another path offering insights into solidification front dynamics involves the employment of transparent model systems of low-melting organics. The model system, succinonitrile-ethanol as an example, displays numerous characteristic properties of metallic alloys and has served as a model substance for studying metallic solidification, Glicksmann (1981). The methods focus on investigation of the interface morphology by observing macro- and microconvective flows by tracers and non-invasive determination of temperature and concentration fields utilizing holography, interferometry and polarimetry (Leonartz et al. 1989).

For the spacelab mission IML-2, which will be launched in 1994, two solidification experiments using model substances are in preparation. Whereas the first experiment using a Bridgman-type furnace inserted into the thermostat of the critical point facility (CPF) focus on the investigation of the purely diffusiv controlled transition from planar to cellular growth morphology and the development of the cell pattern the second experiment will be performed in a power-down configuration on a slow rotating centrifuge called NIZEMI (Fig. 15).

This facility originally was devoted solely to biological samples. However, due to the low melting point of the transparent model substances it was possible to design an experimental set-up providing a controlled temperatur gradient within a glass-cuvette. Thus it shall be possible to observe directional solidification processes at defined g-levels and orientation between 10^{-4} g and 1 g.

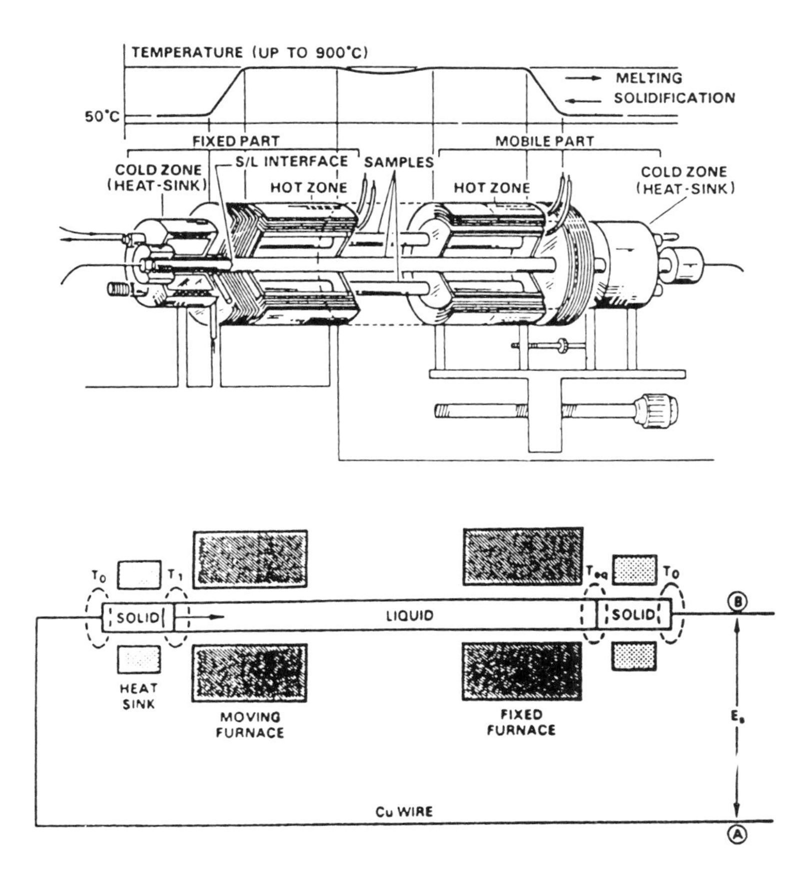

Figure 14: The purpose of MEPHISTO, France, is to perform fundamental studies on the effect of growth parameters and convection on the growth mechanisms and characteristics of the solidified sample. (a) The system consists of a mirror-symmetrical arrangement of two Bridgman-type furnaces in which the center section of a long sample can be partially remelted. (b) By keeping one of the solid-liquid interfaces stationary and by remelting or solidifying at the opposite interface, an electromotive force is generated between the cold ends (Seebeck voltage). From this voltage the actual temperature of the non-stationary solid-liquid interface can be determined with high resolution (ESA, 1989).

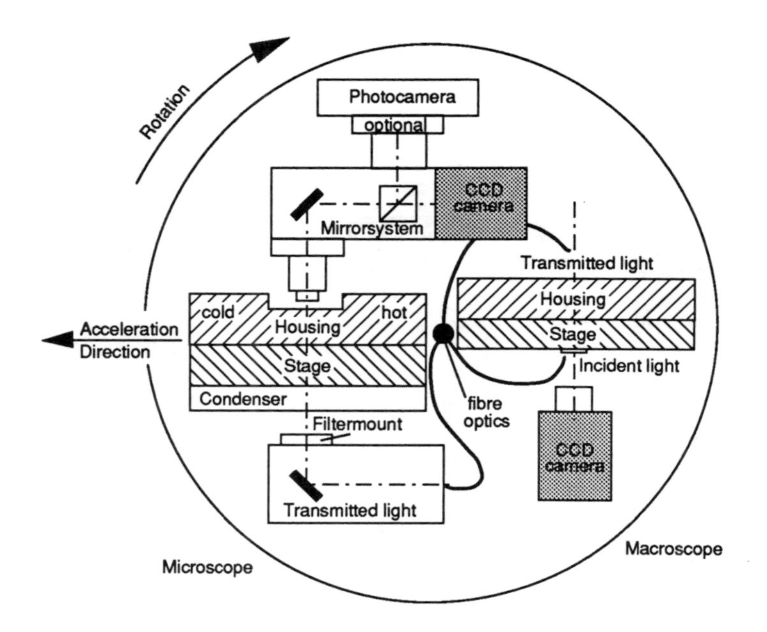

Figure 15: Top view sketch of NIZEMI's rotating platform, Leonartz et al. (1992)

4. Residual accelerations or low gravity modulations.

As already mentioned in chapter 2 the experiences gained in the spacelab flights of the past have led to the realization that the residual acceleration environment on board an orbiting spacecraft is not as low or as steady as would be desirable for certain classes of experiments. The sources of the residual acceleration include crew motions, mechanical vibrations (pumps, motors, etc.), spacecraft maneuvers and basic attitude motion, atmospheric drag and the earth's gravity gradient. The accelerations are characterized by temporal variations in both magnitude and orientation. Soon it became clear that in particular with the low background gravity level, even the very small accelerations may cause discrepancies between the expected and the obtained experimental results.

Recently Alexander (1990) gave a comprehensive review on this topic focussing in particular the mathematical analysis of the problem. Three kinds of research activities of different purposes were induced. Hamacher et al. (1987) and Trappen (1987) started during the spacelab mission D1 with an extensive sampling of accelerometer data by a set of mostly 3-axis sensitive acceleration measurement devices located at different elements of the payload (cf. Fig. 16).

The aim of this work and the subsequent analysis of the data were first to provide the experimenters with detailed informations about the acceleration level during the performance time of their experiment. Secondly to identify the most important sources of low-gravity disturbances and to quantify their impact on the residual gravity level and finally to derive the eigenfrequencies and the transfer functions of this special carrier spacelab D1 to be able afterwards to extrapolate the measured micro-accelerations to any point of the payload. However, at this time hardly quantitativ estimations were made by the experimenters as to the sensitivity of their experiment to residual accelerations.

Due to the beginning of the activities concerning the development of new orbiting platforms following spacelab Tiby and Langbein (1984) made the first attempt to determine allowable g-levels for microgravity payloads. Their main findings are summarized in Figure 17 which shows a sensitivity to residual accelerations decreasing with the frequency being higher than 10^{-2} Hz. The method of an order of magnitude (OM) analysis used by Langbein was also applied by Rouzaud et al. (1985). The advantage of the OM approach is that a great deal of information can be obtained with little computational effort. The disadvantages are related to the fact that the claim of characteristic reference quantities for length, velocity, time etc. are not, in general, known a priori. In addition, the statement of convective motion due to residual acceleration is related only to the general fact of axial macrosegregation. No quantitativ estimation of lateral non-uniformity in composition can be achieved.

Finally, analyses are generally restricted to the examination of a single component disturbance of e. g. sinusoidal in shape. In reality, low-gravity disturbances tend to be associated with more than one frequency. Given that a system may respond to a multi-frequency disturbance in an 'additive' way, tolerance curves such as depicted in Figure 17 may underestimate the response to a given low-gravity disturbance.

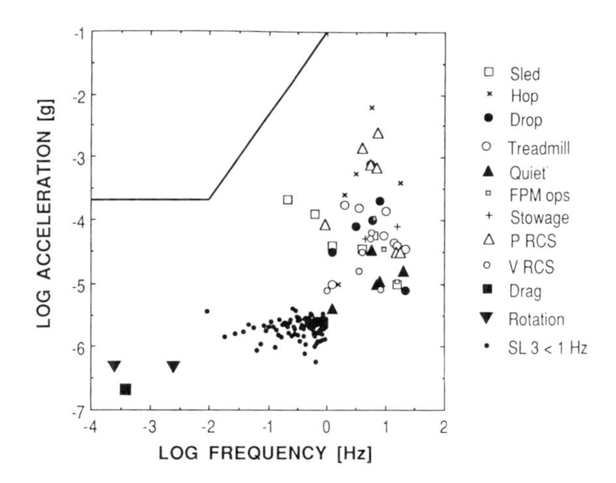

Figure 16: Accelerations measured on orbiting spacecraft. Data courtesy of H. Hamacher, DLR and M. J. B. Rogers, University of Alabama in Huntsville. The Spacelab 3 (SL-3) data are restricted to measured frequencies between 10^{-3} and 1 Hz. The points corresponding to Sled, Hop, Drop, Treadmill, Quiet, FMP ops and Stowage, refer to activities and experiments on the D1 mission. FMP ops stands for Fluid Physics Module operations, P RCS and V RCS refer to the primary and vernier thrusters. The drag and rotation entries correspond to accelerations arising from slow variations in atmospheric drag during an orbit and attitude changes involving rotation, Rosenberger (1990).

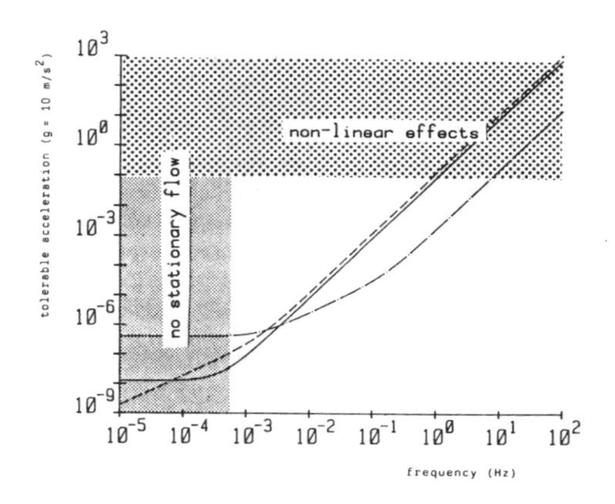

Figure 17: Tolerable residual accelerations in: (a) fluid physics experiment involving a temperature gradient (full line) (b) a crystal growth experiment by the travelling heater method (dash-dot line) (c) a thermodiffusion experiment (dashed line), Langbein (1985).

Direct numerical simulation have been used to get more knowledge of the effects of low-gravity on the transport of heat and momentum. For the case of directional solidification a number of articles are reviewed in the paper of Alexander (1990). By this method the impact of residual accelerations on the buoyancy-driven fluid motion of the melt can directly be coupled to the resulting solute redistribution. It turned out that the specific boundary conditions, thermo-physical properties of the melt, growth rate and sample size are all found to play a role in the determination of the experiment sensitivity. With respect to a steady component of the residual acceleration the worst case, not unexpected, appears to be when the acceleration vector is parallel to the crystal interface as in vertical Bridgman growth on earth.

Table 3 summarizes the results of Alexander et al. (1989) obtained for steady acceleration in case of directional solidifications from diluted gallium-doped germanium.

The directional solidification process also turned out to be extremely sensitive to transient disturbances. The response of the system to a $3 \cdot 10^{-3}$ g impulse of one second duration acting parallel to the interface of a growing crystal in the particular boundary conditions choosen produces a response in the solute field which last for nearly 2 000 s. The response of the solute field to oscillatory accelerations varies from no response at all (at frequencies above 1 Hz with amplitudes below 10^{-3} g) to a significant response at 10^{-3} Hz at amplitudes on the order of 10^{-6} g. Also the effect of accelerations of steady and periodic contributions were investigated in several combinations. However, for frequencies greater than 10^{-2} Hz, there were no discernable effects on the solute fields in the particular boundary conditions.

The third kind of research activities mentioned above as evoked by the fact of log-gravity disturbancies has to be seen in the revived interest in studying convection subject to time dependent modulation in general.

Murray et al. (1991) and Wheeler et al. (1991) have investigated the effect of gravity modulation of high frequencies (1 - 10 Hz) on the onset of solutal convection during directional solidification based on linear stability. They found that a stable base state can be destabilized due to modulations, while an unstable state can be stabilized. The flow and solute disturbance fields show both synchronous and subharmonic temporal response to the driving sinusoidal modulation.

residual acceleration magnitude [g]		orientation N	e_g	ampoule width [cm] 1, growth rate [mm s^{-1}] 6.5	ampoule width [cm] 1, growth rate [mm s^{-1}] 0.65	ampoule width [cm] 0.5, growth rate [mm s^{-1}] 6.5	ampoule width [cm] 2.0, growth rate [mm s^{-1}] 6.5
A)	10^{-4}	↑	←	80			
	10^{-5}		←	92.7	11.9	12.0	
			↙	70.9	11.3		
			↓	6.4	0.95		
	$5 \cdot 10^{-6}$		↓	3.2			
			↙	39			
			←	54.2			
	10^{-6}		←	11.3	2.0		
			↙	8.0			
			↓	0.7	0.0		
B)	10^{-5}		←	22.6			64.5
	10^{-6}		←	2.3			

Table 3: Compositional non-uniformity (in %) for Ge:Ga (Alexander, 1989)

5. Conclusions

It has been recognized for some time that the low-gravity acceleration environment associated with a spacecraft in low earth orbit offers an opportunity to study certain physical processes which are difficult to investigate under the gravitational acceleration experienced at the earth's surface. However, whereas the low-gravity level can potentially eliminate buoyancy-driven convection, the occurance of time-dependent local acceleration due to orbital maneuvers and inherent mechanical vibrations may itself induce buoyant convection.

The impact of residual accelerations is by no means completly understood. Nevertheless, on the basis of our current understanding of experiment sensitivity it has to be required that, in general, experimenters should be concerned about the effect of residual acceleration and that careful modelling should be included as part of the experiment program.

On the other hand, a direct verification of the numerical models by comparison with an experiment was not possible so far. Due to the complexity of the g-jitter on a real spacecraft and the not awareness of the experimenters concerning this problem in the past very little experimental data are available. Thus ESA distributed in March '91 a 'call for proposals' related to defined g-level sounding rocket flights particularly to enable experimenters to evaluate the response and sensitivity of different types of experiments to at first single frequency oscillations or defined pulses. A set of

A technical feasibility study has been conducted to analyze the following model missions:

Mission A:	constant g-levels in the payload with the levels: 0.1 g, 0.03 g, 0.01 g, 0.003 g, 0.001 g, 0.0003 g, 0.0001 g
Mission B:	Sinusiodal g-levels in the payload, with amplitudes of: 0.1 g, 0.01 g, 0.001 g, 0.0001 g and frequencies of: 1 Hz, 10 Hz, 100 Hz
Mission C:	Pulsed accelerations, with g-levels of: 0.1 g, 0.01 g, 0.001 g, 0.0001g and pulse lengths of: 10 ms, 100 ms, 1000 ms

These g-variations are technically feasible and one could conceive a wide variety of mission profiles depending on the proposals received

Table 4: Defined G-level Sounding Rocket Flights (ESA, 1991)

model missions were defined (cf. Table 4) basing on a feasibility study on defined g-level control systems for TEXUS. However, the number of proposals submitted to this issue was such as low that this kind of missions for the time being was post-poned.
Another approach to get at least experimental data comparable to numerical accelerations related to the impact of steady residual acceleration in a low-g environment could be the further use of the slow rotating centrifuge NIZEMI e. g. for directional solidification experiments with transparent model substances as in the Leonartz experiment described in chapter 3.
In general, the use of scaling, mathematical modelling and the results of experiments in a complementary way will be the one and only way to complete our understanding of the response of systems to residual acceleration.

6. References

Alexander, J. I. D. (1990) 'Low-Gravity Experiment Sensitivity to Residual Accelerations: A Review', Microgravity Sci. Technol. 3, 52

Alexander, J. I. D., Ovazzani, J., Rosenverger, F. (1989) 'Analysis of the low Gravity Tolerance of Bridgman-Stockbarger Crystal Growth', J. Cryst. Growth 97, 285 - 302

Billia, B. Jamgotchian, H., Favier, J. J., Camel, D. (1987) 'Cellular Solidification of Pb-Tl Alloys during the D1-WL-GHF 0S Experiment'. in P. R. Sahm, R. Jansen, M. H. Keller (eds.), Scientific Result of the German Mission D1, WPF, Köln, 230 - 236

Camel, D. , Favier, J. J., Dupovy, M. D., Le Maguet, R. (1987) 'Microgravity and Low-Rate Dendritic Solidification', in: P. R. Sahm, R. Jansen, M. H. Keller, Scientific Results of the german Spacelab Mission D1, WPF, Köln , 236-245

Camel, D. Favier, J. J. (1984) 'Thermal Convection on Longitudinal Macrosegregation in Horizontal Bridgman Crystal Growth - I. Order of Magnitude Analysis', J. Cryst. Growth 64, 42 - 56

Camel, D., Favier, J. J. (1986) 'Scaling Analysis of Convective Solute Transport and Segregation in Bridgman Crystal Growth from the Doped Melt', J. Physique 47, 1001

Coriell, S. R., Cordes, M. R., Boettinger, W. J., Sekerka, R. F. (1980) 'Convective and Interfacial Instabilities during Unidirectional Solidification of a Binary Alloy', J. Cryst. Growth 49, 13 - 28

Coriell, S. R., Cordes, M. R., Boettinger, W. J., Sekerka, R. F. (1981) 'Effect of Gravity on Coupled Convective and Interfacial Instabilities during Directional Solidification', Adv. Space Res. 1, 5 - 11

Eisa, G. F., Wilcox, W. R., Busch, G. (1986) 'Effect of Convection on the Microstructure of the MnBi/Bi Eutectic', J. Crystal Growth 78, 159

ESA (1989) 'Experiment Facilities for Materials and Fluid Sciences embarked on Spacelab', ESA SP-1120

ESA (1991) 'Call For Experiment Proposals: Materials and Fluid Science Experiments Using Sounding Rockets', March 1991

ESA (1992)'The High Temperature Materials Processing Laboratory_HTMPL/Phase A Study, 'Final Report of ESA Contract No. 9865/90/F/ BZ(SC), Dornier, Friedrichshafen

Favier, J. J., Camel, D. (1983) 'Order of Magnitude Analysis of Solute Macrosegregation in Crystal Growth from the Melt', ESA SP-191, 295 - 300

Favier, J. J., Goer, J. de (1984) 'Directional Solidification of Eutectic Alloys', ESA SP-222, 127

Feuerbacher, B., Hamacher, H, Jilg, R. (1987) 'Compatibility of Microgravity Experiments with Spacecraft Disturbances', DFVLR-Report 33 - 87/6, Köln

Feuerbacher, B., Naumann, R., Hamacher, H. (1986) Materials Science in Space, Springer, New York

Glicksmann, M. E. (1981) 'Dendrite Growth - Work on Model Systems', in: P. R. Sahm (ed.), Workshop - Erstarrungsfrontdynamik, Gießerei-Institut RWTH, Aachen, 79 - 84

Hamacher, H., Jilg, R., Mehrbold, U. (1987) 'Analysis of microgravity measurements performed during D1', ESA SP-256, 413 - 420

Jackson, K. A., Hunt, J. D. (1966) 'Lamellar and Rod Eutectic Growth', Trans. Metall. Soc. AIME 236, 1129

Krupp, H, Ruppel, W. (1986) 'Bemannte Raumfahrt?', Phys. Bl. 42, 326.

Kurz, W., Fischer, F. J. (1986) Fundamentals of Solidification, Trans Tech Publications, Switzerland, 54

Langbein, D. (1985) in: R. Monti, D. Langbein, J. J. Favier, 'Influences of Residual Accelerations on Fluid Physics and Materials Science Experiments' (1987), H. U. Walter (ed.), Fluid Sciences and Materials Science in Space, Springer, Berlin.

Leonartz, K., Neumann, G., Schmitz, G. J. (1989) 'Micro-g experiment PIETA - Experimental Technique', ESA SP-295, 661 - 664

Leonartz, K., Schmeling, H., Rex, S., Coriell, S. R. (1992) 'Materials Science in the Biological Spacelab Multi User Facility NIZEMI', paper to the 4th Int. Symp. Experimental Methods for Microgravity Mat. Sci. Res., San Diego, CA, 1. - 5.3.92

Müller, G., Kyr P.(1987) 'Directional Solidification of the InSb-NiSb Eutectic', in P. R. Sahm, R. Jansen, M. H. Keller (eds.), Scientific Results of the German Spacelab Mission D1, WPF, Köln, 246 - 259

Murray B. T., Coriell, S. R., McFadden, G. B. (1991) 'The Effect of Gravity Modulations on Solutal Convection during Directional Solidification', J. Cryst. Growth 110, 713.

OFTA (1990),Industrial Applications of Materials Processing in Space, ARAGO 9, Masson Publ., Paris, 24

Pirich, R. G., Larson, D. J., Busch, G. (1980), in: AIAA 18th Aerospace Science Meeting, jan 1980, paper 80 - 0119

Rex, S., Sahm, P. R. (1987) 'Thermosolutal Convection and Macrosegregation', Solidification Processing 1987, Institute of Metals, London, 102 - 105

Rosenberger, F., Alexander, J. I. D., Nadarajah, A. Ovazzani, J. (1990) 'Influence of Residual Gravity on Crystal Growth Processing', Microgravity Sci. Technol. 3, 162

Sahm, P. R., Keller, M. H. (1991) 'Solidification Processing under Microgravity', in R. W. Cahn, P. Haasen, E. J. Kramer (eds.), Material Science and Technology, vol. 15, Weinheim, Stuttgart, 539 - 582

Tensi, H. M., Schmidt, J., Mackrodt, C. (1989) 'Influence of Microgravity on the Morphology of the Eutectic Volume between the Dendrites and the coarsening of Dendrites', in N. B. Singh, V. Laxmanan, E. W. Collings (eds.), Materials Processing in Space, Trans Tech Publications Ltd. Switzerland, 45 - 64

Tiby, C., Langbein, D. (1984) 'Allowable g-Levels for Microgravity Payloads', Batelle Frankfurt, Final Report of ESA Contract No. 5.504/83/F/FS(Sc)

Tiller, W. A., Jackson, K. A., Rutter, J. W., Chalmers, B. (1953) 'The Redistribution of Solute Atoms During the Solidification of Metals', Acta Met. 1, 428

Trappen, N., Demond, F. J., (1987) 'Post Flight Accelerometer Data Evaluation', in: P. R. Sahm, R. Jansen, M. H. Keller (eds.), Scientific Results of the German Spacelab Mission D1, WPF, Köln

Turner, J. S. (1973) Buoyancy Effects in Fluids, Cambridge Univ. Press, London

Walter, H. U. (1987) Fluid Science and Materials Science in Space, Springer, Berlin

Wheeler, A. A., McFadden, G. B., Murray, B. T., Coriell, S. R. (1991)' Convective Stability in the Rayleight-Bénard and Directional Solidification Problems: High Frequency Gravity Modulations', Phys. Fluids A, 3, 2847

TIME DEPENDENT ASPECTS OF SOLIDIFICATION

V. Pines, M. Zlatkowski, and A. Chait
NASA Lewis Research Center
Cleveland, Ohio 44135
USA

ABSTRACT. This work focuses on time dependent aspects of solidification from the melt. Starting from the basic thermodiffusion model, we review recent findings using both analytical and fully numerical approaches. It is demonstrated that omission of the inherent time dependency of the problem in the static approach and in the quasistationary approximation can lead to serious shortcomings. Time dependent linear stability analyses for a sphere results in a power law growth of small perturbations, and for a needle crystal in a wave propagation along the parabolic interface. The nonlinear development of a plane interface initially perturbed with stochastic perturbations, results in the generation of solitary waves propagating away from cell tip regions along the interface with increasing amplitudes, terminating in the valleys between adjacent cells. The developing morphologies are sensitive to localized perturbations, revealing the existance of long range interaction in the growing structure.

This work concentrates on certain aspects of a solid-liquid interface dynamics during solidification. The dynamics of solidification processes from supercooled melts leading to complex morphological solid-melt interface structures represents a challenging problem of both practical and theoretical importance. The general description of solidification of pure materials is commonly formulated within the framework of continuum non-equilibrium thermodynamics[1]. The basic commonly accepted model which captures the essential features of free solidification of pure materials from a supercooled melt was introduced by Mullins and Sekerka[2]. This model considers the transport of heat in both phases by diffusion. Mass density jump across the interface is neglected, and heat energy generated at the solid-liquid interface is removed by diffusive transport only. Local equilibrium conditions at the interface are assumed. The interface temperature departs from its thermodynamic equilibrium value for a planar interface due to curvature (Gibbs-Thompson effect), but surface energy anisotropy (Herring effect) is neglected.

Since the pioneering work of Mullins and Sekerka[2], many analytical and numerical studies of the basic model have been undertaken. The common simplifications to this model are either the steady-state assumption[3], or the quasistationary approximation[4] to the diffusion equations. However, recent attention has focused on the inherent time dependent nature of the process. Analytical and numerical stability analyses[5,6] of a growing Ivantsov

S. H. Davis et al. (eds.), Interactive Dynamics of Convection and Solidification, 179–181.

needle crystal from a supercooled melt were made in a fully time-dependent context. In these works it was demonstrated that the term neglected in the quasistationary approximation is of vital importance in determining the time-dependent features of the dendritic surface, since the term introduces a branch point in the complex frequency space. A consequence of this branch point is that any perturbations near the tip propagate down along the dendrite interface as waves with increasing amplitudes. This wave propagation process has the effect of stabilizing the tip and its immediate vicinity. The decay of perturbations in the form of travelling waves is found to be a characteristic of geometries with broken translational symmetries along the interface (e.g. Ivanstov's paraboloid). This phenomenon is not found in geometries where the solid-liquid interface possesses translational symmetry (e.g. a plane, a sphere, or a circular cylinder). In Ref. 7, the basic thermodiffusion model[2] was directly integrated numerically from arbitrary initial conditions with an initially spherical solid nucleus submerged in a supercooled melt. The emergence of a dominant spatial mode from arbitrary initial conditions, including stochastic noise, was found to correspond to the most unstable mode determined from linear stability analysis of the time-dependent problem. Instabilities initiated by localized defects, analogous to point and line defects in crystal lattices, were found to grow and spread in a wave-like manner, a phenomenon observed in experiments[8]. The formation of a pre-dendritic growth stage was characterized by the natural establishment of constant tip radii and solidification rates[7].

In a recent study[9] we demonstrated that when the nonlinear development of instabilities breaks the translational symmetry along the interface a new phenomenon is observed. The growth is characterized by the formation of cell-like pre-dendritic structures and is accompanied by the generation of solitary waves. We believe this process may lead to later development of dendrite side branches. The time evolution of the interface morphology was followed from a planar interface initially perturbed with small amplitude stochastic noise. During the initial growth stage, the short wavelength perturbations decayed due to the Gibbs-Thompson effect. Subsequent nonlinear evolution showed that the morphological *details* of the interface are initial condition sensitive; however, the characteristic size of the pre-dendritic structure and the establishment of constant values of cell tip radii and velocities are independent of the initial conditions. The results were obtained from fully time dependent two-dimensional nonlinear numerical simulations within the framework of the basic thermodiffusion model[2]. The details of the model and numerical solution methodology are presented elsewhere[7].

Subsequent growth from the same initial conditions was performed, but a small perturbation was placed during the middle of the growth on a tip of a single cell in the morphology, and was allowed to naturally develop in time. A point-by-point difference between the perturbed and unperturbed interface positions at each time step revealed a solitary wave generation phenomenon. These solitary waves propagate along the interface, experiencing both dispersion and growth during their evolution. The structural details of the solitary waves were found to be dependent on the initial perturbations, but their existance was independent of the particular type of perturbation used. A point-by-point differencing between successive time interface positions of an unperturbed interface has also shown the natural development of solitary waves over the entire interface. These waves are apparently a phenomenon correlated with the translational symmetry breakdown and the subsequent formation of constant values of morphology features.

These findings, augmented by subsequent studies reported elsewhere and by recent independent experimental works, clearly point out the need for retaining the complete time dependent terms in the mathematical models used to describe solidification phenomena. We believe that many of the physical phenomena exhibited during the solidification process are fundamentally time dependent in nature, and that the formation of constant operating conditions is a manifestation of a delicate dynamic equilibrium which is the result of, and is accompanied by, time dependent aspects.

This work is sponsored by NASA's Microgravity Science and Application Program.

REFERENCES

1. S. R. de Groot and P. Mazur, Non-Equilibrium Thermodynamics, North Holland P.C., Amsterdam-London (1969).
2. W. W. Mullins and R. F. Sekerka, J. Appl. Phys., 34, p. 323 (1963).
3. D. A. Kessler, J. Koplik and H. Levine, Adv. Phys., 37, p. 255 (1989).
4. J. S. Langer, Rev. Mod. Phys., 52, p. 1 (1980).
5. V. Pines and P. L. Taylor, Phys. Rev. A., 41(2), p. 1006 (1990).
6. V. Pines, P. L. Taylor and M. A. Zlatkovski (M. Zlatkowski), Phys. Rev. A., 41(2), p. 1021 (1990).
7. V. Pines, M. Zlatkowski and A. Chait, Phys. Rev. A., 42, p. 6129 (1990); Phys. Rev. A., 42, p. 6137 (1990).
8. R. J. Schaefer and M. E. Glicksman, Met. Trans., 1, p. 1973 (1970).
9. V. Pines, M. Zlatkowski, and A. Chait, submitted to Phys. Rev. Let.

OPTIMIZATION APPLIED TO SOLIDIFICATION PROCESSES

Jonathan A. Dantzig and Daniel A. Tortorelli
Department of Mechanical and Industrial Engineering
University of Illinois
1206 W. Green Street
Urbana, Illinois 61801 USA

ABSTRACT. An optimal design algorithm is presented for the analysis of general solidification processes, and is demonstrated for the growth of GaAs crystals in a Bridgman furnace.

1. Introduction

An optimal design algorithm is presented for the analysis of general solidification processes, and is demonstrated for the growth of GaAs crystals in a Bridgman furnace. The system is optimal in the sense that the prespecified temperature distribution in the solidifying materials is obtained to maximize product quality. The optimization uses traditional numerical programming techniques which require the evaluation of cost and constraint functions and their sensitivities. The finite element method is incorporated to analyze the crystal solidification problem, evaluate the cost and constraint functions, and compute the sensitivities. These techniques are demonstrated in the crystal growth application by determining an optimal furnace wall temperature distribution to obtain the desired temperature profile in the crystal, and hence to maximize the crystal's quality.

We have chosen to adapt an existing commercially available finite element program, FIDAP [1], to compute of the sensitivities, rather than develop a new code. The explicit sensitivities are computed analytically by the adjoint technique[2], which has been applied to nonlinear transient conduction problems by Tortorelli *et. al.* [3]. Large computational savings and accurate calculations are realized by utilizing an explicit approach as opposed to the costly and sometimes unreliable finite difference method [3, 4].

The implicit dependence of the objective on the temperature can be resolved by solving the following adjoint problem: Find that value of λ for which

$$-\frac{\partial G}{\partial T} = \frac{\partial R(T, \mathbf{b}, \lambda)}{\partial T}\frac{\partial T}{\partial \mathbf{b}} \tag{1}$$

for all admissible $\frac{\partial T}{\partial \mathbf{b}}$, where R is the residual as used in finite element analysis. This equation is linear in λ, and is the adjoint operator for the incremental problem. This allows us to solve the adjoint problem efficiently when the finite element method is used.

After solving the original problem with Newton-Raphson iteration, we compute the adjoint load vector $(\frac{\partial G}{\partial T})$ and perform a back substitution on the transpose (adjoint) of the decomposed stiffness matrix to evaluate λ. Once λ is determined, the sensitivities are obtained directly. The efficiency of this method lies in the fact that a single back-substitution into the already decomposed stiffness matrix yields all of the components of the sensitivity vector. In general, the solution of the

S. H. Davis et al. (eds.), Interactive Dynamics of Convection and Solidification, 183–185.

primal problem requires several Newton-Raphson iterations. Hence, the added cost of evaluatin the sensitivities is relatively small.

1.1. APPLICATION TO BRIDGMAN CRYSTAL GROWTH PROCESS

When crystals for electronics applications are grown using the Bridgman process, the finished bul crystals are sliced into thin wafers perpendicular to the growth direction. Fluid flow in the me during solidification can interact with the solute field near the crystal-melt interface to adversel affect the chemical composition of the crystal.[5] The primary means for controlling the convectiv flow is to control the shape of the crystal-melt interface, which may be accomplished by definin appropriate process parameters. A schematic model of the Bridgman furnace configuration i shown in Figure 1.

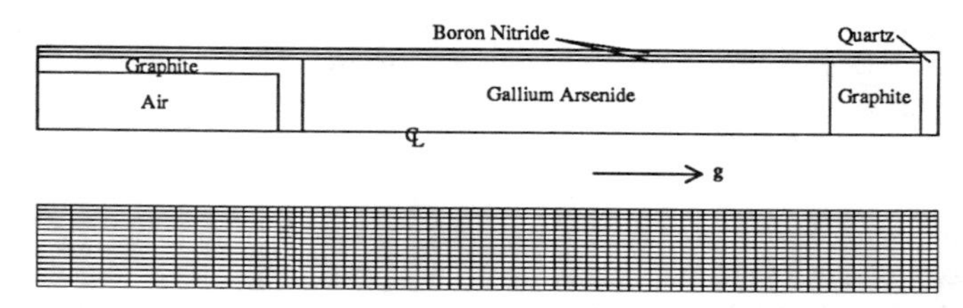

Fig. 1: Schematic view of the model for the proposed experiment to grow GaAs crystals, and corresponding finite element mesh containing 1216 nodes and 1230 elements.

The commercial code FIDAP[1], with modifications to enable the design sensitivities to b calculated, was used for the analysis. The container and melt were modeled using four–noded linea isoparametric elements, whereas the presence of the furnace wall was represented by a specifie temperature distribution exchanging heat by radiation with the exterior surface of the ampoule Further details of the radiation calculation are given in the reference [6].

The desired temperature profile, $\overline{T}(z)$, to be attained in the ampoule was specified on both th center-line and the outer radius of the crystal (inner radius of the ampoule). The objective functio was then defined as the error between the desired and computed temperatures at N discrete points

$$G = \sum_{i=1}^{N} \left(T_i - \overline{T}_i\right)^2 \qquad (2$$

Thus, G represents the function to be minimized.

The furnace was modeled using ten separate heating zones to span the entire length o the furnace. The eleven specified wall temperatures represent the design parameters, and th intermediate wall temperatures were determined *via* linear interpolation. The results for this cas are shown in Figure 2, illustrate that one may attain the final objective.

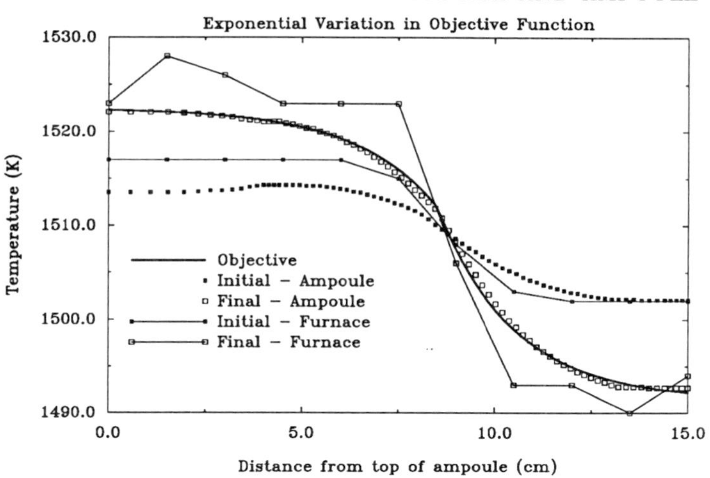

Fig. 2: Optimal furnace wall and ampoule temperature profiles for exponential variation in the objective function

Conclusions and Acknowledgment

The results indicate the practicality of optimal process design and the utility of the sensitivity analysis for this class of problems. The optimal solution can be found with little user intervention. Indeed, the only work required beyond that for the normal analysis is the definition of the design variables and objective function.

The authors wish to thank NASA, which has supported this work under grant NASA NAGW-683 and NAG3–1286, and the National Center for Supercomputing Applications for partial support of one of the investigators.

References

[1] M. S. Engelman. *FIDAP Theoretical Manual.* Fluid Dynamics International, Evanston, IL, 1987.

[2] E. J. Haug, K. Choi, and V. Komkov. *Design Sensitivity Analysis of Elastic Mechanical Systems.* Academic Press, New York, 1986.

[3] D. A. Tortorelli, R. B. Haber, and S. C.-Y. Lu. Design Sensitivity Analysis for Nonlinear Transient Thermal Systems. *Computer Methods in Applied Mechanics and Engineering*, 75:61–78, 1990.

[4] D. Tortorelli. Sensitivity Analysis for Nonlinear Constrained Elastostatic Systems. In S. Saigal and S. Mukherjee, editors, *Sensitivity Analysis and Optimization with Numerical Methods, AMD Vol. 115*, pages 115–126, 1990.

[5] C. J. Chang and R. A. Brown. Radial segregation induced by natural convection and the melt/solid interface shape in vertical Bridgman growth. *J. Crystal Growth*, 63:343, 1983.

[6] J. A. Dantzig and D. A. Tortorelli. Optimal design for solidification processes. In G. S. Dulikravich, editor, *Third International Conference on Inverse Design Concepts and Optimization in Engineering Design*, pages 213–226, 1991.

NUMERICALLY SIMULATED TEMPERATURE FIELDS IN A GRADIENT FURNACE MODELLING THE MICROGRAVITY ENVIRONMENT

G. ZIMMERMANN, J. OTTEN, N. HOFMANN
ACCESS e. V.
Intzestr. 5
5100 Aachen
Germany

Abstract: In preparation of the German Spacelab Mission D-2 samples of CuMn are directionally solidified with a planar solid-liquid interface in the gradient furnace GFQ. The temperature distribution in the sample is numerically simulated with the finite element solver CASTS. In microgravity the heat transfer between the furnace and the sample is reduced because of less convection. Therefore the position of the solid-liquid interface shifts towards the heating zone. To compensate this, the gas-mixture in the furnace and the furnace temperatures can be varied. As a result an increase of the heater temperatures can quite well reproduce earth conditions, which are necessary to get defined solidification parameters.

1. Introduction

Directional solidification of metallic alloys is often done in gradient furnaces using the Bridgman-Stockbarger technique. It allows a quasi one-dimensional heat flux along the sample axis in the region of the solid-liquid interface, combined with a substantial temperature gradient.

Besides concentration dependent effects like thermosolutal convection and segregation in the melt the temperature field in the furnace is decisive for the solidification behaviour. Experimentally the temperature in the sample can be detected at defined points by thermocouples. To optimize process parameters and avoid expensive experiments the numerical simulation of the temperature field in the sample is a very helpful tool.

2. Numerical Simulation

The system to be simulated consists of the Dornier gradient furnace GFQ [1] and a CuMn sample. One half of the cross-section of the axisymmetrical set-up and some relevant distances are shown in Figure 1. The furnace consists of two heaters and a cooling zone, separated by an adiabatic zone. The quenching nozzle is used to stop the controlled directional solidification immediately and to fix the position and morphology of the solid-liquid interface. The CuMn sample of 165 mm length is sealed in a MoRe tube of 10 mm outer diameter. For temperature measurements an Al_2O_3 tube including six thermocouples is put in the center hole of the sample. During solidification the furnace is moved along the fixed sample with defined velocities.

To simulate the temperature distribution in the furnace numerically the 3D-FEM program CASTS [2] is used. Input data are the geometry and the temperatures of the furnace, as well as the geometry and temperature dependent material properties of the sample, which consists of nine different materials. The heat transfer between the furnace and the sample is produced by radiation and heat conduction by a gas mixture of He and Ar in the furnace. The grid chosen for the simulation has 17 points in radial direction and is equidistant in axial direction with distance of 0.5 mm.

As a result we attain the time-dependent temperature field in the sample at different relative positions of furnace and sample and at various solidification velocities.

S. H. Davis et al. (eds.), Interactive Dynamics of Convection and Solidification, 187–190.

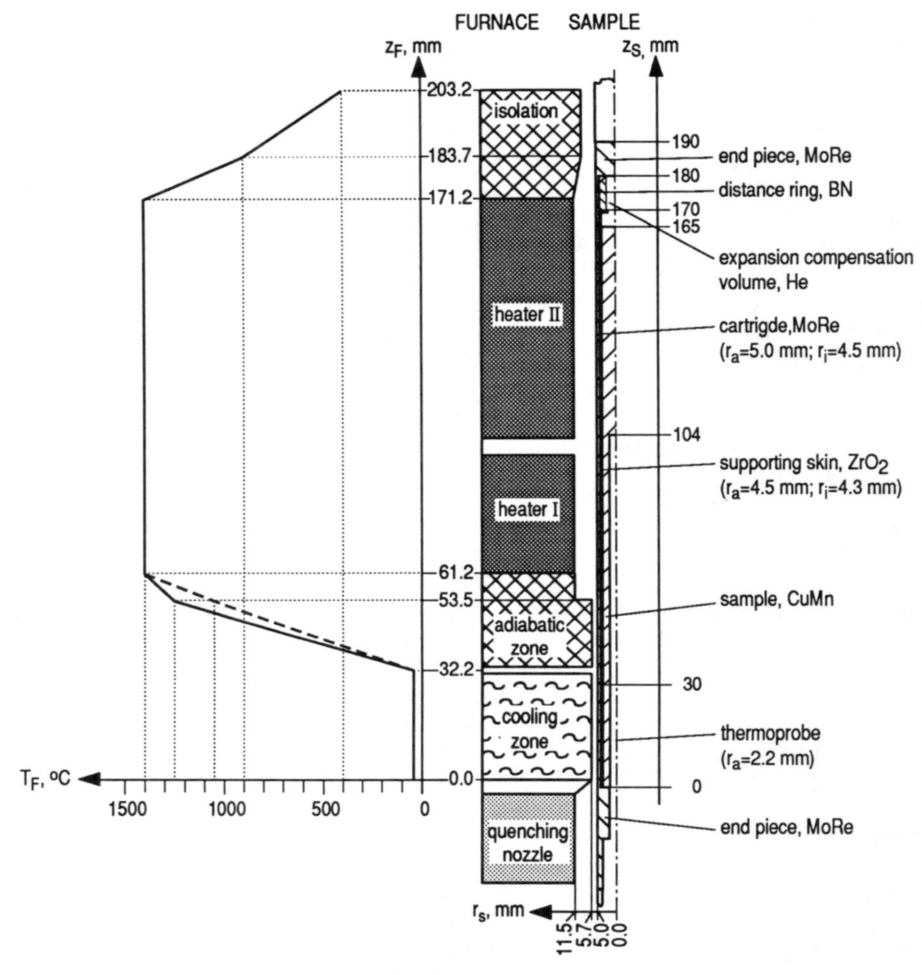

Figure 1: Cross-section of the set-up of the system furnace - sample with relevant measures and materials and the furnace temperature profiles

3. Results and Discussion

3.1. VALIDATION OF THE NUMERICAL RESULTS BY EXPERIMENT

To check the quality of the numerical calculations a solidification experiment is simulated. A CuMn sample is directionally solidified with three different velocities. The setup with relevant measures and materials is shown in Figure 1. The gas mixture in the furnace consists of 35 mbar He and 55 mbar Ar. The heat transfer between the furnace and the sample is simulated by radiation and heat conduction by the gas mixture. The temperature profile at the inner surface of the furnace is given by the solid curve.

First calculations show that in the cooling and adiabatic zone radial temperature differences between the inner surface of the furnace and the surface of the sample up to 200 °C exist. Despite the small gap of 0.7 mm between cooling zone and sample, convective heat transport has to be taken into consideration. An estimation based on convection in a vertical gap leads to an increase in the heat conduction in the region of the cooling zone by a factor of 1.75 for the simulations.

The temperature difference between the simulated temperature profile in the sample and measured values is for all cases lower than 50 °C. Because of typical temperature gradients in the sample of 250 K/cm, this means a maximum discrepancy of 0.2 mm for the actual position of the solid-liquid interface between experiment and simulation.

In conclusion, in spite of a quite simple furnace temperature profile, the numerically simulated temperature field in the sample corresponds with the experimental values.

3.2. MODELLING OF THE TEMPERATURE FIELD UNDER MICROGRAVITY CONDITIONS

Solidification experiments carried out with an AlMg alloy having the same sample design in the GFQ furnace during the German D1-Mission by Rex and Sahm [3] show that in microgravity the heat transfer in the furnace is changed. Using the same solidification parameters, gas mixture, heater and cooler temperatures, the position of the solid-liquid interface at the beginning of the solidification process is 2.7 mm higher than on earth. The reason for this is the reduced convective heat transport from the heater zone of the furnace into the adiabatic region.

This effect is expected for the experiments with CuMn samples during the German D2-mission, too. Modelling the heat transfer under reduced gravity we made two assumptions: First the heat-transfer in the cooling zone is produced by radiation and pure heat conduction. Second the furnace temperature in the adiabatic region is slightly decreased (see dashed line in Figure 1). Curves a and b of Figure 2 show the temperature profile in the sample at the beginning of the directional solidification. Under earth conditions (curve a) the temperature is higher because of the additional convective heat flux in the furnace. The changes in the slopes of the curves are caused by different heat conductivities of CuMn in the solid and liquid phase. At the melting point of the CuMn alloy of 885 °C the shift of the solid-liquid interface is 2.6 mm and comparable to the effect in the AlMg sample. Under simulated microgravity conditions the solidification length is 0.8 mm shorter and the quenched phase boundary is about 1.8 mm higher.

It is important to know the position of the solid-liquid interface relative to the adiabatic zone. If the interface moves too much to the heater region the isothermes in the sample are curved because of a non-axial heat flux and the occurance of radial segregation.

To modify the temperature field in the sample there are only two possibilities: the variation of the gas mixture and changes in the heater temperatures.

Curve c in Figure 2 represents the situation under microgravity but with an increase in the heater temperature of 100 °C. The more effective heat transport in the heating zone shifts the phase boundary to the cooling zone. The differences in the positions of the solid-liquid interface at the beginning and at the end of the simulated solidification process are less than 0.4 mm. The temperature gradient is slightly increased and the solidification length is reduced by about 1.6 %.

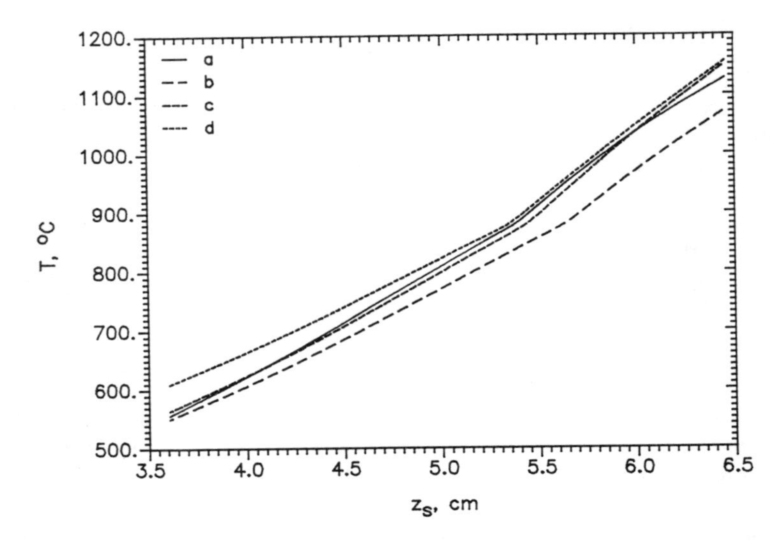

Fig. 2: Simulated steady state temperature profiles in the CuMn sample at the beginning of the solidification experiment: a) 1 g conditions; b) "μg", conditions; c) "μg" with an increase in heater temperature of 100 °C; d) case c) and half heat conductivity of the gas.

Lowering the thermal conductivity of the gas mixture reduces the cooling power. Curve d in Figure 2 shows the situation like curve c but with half the value of the heat conductivity of the gas. As a result both positions of the solid-liquid interface at the beginning and at the end of the solidification are shifted to the cooling zone. The less effective heat transport due to conduction reduces the temperature gradient and the solidification length by about 7.3 %.

4. Conclusions

Using the finite element solver CASTS it is possible to simulate the temperature field during a certain solidification experiment numerically with good accuracy. The changes in thermal behaviour under microgravity can be compensated by an increase of the heater temperatures of about 100 °C and not by reducing the heat conductivity in the furnace. Because of an upper limit in the heater temperature of the GFQ furnace of 1 500 °C, earth experiments had to be carried out with maximum heater temperature of 1 400 °C.

5. References

[1] Zell, U. (1991) 'GFQ Manual', Dornier, Document No. MA-2057-0200 DS / 01

[2] Hediger, F. and Hofmann, N. (1991) 'Process simulation for directionally solidified turbine blades of complex shape' in M. Rappaz, M. R. Ozgu, K. W. Mahin (eds.), Modeling of Casting, Welding and Advanced Solidification Processes V, The Minerals, Metals and Materials Society, pp. 611 - 619

[3] Rex, S., Sahm, P. R. (1987), 'Thermosolutal Convection and Macrosegregation, Results of Directional Solidification of Al-Mg Alloy under Reduced Gravity', Solidification Processing, Sheffield, UK, 102 - 105

EFFECT OF SOLIDIFICATION MORPHOLOGY ON THE MACROSCOPIC BEHAVIOR OF SOLIDIFICATION SYSTEMS

V. R. VOLLER
Department of Civil and Mineral Engineering
University of Minnesota
Minneapolis, USA

ABSTRACT. One domain equations modelling the heat and mass transfer processes that occur in the solidification of a binary alloy are presented. Using an ammonium chloride solution the effect of the morphology of the solid+liquid mushy zone on macrosegregation is investigated.

1. INTRODUCTION

In macroscopic modelling of solidification systems an important consideration is the nature of the two phase solid+liquid mushy zone, see Figure 1. In general the solid component in the mushy zone will consist of both a fixed columnar dendritic structure (moving with a prescribed velocity) and/or equiaxed grains which can flow with the fluid. Previous models of solidification systems (Bennon & Incropera, 1987; Beckermann & Viskanta, 1988; Voller et al., 1989) only consider columnar dendritic systems and do not allow for a relative movement of the solid with respect to the liquid. In free equiaxed systems, however, a relative movement of the solid can take place. A relative movement that can change the nature of the heat and solute transport in the system (the so called macrosegregation process). The aim in this paper is to investigate the effects of relative solid movement on predictions of macrosegregation.

2. THE GOVERNING EQUATIONS

Following Voller et al. (1989) the governing macrosegregation equations can be written in the generic form

$$\frac{\partial(\rho\phi)}{\partial t} + \nabla.(\rho \mathbf{u}\phi) = \nabla.(\Gamma\nabla\phi) + S_{term} \tag{1}$$

where $\mathbf{u} = (1-g)\,\mathbf{u_s} + g\,\mathbf{u_l}$ is the systems velocity, g is the local liquid fraction and the subscripts s and l refer to the solid and liquid phases respectively. On appropriate definition of the transport coefficient, Γ

S. H. Davis et al. (eds.), Interactive Dynamics of Convection and Solidification, 191–193.

Equ.(1) can represent the general Navier-Stokes equations on setting ϕ to; 1 (mass continuity), u and v (x and y momentum), T (energy conservation), and C (solute conservation). In arriving at a full description of the macrosegregation process constitutive equations relating the temperature, T, and solute concentration, C and the liquid and solid velocities are also required.

A key feature in the specific governing equations derived from Equ.(1) will be the source term, S_{term}. In the u and v momentum equations this term will have two components; (i) a Boussinesq source term, accounting for both thermal and solutal bouancies and (ii) a drag term $\mathbf{F_s} \propto (\mathbf{u_s} - \mathbf{u_l})$, which accounts for the interphase slip between the solid and liquid. Attention should also be directed to the source terms in the energy and solute equations which include a convective component to account for the transport of latent heat and solute due to relative solid movement.

The focus of the current work will be an investigation of a constitutive equation used to relate the solid and liquid velocities in the two phase mushy zone, viz., the "consolidation" relationship (Flood et al., 1991)

$$\mathbf{u_s} = \left[\frac{g - g_{cut}}{1 - g_{cut}}\right]\mathbf{u_l} \tag{2}$$

where a static casting has been assumed and g_{cut} is a given liquid fraction value. In using this relationship, as the liquid fraction decreases so does the solid velocity; from a value equal to the liquid velocity at g=1 to a value of zero at g_{cut}.

3. PRELIMINARY RESULTS

The use of the consolidation factor is examined on calculating the solute field in an ammonium chloride solution solidifying in a square box (25mmx25mm) with one face fixed at a cold temperature. The appropriate thermal and physical data is taken from Voller et al. (1989). The Results (calculated on a 15x15 control volume grid) are shown in terms of the solute contours at 0.095, 0.1 (nominal composition) and 0.105 at time t=1000 seconds (when solidification is about 2/3 rds complete). Figure 2. shows three solute fields corresponding to g_{cut} values of 1.0 (no solid movement), 0.5, and 0.1 respectively. As the solid movement is enhanced these results reveal two features, (i) the scale of the macrosegregation is reduced, and (ii) the width of the mushy zone (indicated by dashed lines at g=0.25 and g=0.75) thickens

4. CONCLUSIONS

The work in this paper represents an attempt to account for solid movement in a macrosegregation model. The results clearly show that solid movement will effect the resulting macrosegregation patterns.

5. REFERENCES

Bennon, W.D., and Incropera, F.P., "The Evolution of Macrosegregation in Statically Cast Binary Ingots", Metal. Trans. B, 18B (1987), 611-612.

Beckermann, C., and Viskanta, R., "Double-diffusive Convection during Dendritic Solidification of a Binary Mixture," PCH, 10 (1988),195-213.

Voller, V.R., Brent, A.D., and Prakash, C., "The Modelling of Heat, Mass and Solute Transport in Solidification Systems", Int. J. Heat Mass Transf., 32 (1989), 1719-1731.

Flood, S.C., Katgerman, L., and Voller, V.R., " The Calculation of Macrosegregation and Heat and Fluid Flow in the DC casting of Aluminum Alloys" in Modelling of the Casting, Welding and Advanced Solidification Process -V", TMS (1991), 683-690.

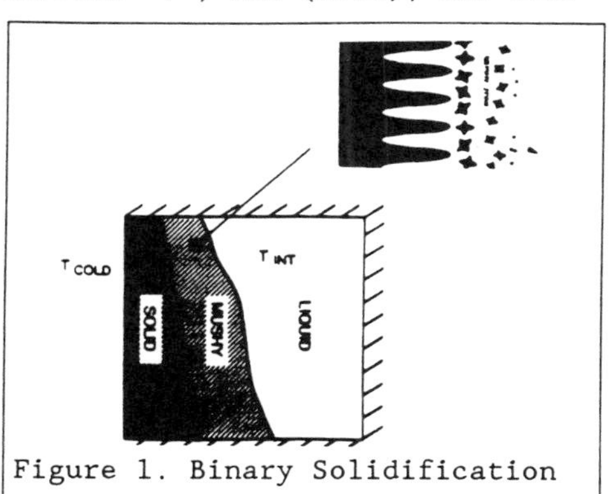

Figure 1. Binary Solidification

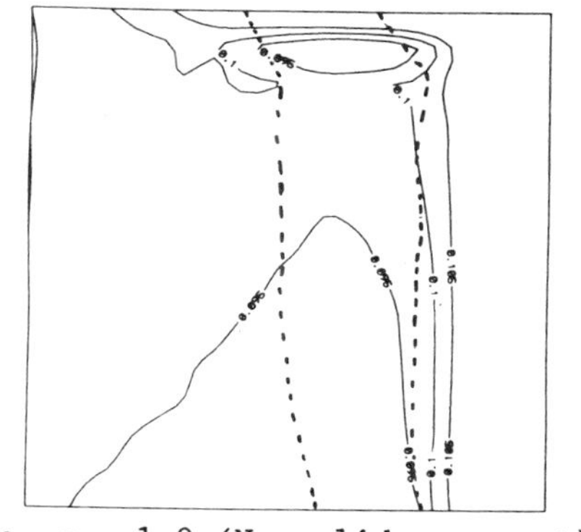

2a. g_{cut}=1.0 (No solid movement)

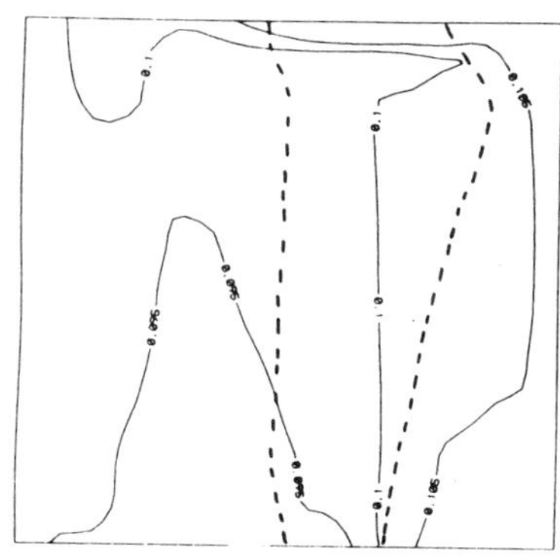

2b: g_{cut}=0.5

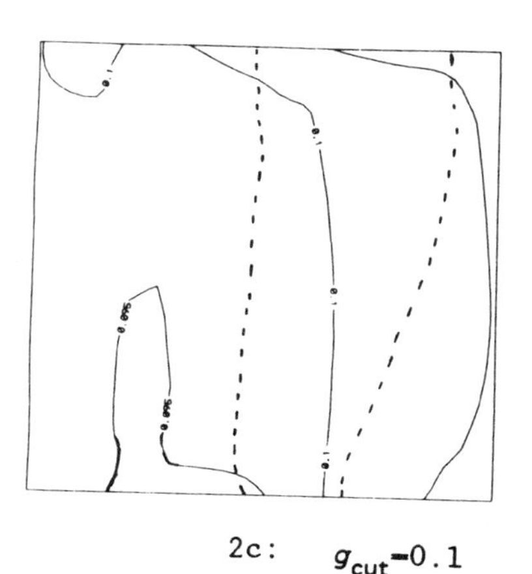

2c: g_{cut}=0.1

Figure 2. Solutal Fields (Concentration Contours 0.095-0.105) at t=1000 s.

Simulation of the Columnar Dendritic Solidification of a Pb-Sn Alloy

M. C. SCHNEIDER AND C. BECKERMANN
Department of Mechanical Engineering
The University of Iowa
Iowa City, IA 52242

ABSTRACT. This study reports on numerical simulations of the columnar dendritic solidification of a Pb-20 wt.% Sn alloy in a rectangular cavity cooled from the side. Noteworthy features of the model include the detailed consideration of the liquid, α- and γ- phases, the coupling of convection to microstructural characteristics (e.g., anisotropy of the dendrites, dendrite arm spacings) and fully temperature and concentration dependent enthalpies and densities. The effects of flow due to both thermal/solutal buoyancy forces and solid/liquid density differences on the resulting solid structure and on macrosegregation are investigated.

1. Introduction

During the columnar dendritic solidification of an alloy, convective flow in both the liquid and mushy regions has a significant effect on the final composition and structure of the cast product (e.g., causing macrosegregation). A model based on volume averaging of the conservation equations has recently been developed (Schneider, 1991). This model is a generalization of models used by Bennon and Incropera (1987), Beckermann and Viskanta (1988), Voller, et al. (1989) and others, and includes several new features. For instance, both the temperature and concentration dependence of the liquid, α- and γ- phase enthalpies are considered (Schneider and Beckermann, 1991). The model also allows for both thermal/solutal buoyancy driven flow and flow due to solid/liquid density differences. By allowing the permeability in the columnar dendritic mushy zone to be anisotropic and dependent on the dendrite arm spacings, convection in the mushy zone is coupled to microstructural characteristics. In this study numerical simulations are presented for a Pb-20 wt.% Sn alloy cooled from the side, and the effects of convection on the solid structure and on macrosegregation are investigated.

2. Model Description

Due to space limitations, only a few details of the solidification model are presented here. A complete description is available in Schneider (1991). In deriving the model equations, the following assumptions are used: rigid and stationary solid phases, local (i.e., within an averaging volume) thermal equilibrium, locally well-mixed liquid, no microscopic or macroscopic solid species diffusion and no dispersive fluxes. The volume averaged mass, energy and species conservation equations for the liquid, α- and γ- phases are summed to obtain mixture conservation equations, with the solid species equations also used individually. In order to account for the interfacial friction of the liquid in the mushy zone, the liquid momentum equation is considered separately. The permeability in the mushy zone is assumed to be anisotropic and is modeled based on the Blake-Kozeny equation. The permeability normal to the primary dendrite arms is given by (Poirier, 1987)

$$K_{\eta\eta} = 1.73\text{x}10^{3}(d_1/d_2)^{1.09}d_2^2\varepsilon_\ell^3/(1 - \varepsilon_\ell)^{0.749} \tag{1}$$

and the permeability parallel to the primary dendrite arms is given by

$$K_{\xi\xi} = [4.53\text{x}10^{-4} + 4.02\text{x}10^{-6}(\varepsilon_\ell + 0.1)^{-5}]\, d_1^2\varepsilon_\ell^3/(1 - \varepsilon_\ell) \tag{2}$$

In the above equations ε_ℓ is the liquid volume fraction, and d_1 and d_2 are the primary and secondary arm spacings, respectively. The primary dendrite arm spacings are calculated from an experimental correlation (Klaren, et al., 1980) valid for lead-tin alloys

S. H. Davis et al. (eds.), Interactive Dynamics of Convection and Solidification, 195–197.

$$d_1 = 22.62 \times 10^{-6} G^{-0.33} V^{-0.45} \tag{3}$$

where G is the local temperature gradient and V is the dendrite growth velocity. The secondary dendrite arm spacings are calculated from a coarsening model (Roosz, et al., 1986) where

$$d_2^3 = d_2(0)^3 + 35 \int_0^t \Gamma D_\ell / [m_\ell(\kappa - 1)\langle C_\ell \rangle^\ell] \, dt \tag{4}$$

where D_ℓ is the liquid mass diffusivity, m_ℓ is the liquidus slope, κ is the segregation coefficient, $\langle C_\ell \rangle^\ell$ is the average liquid concentration, Γ is the surface energy and $d_2(0)$ is the initial secondary arm spacing. The dendrites are assumed to grow in the direction opposite that of the heat flow. All of the phase properties vary with both temperature and concentration (Thresh and Crawley, 1970; Poirier, 1988; Poirier and Nandapurkar, 1988; Poirier, et al., 1991). The model equations were solved using a control volume based finite difference scheme with the SIMPLER algorithm (Patankar, 1980; Schneider, 1991). In this study, two simulations were carried out using the system shown in Figure 1. In Case S1 the effects of flow due to only solid/liquid density differences (contraction) were investigated, while in Case S2 both thermal/solutal buoyancy driven flow and contraction flow were considered. The calculations were performed on a 25 x 30 grid that was biased near the walls.

3. Results and Conclusions

A comparison of the mixture concentration patterns in Figures 2 and 3 shows the importance of considering both volume contraction and thermal/solutal buoyancy driven flow in trying to predict macrosegregation. The solute rich and solute poor regions at the top and bottom of the cavity, respectively, in Figure 3 are obviously caused by buoyancy driven flow. The solute rich region at the left wall in both Figures 2 and 3, on the other hand, is due to suction of relatively high concentration liquid through the mushy zone. In addition, the anisotropic nature of the permeability seems to have little influence on the flow. Figure 4 shows that the primary dendrite arm spacings are similar for both Cases S1 and S2, indicating an insensitivity of the spacings to the fluid motion. Finally, the secondary dendrite arm spacings for

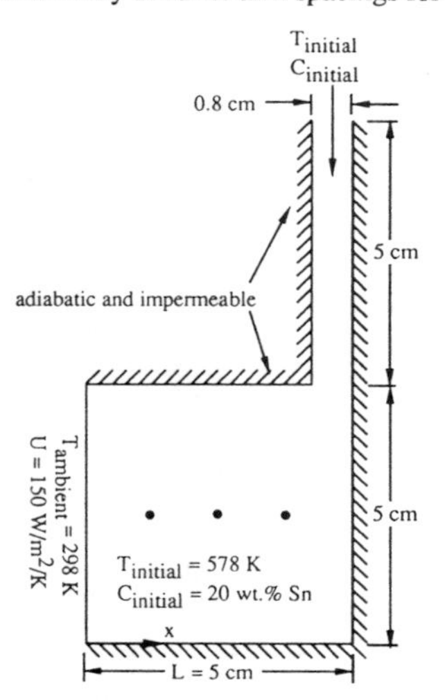

Figure 1. System and boundary conditions used in simulations.

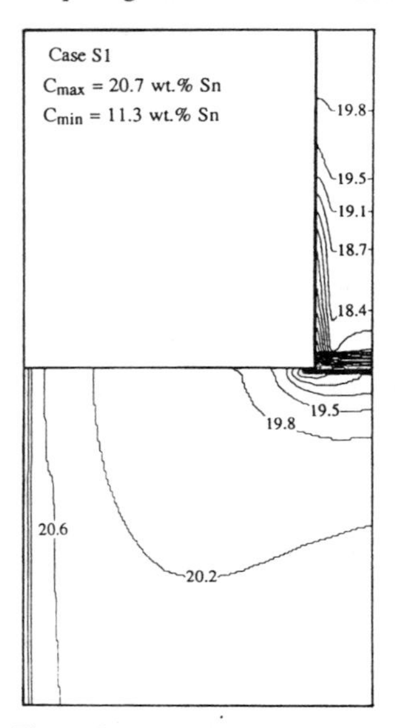

Figure 2. Mixture concentration isoplets for case S1 at 950 sec.

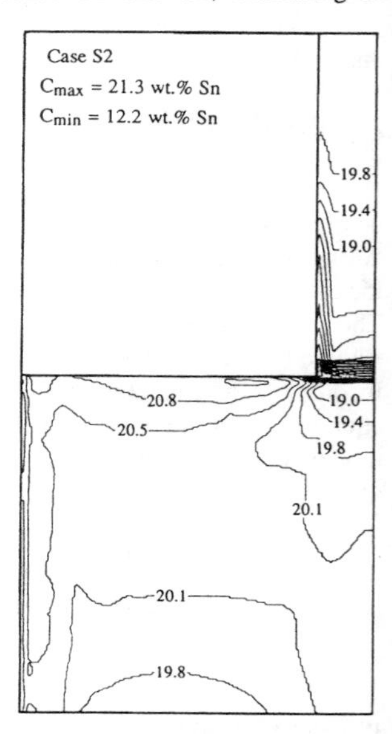

Figure 3. Mixture concentration isoplets for case S2 at 950 sec.

Case S1 are virtually identical to those shown in Figure 5 for Case S2, again illustrating the insensitivity of the arm spacings to the fluid flow.

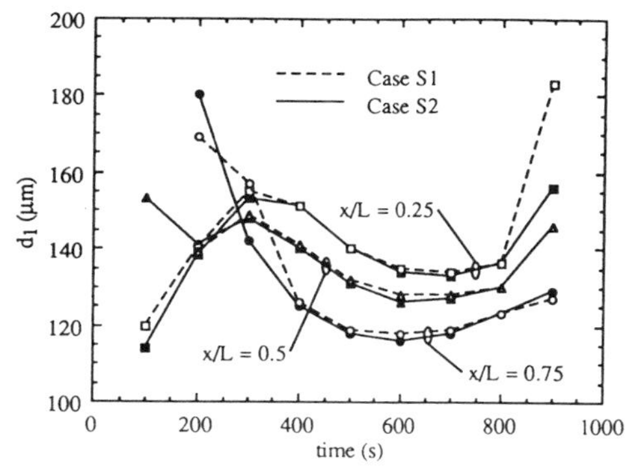

Figure 4. Variation of the primary dendrite arm spacings with time.

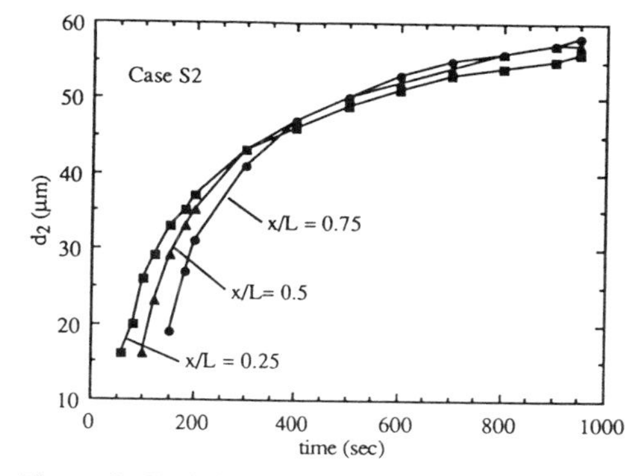

Figure 5. Variation of the secondary dendrite arm spacings with time for Case S2.

4. Acknowledgments

The work reported in this paper was supported by the National Science Foundation under Grant Nos. CBT-8808888 and CTS-8957149. The authors thank C. Fan, B. Feller and J. Ni for their assistance.

5. References

Beckermann, C. and Viskanta, R., 1988, "Double-Diffusive Convection during Dendritic Solidification of a Binary Mixture," *PhysicoChemical Hydrodynamics*, **10**, 193-213.

Bennon, W.D. and Incropera, F.P., 1987, "A Continuum Model for Momentum, Heat and Solute Transfer in Binary Solid-Liquid Phase Change Systems - I. Model Formulation," *Int. J. Heat Mass Transfer*, **30**, 2626-2170.

Klaren, C.M., Verhoeven, J.D. and Trivedi, R., "Primary Dendrite Spacing of Lead Dendrites in Pb-Sn and Pb-Au Alloys," *Metall. Trans. A*, **11A**, 1853-1861.

Patankar, S.V., *Numerical Heat Transfer and Fluid Flow*, McGraw-Hill, New York, 1980.

Poirier, D.R., 1987, "Permeability for Flow of Interdendritic Liquid in Columnar-Dendritic Alloys," *Metall. Trans. B*, **18B**, 245-255.

Poirier, D.R., 1988, "Densities of Pb-Sn Alloys during Solidification," *Metall. Trans. B*, **18B**, 245-255.

Poirier, D.R and Nandapurkar, P., 1988, "Enthalpies of a Binary Alloy during Solidification," *Metall. Trans. A*, **19A**, 3057-3061.

Poirier, D.R., Nandapurkar, P.J. and Ganesan, S., 1991, "The Energy Equation for Dendritic Solidification," *Metall. Trans. B*, **22B**, 889-900.

Roosz, A., Halder, E. and Exner, H.E., 1986, "Numerical Calculation of Microsegregation in Coarsened Dendritic Microstructures," *Mat. Sci. and Tech.*, **2**, 1149-1155.

Schneider, M.C., 1991, "Modeling the Columnar Dendritic Solidification of Lead-Tin Alloys," M.S. Thesis, The University of Iowa, Iowa City, IA.

Schneider, M.C. and Beckermann, C., "Effects of Simplified Enthalpy Relations on the Prediction of Heat Transfer during Solidification of a Lead-Tin Alloy," *Appl. Math. Modelling*, **15**, 596-605.

Thresh, H.R. and Crawley, A.F., 1970, "The Viscosities of Lead, Tin and Pb-Sn Alloys," *Metall. Trans.*, **1**, 1531,1535.

Voller, V.R., Brent, A.D. and Prakash, C., 1989, "The Modeling of Heat, Mass and Solute Transport in Solidification Systems," *Int. J. Heat Mass Transfer*, **32**, 1719-1731.

SCALE ANALYSIS AND NUMERICAL SIMULATION OF SOLIDIFICATION OF AN ALLOY COOLED AT A VERTICAL BOUNDARY

GUSTAV AMBERG
Department of Hydromechanics
Royal Institute of Technology
S-100 44 Stockholm, Sweden
e-mail: gustav@hydro.kth.se

In the last few years, significant progress has been made in the mathematical modelling of binary solidification. Several papers reporting solidification simulations based on first principles have appeared (Bennon & Incropera 1987, Voller et.al. 1989, Amberg 1991), and there are also a few comparisons between experiments and simulations (Christenson et.al. 1989, Shahani et.al. 1992). These simulations are fairly successful in predicting the expected qualitative behaviour for solidification due to cooling at a vertical wall, and in some cases also agree quantitatively.

However, the qualitative understanding of the solidification process is still limited. Consequently there are no established scaling laws for such fundamental properties as the time for complete solidification, or the degree of macrosegregation. In this talk such laws are derived for the restricted case of a low Prandtl number binary system, solidifying due to constant heat flux cooling at a vertical wall.

By a systematic classification of different possibilities, and estimates of the order of magnitudes of terms in the nondimensional governing equations, a parameter map is derived. This contains fourteen different parameter ranges, each of which signifies a particular set of qualitative properties of the cooling process. Among others, The 'long' and 'short freezing range' modes of solidification (Ruddle 1970) are recovered. The former is characterized by a mush that fills the entire mold, while in the latter the mush forms a front separating solid and liquid regions. In each case, quantitative estimates for solidification time, degree of undercooling and segregation, are derived. The qualitative and quantitative estimates are confirmed and illustrated by timedependent numerical simulation of typical solidification histories.

In order to check predictions in the long freezing range, slow solidification of an iron - 1% carbon system with 5°C initial superheat filling a 10 x 10 cm mold was simulated. This corresponds to a particular parameter range which is identified by the inequalities $Bo^{1/4} St_q/(St_s A) > 1$; $St_s > St_q > St_m$, where St_s, St_q, St_m are Stefan numbers ($L/(C\Delta T)$, L latent, C specific heat) based on three different typical temperatures $\Delta T_s = T_i - T_L(c_i)$, qB/k, and $\Delta T_m = T_L(c_i) - T_S(c_i)$. ΔT_s, the difference between the initial melt temperature and the initial liquidus temperature is the characteristic temperature difference for the liquid region. qB/k (q is heat flux, B half width and k heat conductivity) is typical of the entire mold, and ΔT_m, the difference between the initial liquidus and solidus temperatures, is characteristic of the

S. H. Davis et al. (eds.), Interactive Dynamics of Convection and Solidification, 199–201.

mush region. Bo is a Boussinesq number for the thermal convection in the liquid region $Bo = \alpha g \Delta T_S H^3/\kappa^2$.

A number of specific quantitative predictions can be made. Firstly it is found that the initial temperature is too high (or cooling rate too slow) for any mush to form initially. By simple considerations of the initial free convection, a mushy zone is predicted to appear first at t = 0.19 (time nondimensional with B^2/κ). Figure 1a shows the simulated velocity field and the position of the mush front at approximately this instant. As seen there, a mush is just appearing and covers about half the cooled wall, in good agreement with the estimate. After the appearance of the mush, the liquid region experiences the mush front as a boundary of approximately constant temperature, and is consequently cooled towards that temperature. From a simple model of the convective heat transfer in the liquid region, it can be predicted

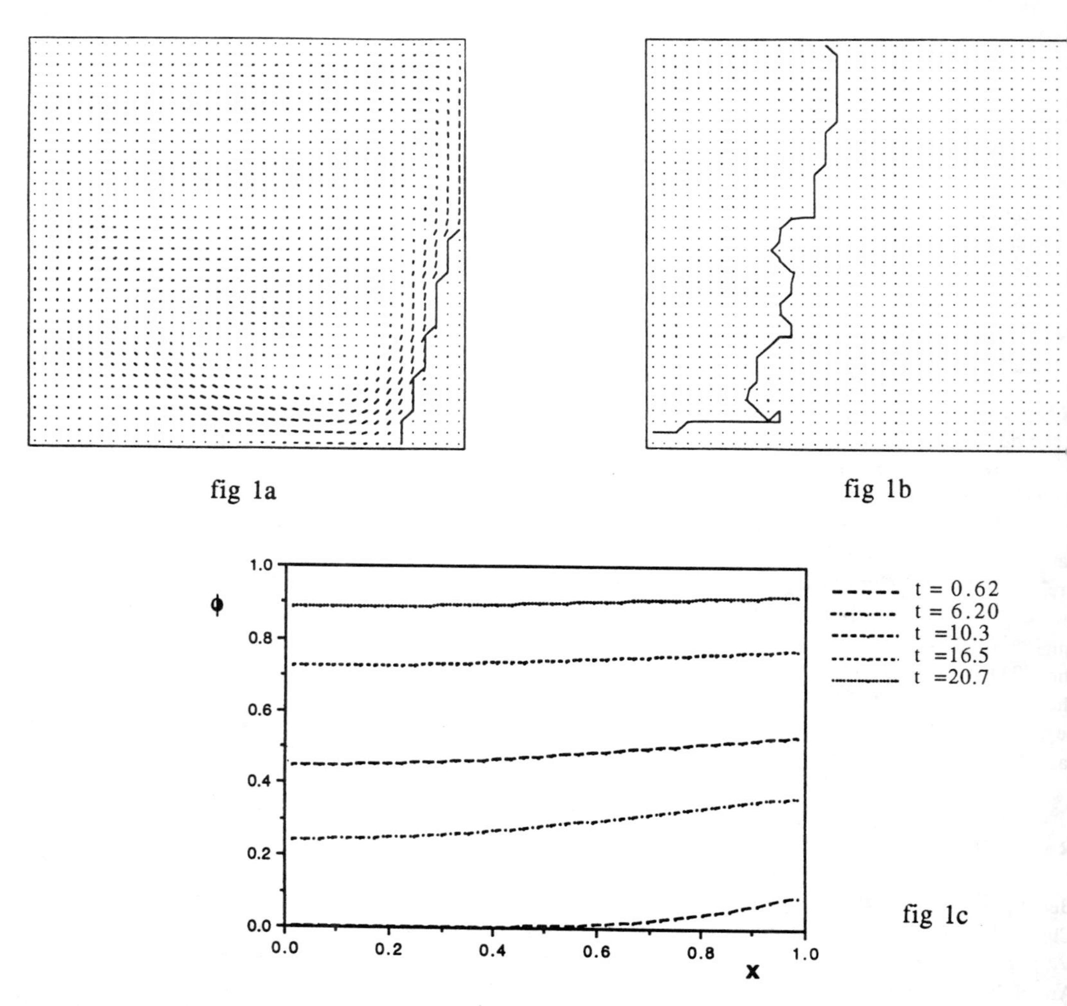

Figure 1. Simulated results for an example of the 'long freezing range', a) velocity field and mush front at t=0.20, b) at t=0.52, c) volume fraction solid vs. horizontal coordinate x, at midheight, for later times.

that convection ceases after t ≈ 0.53, corresponding to the simulated results in figure 1b. As seen there, there is no visible convection, and the mush front is becoming irregular. This is a situation where in reality undercooling of the liquid region is expected, in the simulation this will appear as a very thin mush spanning the entire cavity. The subsequent solidification now proceeds by a gradual thickening of the mush, i.e. a rather uniform increase of the solid fraction. Figure 1c shows the simulated solid fraction vs the horizontal coordinate x, taken at midheight, y=H/2. Indeed the variation of solid fraction over the mold is modest. Finally, the time for complete solidification is 22.9 from the simulation, compared to the predicted 18.1.

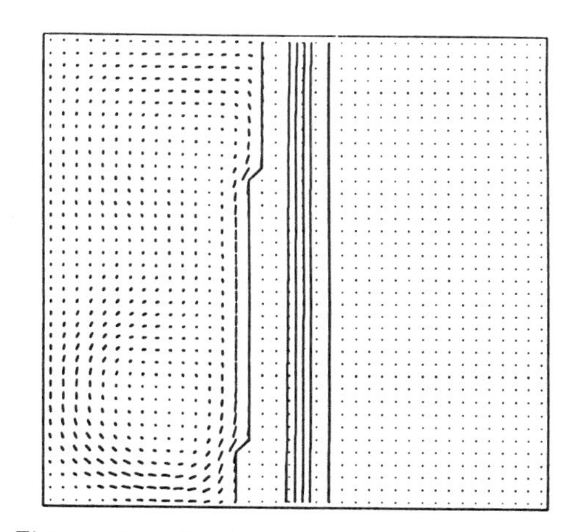

Figure 2. Simulated velocity arrows and contours of solid fraction (levels are 0.00001, 0.2, 0.4 ... 0.8 0.999) for the 'short freezing range', at time t=0.034.

As an example of the short freezing range, a case corresponding to very rapid cooling was investigated. The particular case was designated by the inequalities $Bo^{1/4} St_q/(St_s A) <1$, $St_s > St_m > St_q$. The major prediction in this case is that the mush here forms a very thin front separating the liquid and solid regions. Figure 2 shows the simulated velocity field and levels of solid fraction at nondimensional time 0.034, about one third of the predicted solidification time 0.091. There is indeed a front separating the liquid and solid regions, even though this is considerably wider than predicted. The simulated solidification time agrees quite well with the predicted.

Several other possible parameter ranges have been investigated, and the agreement between estimates and simulations above are typical. The above two cases are the most important qualitatively, even though each of the two divides further in subcategories giving different quantitative estimates. Also more quantitative predictions than those presented here can be made. In particular, an estimate of the final macrosegregation has been worked out for each case, showing it to be proportional to the mush Rayleigh number and a factor that differs in different ranges. In the cases above, segregation was predicted to be fairly large in the first, and completely negligible in the last, which is roughly confirmed by the simulated results.

References

Bennon W.D. & Incropera F.P. *Int. J.Heat Mass Transfer*, **30**, pp2171-2187, 1987.
Christenson M.S., Bennon W.D. & Incropera F.P. *Int. J.Heat Mass Tr.*, **32**, pp69-79, 1989.
Voller V.R., Brent A.D. & Prakash C. *Int. J.Heat Mass Tr.*, **32**, pp1719-1731, 1989.
Amberg G. *Int. J.Heat Mass Tr.*, **34**, pp217-227, 1991
Shahani H., Amberg G. & Fredriksson, H. *Met. Trans.* in press, 1992.
Ruddle, R.W. *Applied Science in the Casting of Metals*, ed K.Strauss. Pergamon Press 1970.

CONVECTIVE EFFECTS ON SOLIDIFICATION GRAIN STRUCTURE

F. DURAND
Grenoble Polytechnic Institute
Madylam, ENSHMG B.P. 95
38402 SAINT MARTIN D'HERES (France)

ABSTRACT. Blooms or billets made by continuous casting, or shaped castings from foundry, or even pieces treated by surface remelting, more generally all solidification products are agglomerates of crystals called solidification grains. Solidification grain structure includes informations on the space distribution of the grains, their size, shape, orientation etc... This structure is strongly influenced by convection in the liquid during solidification. According to observations on solidifification of transparent material in ingot molds, grain structure results from successive stages : filling, start of the natural convection, growth of the columnar crust, formation of equiaxed crystals, sedimentation regime, percolation regime. Experiments on Al-Cu 2 pct alloys solidified in the presence of rotative stirring give informations about the effect of stirring on the relative extension of the columnar crust, on the size of the equiaxed crystals, on the variation of temperature with time, including recalescence effects. The observations are discussed with respect to heat transfer phenomena at macro-scale, nucleation and growth of equiaxed crystals at micro-scale.

1. Grain structure, concept and importance

Most of the solidifications products are aggregates of crystals, called solidification grains. The concept of solidification grain structure involves first the space distribution of the crystals resulting from solidification. More precisely the concept contrives different types of information about their shape (are they oriented as in the columnar zone, or non-oriented as in the equiaxed zone ?), about their size, and about the associated defects : grain coarseness , porosities and voids, meso-and macro-segregations, etc... This concept applies to all kinds of solidification products, taking into account a large variety of size, geometry, and solidification conditions.

Most of metallic materials undergo solidification as the first stage of their life as materials.

- Slabs, billets produced by continuous casting, or ingots cast into molds, will be submitted to plastic forming by rolling or forging or extrusion. The geometry is simple and massive. The main parameters are data on the solid-liquid equilibrium, solid and liquid thermal properties, superheat defined as the difference between the initial temperature and the equilibrium temperature, and heat exchange conditions with ambience. Due to the massiveness of the product, the cooling rate is low, typically 0.1 to 1 Ks^{-1}. Temperature field in the liquid is strongly influenced by convection .
- Shaped castings receive their final geometry from the solidification stage. Since thousands of years, foundrymen make such castings, and some of these castings are among the finest works of art.

Presently, shaped castings are basic products for many industrial activities. Nowadays masterpieces are multivalve heads for automotive engines, casings for aeronautical turbine engines, bodies for cash machines... The required geometry is more and more complex, combining thin elongated veils connected to massive parts. In the thinner parts, the solidification time can be very short, less than a second. In such cases, there is a strong interaction between solidification and filling flow.

- The solidification time is much shorter in the products resulting from rapid solidification, typically powder particles (200 microns in diameter) or thin ribbons (less than 50 microns in thickness) or products treated by surface remelting. The geometry must ensure a very high cooling rate, 10^4 Ks^{-1} or more.

S. H. Davis et al. (eds.), Interactive Dynamics of Convection and Solidification, 203–215.

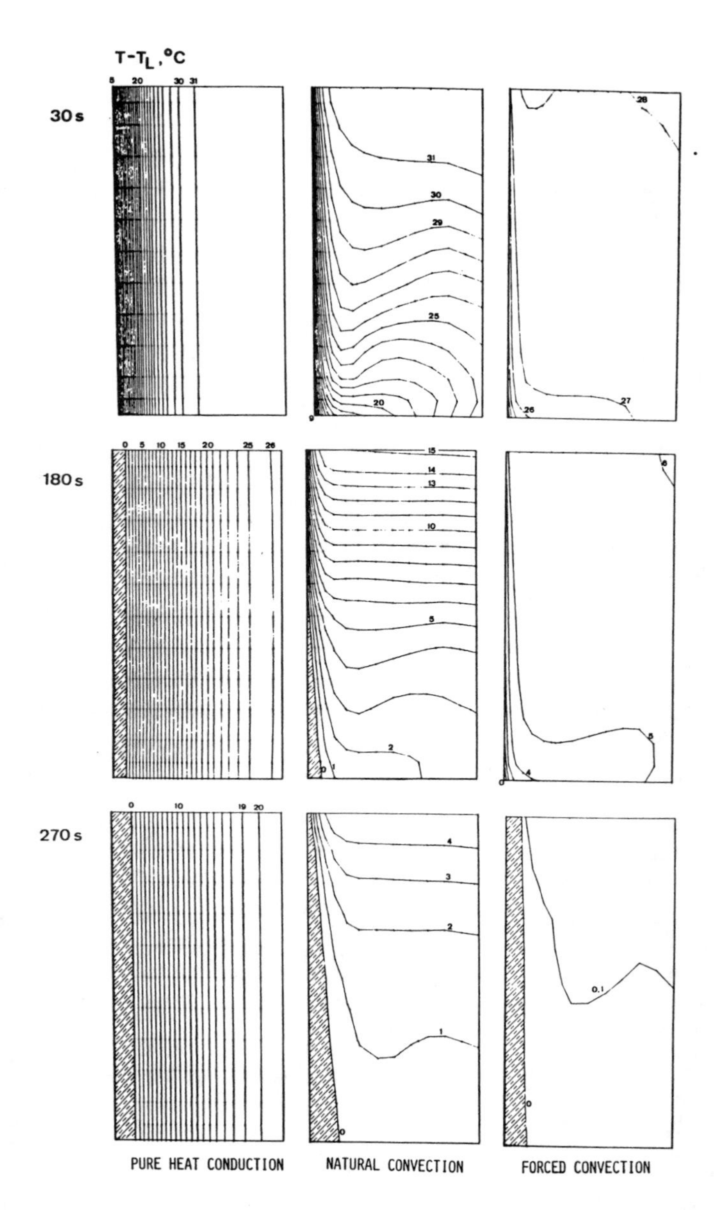

FIg. 1 - Temperature distribution in a rectangular ingot mold full of aluminium, at 30 s, 180 s and 270 s after pouring. Height 300 mm, with 200 mm, thickness 70 mm. The ingot mould is cooled through the wall on the left only . Liquid in forced convection (fig on the right), or natural convection (middle), or conduction only (left). From Meyer [1984].

- A welded junction may be considered as a solidification product, at least if fusion and solidification are the basic phenomena making the junction.

Most examples of solidification products are metallic materials. However the foundry techniques are industrially applied to cast ceramics. The concept of solidification grain structure may be successfully applied to crystalline aggregates resulting from solidification processes in geological systems, although solid state transformations make the identification of the structure difficult.

2. Process parameters of grain structure

They may be classified into 3 groups :
a) Chemical composition
For similar thermal and convective conditions, the extent of the equiaxed structure and the number of equiaxed grains per unit volume increase with the solute concentration. According to empirical observations, "good alloys" for solidification grain structure usually have a relatively large freezing range. As indicated below, this effect can be represented by the DT_{MH} parameter, which is the freezing range calculated for the composition of the first formed crystal. However an equiaxed fine grained structure may be obtained even with pure metals if stirred during solidification.
b) Thermal parameters
For given composition and convective conditions, the number of equiaxed grains is higher with higher values of the initial cooling rate. However, such cooling conditions involve high heat flow. Consequently convection is necessary to limit the columnar growth as explained below.
c) Convective state of the liquid
For given heat losses and composition conditions, stirring reduces the columnar zone, favours the equiaxed zone and increases the equiaxed grain number.

Its effect on the temperature distribution was clearly demonstrated by Meyer's numerical simulation [1984] (Fig. 1). Forced convection gives a very uniform temperature distribution, the gradients being confined close to the cooling wall. The liquid adjacent to the wall reaches later the equilibrium temperature, and at that moment the gradients are much lower than in the liquid motionless or in natural convection .

3. Importance of convection on solidification structure

An ice-cream is a solidification product demonstating what can be obtained by convection on solidification structure. One observes that the product is attractive only if its structure is equiaxed and fine grained. This is performed by stirring the melt up to solidification. But if the stirrer does not operate, the product is coarse grained, it crunches under the teeth, it has not any value.

More generally, solidification grain structure is strongly influenced by convection conditions during solidification. The discussion of this observation takes into account phenomena interacting at different scales. The number of equiaxed grains depends on nucleation and growth phenomena (micro-scale). The key parameter is the supercooling, that is the difference between the melt temperature and the solid-liquid equilibrium temperature. These phenomena are governed by heat losses at the scale of the product (macro-scale). The shape and space distribution of crystals are controlled by heat and solute transport phenomena at the solid-liquid interface (micro-scale), which depend strongly on the convective state of the liquid. The next stage of discussion is to discriminate in the flow structure different components able to act at the relevant scale of crystallization mechanisms. According to the considered structure or defect, a large variety of geometries, scales, thermal and flow conditions are to be taken into account. Heat and mass transport are coupled. Consequently, the interpretation is specific of the considered case. It is important to keep in mind the experimental facts in their very complexity.

In the following, experimental observations will be presented in order to highlight the main parameters, then simple theoretical models will be used to explain the observed tendencies.

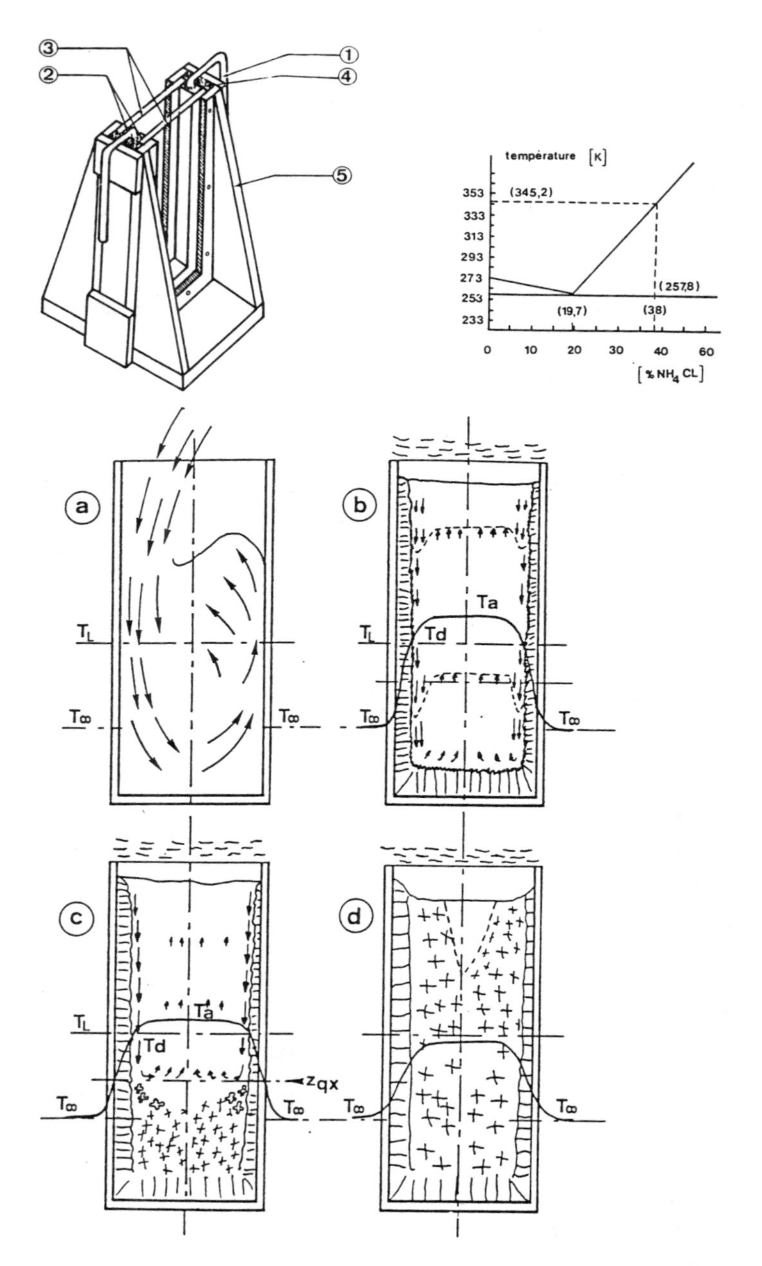

Fig. 2 - Solidification experiment on a transparent material, here ammonium chloride in solution in water. Ingot mold, height 180 mm, width 60 mm, thickness 25 mm. Corresponding phase diagram. a) filling stage, b) natural convection sets up and columnar crystals grow in crust, c) equiaxed crystals are observed in front of the lower part of the vertical wall, d) sedimentation and percolation regime. From Witzke [1979].

4. Formation of the grain structure, a kinematic view

When considering the grain structure of a solidified product, the successive stages of the process are to be taken into account . Some informations emerge from the critical examination of the macrostructure itself. In fact the most suggestive ideas come from solidification experiments on transparent materials. Five stages can be distinguished. They were particularly clear in a series of experiments performed by Witzke [1979] (Fig.2). The material was a solution of ammonium chloride in water, the liquidus temperature being 72°C. It was cast in a small rectangular ingot mold . The large lateral walls were transparent and insulating, the liquid was cooled by the small lateral walls and the bottom, made of a copper tube with rectangular section. In most cases, the mold walls were preheated by hot water flowing in the tube before casting.

The five stages were the following :

a) The filling stage itself

The movement due to filling was damped in a few seconds. Equiaxed crystals could be observed immediately at this moment if the mould walls were cold, but no crystal is formed if the tube constituting the walls was preheated.

b) Onset of natural convection

One can see clearly how liquid layers are accelerated along the vertical cold wall, with a recirculation in the central part of the liquid.

c) Formation of a columnar crust

Columnar crystals form in a continuous crust, which grows progressively.

d) Formation of equiaxed crystals

The first equiaxed crystals can be discerned in the very lower part of the accelerated layer, clearly the coldest part of the liquid. The flow carries the small crystals away in the central part of the liquid, which is not so cold, and where they partly remelt, at least at the first time. After a while, in all the lower central part of the liquid the equiaxed crystal stop remelting. Since then, two zones can be distinguished in the liquid, separated by an horizontal limit. In the lower zone, equiaxed crystals are sedimenting, their fall promotes a specific recirculating flow. In the upper zone, no crystal is discernible, and the natural convection vortices continue to operate. The limit between the two zones progressively moves to the free surface, on account of superheat extraction.

e) Sedimentation and percolation.

When the limit between the 2 zones reaches the free surface, equiaxed crystals are either floating or sedimenting in the whole liquid, at least in the upper part. (sedimentation regime).

In the lower part, the solid fraction is higher, crystals join one to another in a rigid skeleton, through which the residual liquid may percolate (percolation regime).

According to the observations, the columnar growth rate progessively slows down, until it stops.

Concerning solidification of metallic ingots, experimental facts let suppose that the stages mentionned above exist also, keeping in mind that each stage results from a specific space distribution of isotherms. In particular, if superheat is low, equiaxed crystals can form very early in the liquid. However solidification in ingots is typically a time-dependent phenomenon. Natural convection strongly influences isotherm space distribution in the liquid. Moreover, crystal formation itself promotes a specific convective flow. Consequently the interpretation of ingot solidification is far from straighforward. Comparatively, continuous casting which is a steady-state process, allows easier interpretation, although most observed defects, as for instance meso- and macro-segregations, involve non-steady phenomena.

5. Solidification in good stirring conditions

The present paragraph is based on a series of experiments on a single alloy composition (Al-2% Cu), solidified at different cooling rates in the presence of stirring [N'Guyen Than Linh 1991]. The ingot was a cylinder with a low heat capacity (Fig. 3). A typical macrostructure is presented on Fig. 4 c and d. The external zone is formed by fine columnar grains, which are inclined at counter-current with respect to the liquid flow. The internal zone is formed by equiaxed crystals, more or less fine and more or less uniformily distributed according to the rotation rate.

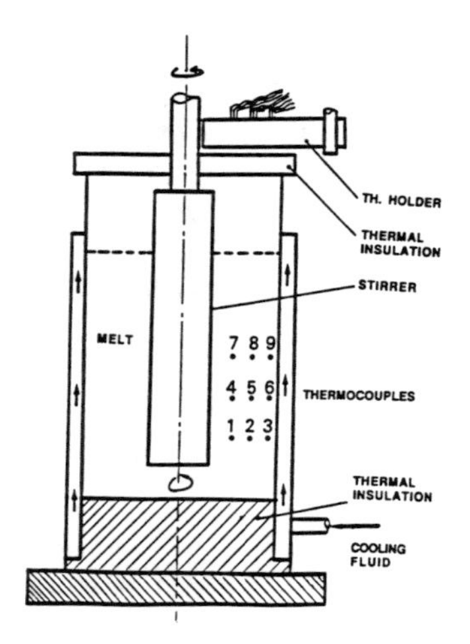

Fig. 3 - Schematic cut of the ingot mould for rotative stirring experiments. Internal diameter 140 mm, height 180 mm.

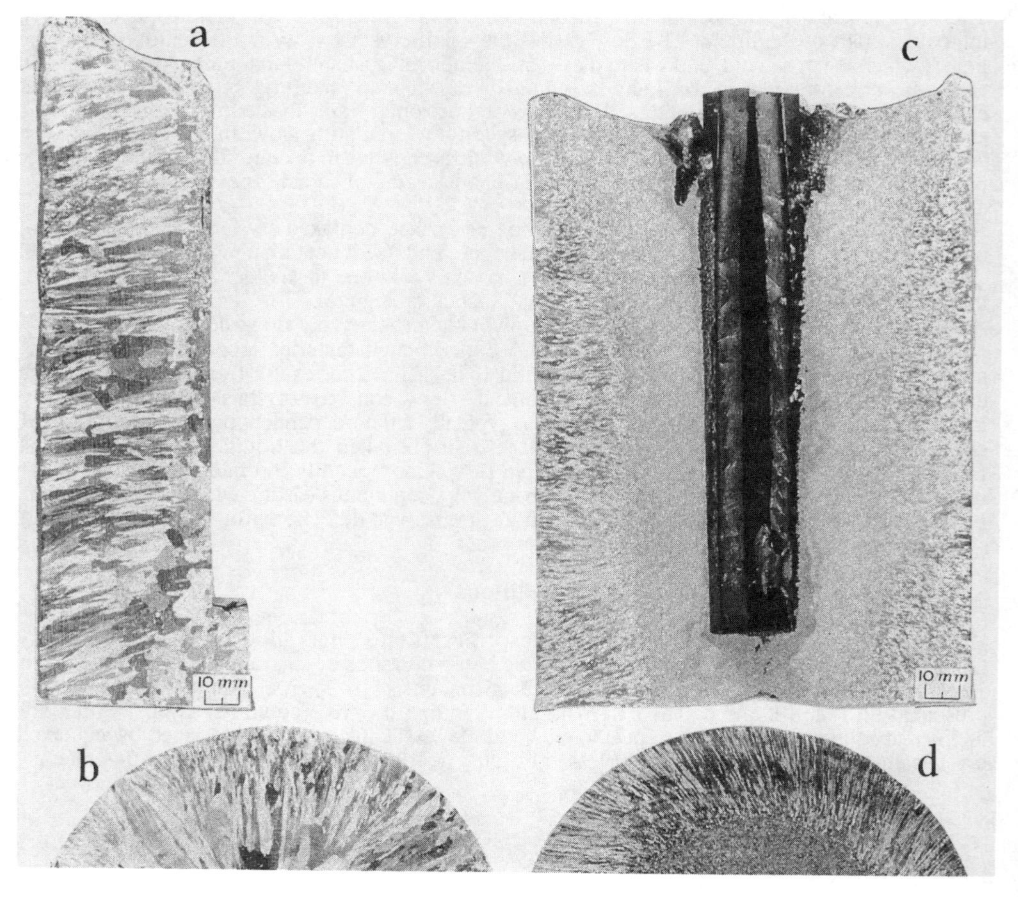

Fig. 4 - Macrographs of 2 ingots cooled by water : a) and b) without stirring, c) and d) stirred at 350 rpm.

Nine thermocouples, numerically recorded, gave lots of informations on the thermal behaviour in solidification conditions. On typical temperature-time curves, several stages could be distinguished (Fig. 5 a and b): cooling of the liquid (segment AB on Fig. 5b), recalescence (BG), pseudo-plateau (GH) and final cooling acceleration (HDE). For further exploitation each curve was summarized by noting the values of the following characteristics: P_L, cooling rate at the liquidus, T_{min} and t_{min}, minimum temperature at the recalescence step and corresponding time, T_{max}, maximum temperature after T_{min}, and D_{pp}, duration of the pseudo-plateau.

In good stirring conditions, all thermocouples display the recalescence step practically at the same time, with the same amplitude and the same duration and also the same slope of the pseudo-plateau (Fig.5c). This result means that the whole liquid is submitted to the same temperature-time history, with an accuracy better than a few tenths of a degree, which is an experimental result enabling valuable zcientific analysis.

The only measurement which varies from one thermocouple to the other is the pseudo-plateau duration, indicating that the solid-liquid front is progressing from the cooled wall toward the inside of the ingot. Concerning this point, fluctuations observed on the temperature curves indicated that the change of slope at the end of the pseudo-plateau can be related to the change in flow regime. During the pseudo-plateau, the thermocouple receives heat flows from a melt with equiaxed crystals growing in suspension in it. After the pseudo-plateau, the thermocouple is embeded in a rigid dendritic skeleton, in which the residual liquid is percolating.

It is worth noting that such experiments gives only indirect information on the supercooling existing during the solidification step. In effect, the thermocouples were not calibrated. Moreover, the thermodynamic equilibrium temperature,which is the reference for supercooling. if it has been measured as it is usually the case for pure metals, is given by measurements performed in a solidification experiment comparable to ours, the conditions of which are difficult to be taken into account.

Our curves give 2 indirect informations on supercooling:

1) The recalescence amplitude $T_{max} - T_{min}$ is a lower limit of the supercooling.

2) In the pseudo-plateau regime, the supercooling remains practically constant, its variation being lower than 1 degree.

The two informations above depend on the stirring intensity.

6. Effect of stirring on experimental features of solidification in ingots

6.1 EFFECT ON GRAIN STRUCTURE

Modification in the ingot structure is summarised by Fig.7, representing the relative extent of the equiaxed zone, and Fig.8, on the grain size of the equiaxed crystals. Stirring decreases the extent of the columnar zone, and decreases the equiaxed grain size. This decrease is very strong when stirring increases from low to moderate value, then this tendency vanishes progressively .

6.2 EFFECT ON TEMPERATURE-TIME CURVES

When rotation rate decreases, recalescence starting time varies from one point to its neighbour, recalescence amplitude increases and moreover its duration can become very long (Fig.8). This variation can be interpreted as follows. The theoretical model shows that equiaxed crystals nucleate in a very narrow temperature range, a few tenths of a degree close to T_{min}. But the thermocouple in M receives not only the latent heat release from the volume element in M, but also the thermal flow coming from the surrounding volume elements in M', M" etc... In good stirring conditions, recalescence occurs practically at the same time in M, M', M", and the recorder signal is very sharp. If stirring is not so good, recalescence starts at different times in M, M', M" so the recorded signal is enlarged and deformed.

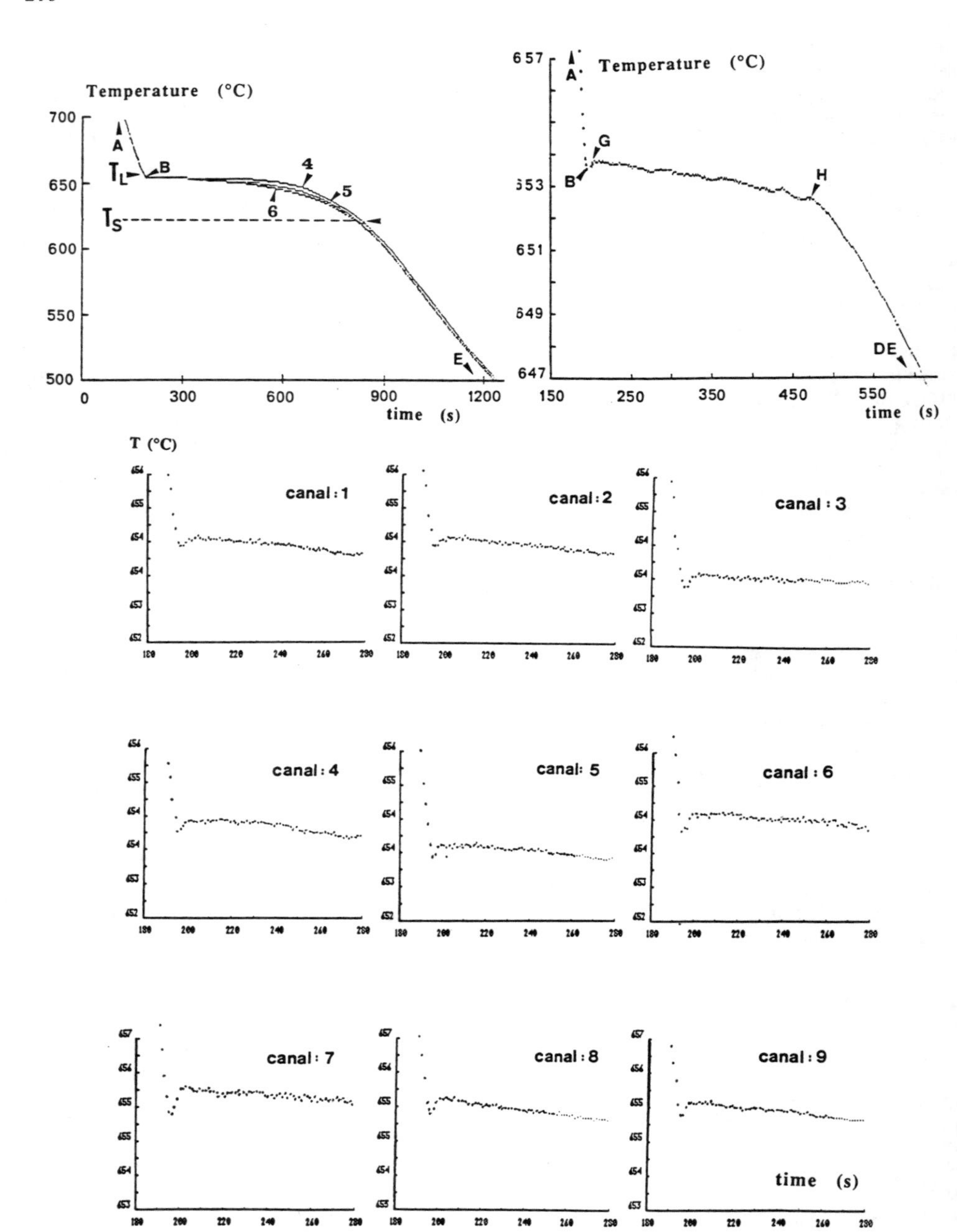

Fig. 5 - Typical temperature-time curves from the ingot cooled by compressed air and stirred at 200 rpm. a) thermocouples placed at 75 mm, r = 34 mm (nb. 4), 49 mm (nb.5) and 64 mm (nb.6).
b) Close view of curve 5.
c) Comparison of the temperature-time curves of the 9 thermocouples, in the ingot cooled by compressed air, stirred at 200 rpm. Numbers refer to the thermocouple position on fig. 3.

The most important effect of stirring is the uniformization of the temperature distribution in the melt, the reduction of temperature gradients, and the delay in the start of the columnar growth, already mentionned in par.2c. The columnar volume fraction is directly lowered, and moreover the equiaxed grain formation may operate in a larger volume and more favourable supercooling conditions.

Stirring increases the cooling rate P_L, but only in a limited range, less than 10 %. This result is logical, because heat transfer resistance in the liquid is only a minor part of the thermal resistance controlling heat losses.

6.3 ORIGIN OF EQUIAXED GRAINS

Concerning the effect of stirring on the decrease in grain size, the interpretation involves a discussion on the origin of equiaxed grains. From the examination of the temperature records during solidification, it appears that supercooling, although not precisely measured, stays at a moderate level, very far from the range associated to homogeneous nucleation. So crystals should form on hypothetical preexisting "nucleating particles". Formation of "redendritic nuclei" in the strong supercooling conditions of the liquid against the mould wall, as proposed by Gender [1926, quoted by Chalmers 1963], by Biloni [1968], Ohno (1968] , possibly explains what is observed in Witzke 'experiment [1979] when the mould walls are cold, but not when they are preheated. In the experiments on aluminium alloys by N'Guyen Thanh Linh, superheat was relatively high and mold thermal capacity very low, so we consider that this mechanism does not operate. Moreover there was no addition of nucleating agent , and titanium and boron were far below the level able to promote equiaxed seeding.

Another assumption is to consider that the liquid can be seeded by dendritic fragments detached from the columnar front. This mechanism could explain the effect of stirring. Obviously it operates in transparent solidifying systems when stirring starts during solidification. However, in this case temperature field is modified, so the dendritic front becomes unstable and tends to remelt. This is not the case of a continuously applied stirring. The turbulent fluctuations able to make such remelting operate at a scale probably much larger than the dendritic structure.

6.4 RESULTS OF THEORETICAL MODELS CONCERNING EQUIAXED GRAINS

The variation of the number of equiaxed grains with solidification conditions was treated recently by a few theoretical models for inoculated nodular irons [Lacaze 1988, following Oldfield] and for aluminium based, inoculated alloys [Desbiolles 1989]. In these models, the number of grains is a function of supercooling with 2 parameters fitted on experimental results.

A model by Maxwell and Hellawell published in 1975 uses directly the capillary law for the nucleation rate dN/dt :

$$\frac{dN}{dt} = B \exp(- \frac{A}{\Delta T^2}) \quad (1)$$

The nucleation law is used to calculate numerically N_i, the number of grains formed at time step i during the supercooling excursion. The melt temperature is assumed to be uniform, its variation wilth time is given by a simplified heat balance between the heat losses (represented by the initial cooling rate P_L) and the latent heat production:

$$dT = - P_L dt + \Delta H_F \, df_s \quad (2)$$

ΔH_F latent heat of fusion per unit volume

In the original model, B is related to the number of nucleating sites, A is a function of the capillary ability of the nucleating substrate, both parameters being fitted.

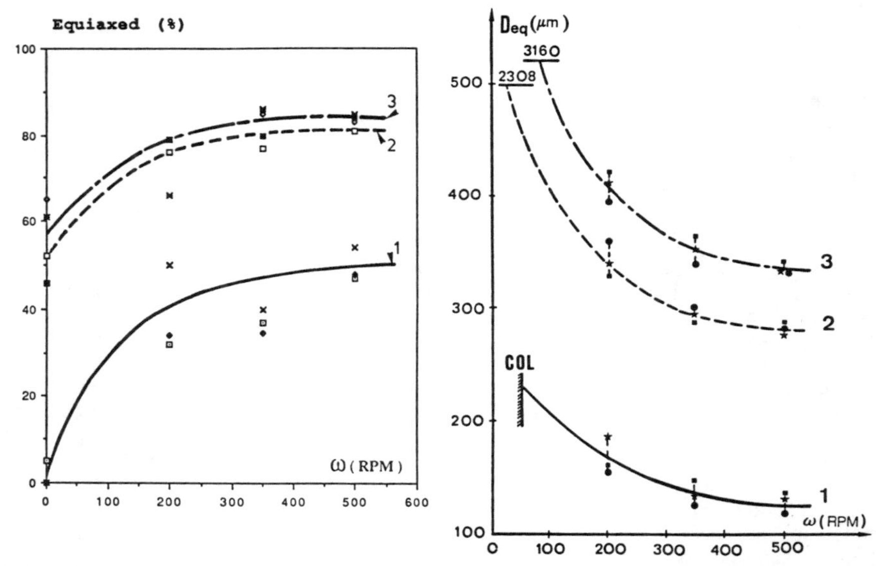

Fig. 6 - Equiaxed proportion as a function of stirring intensity. Ingots cooled by water (1), by compressed air (2), by air in natural convection (3).

Fig. 7 - Grain size (D_{eq}) as a function of stirring. Ingots cooled by water (1), by compressed air (2) and by air in natural convection.

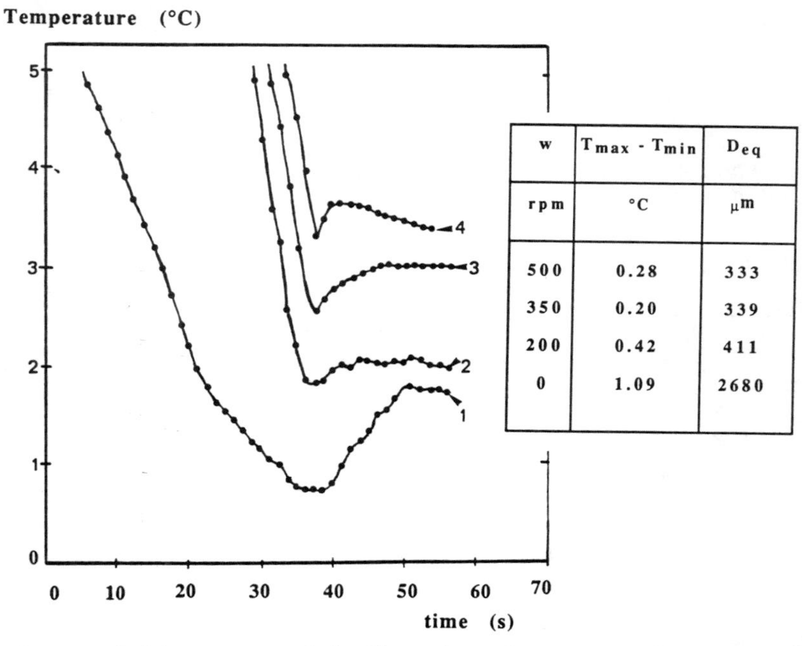

w	Tmax - Tmin	Deq
rpm	°C	μm
500	0.28	333
350	0.20	339
200	0.42	411
0	1.09	2680

Fig. 8 - Ingot cooled by compressed air. Close view of temperature-time curves of the same position in 4 stirring conditions. The curves are shifted, so only the relative variations are to be considered.

We adapted Maxwell-Hellawell model to non-inoculated, multicomponent alloys solidified in stirring conditions [Desnain 1990]. The preexponential term B is calculated from physical properties. A-parameter remains fitted. df_S, production of solid, is a sum on all the classes i of crystals, represented by their number N_i, their present radius R_i and growth velocity v_i:

$$(3) \qquad dfs = \sum_i N_i \, 4\pi \, R_i^2 \, v_i \, dt$$

We demonstrated that in reversible conditions, the production of solid is related to the DT_{MH} parameter which is a numerical estimate of the freezing range for the composition of the first formed crystals:

$$(4) \qquad \frac{dfs}{dT} = \frac{1}{DT_{MH}}$$

$$(5) \qquad DT_{MH} = \sum_j m_j \, C_{bj} \, (k_j - 1)$$

Here the subscript represents the solute element j, C_{bj} is its concentration in the initial liquid, m_j is the corresponding liquidus slope, k_j the partition coefficient.

On fig. 9 is presented a typical temperature-time curve calculated by the model (curve a), together with the corresponding nucleation rate variation (curve b). This latter variation is strongly dependent on supercooling. Curve b depicts the strongly non-linear variation of eq.1. Nucleation begins at DT_n, practically 1K in the considered case, just 0.20 degree higher than DT_{max}. At DT_{max}, the latent heat release begins to exceed the heat losses of the system, the temperature increases. At DT_{ex}, nucleation stops.

When the cooling rate P_L increases, due to modification in the cooling system, or (in a more limited range) to the effect of stirring, DT_n and DT_{ex} do not change(because P_L is not in eq. 1). DT_{max} is only slightly increased (less than 0.10 K), but the number of additional grains is large, due to the form of eq.1. Of course the numerical results depend on the value of A-parameter, which results from trial-and-errors fit in order to give to DT_{max} a reasonnable value. But this numerical behaviour is in agreement with current observations on transparent materials solidified in the presence of stirring : when the first crystal is observed, in a while, millions of crystals are present everywhere in the melt.

Considering the effect of stirring intensity on grain size mentionned in par.4, the present theoretical result can be used to pursue the discussion . Increasing the stirring intensity makes only a minor increase in the initial cooling rate P_L. According to the model, DT_{max} increases very slightly. Considering fig.8 and taking the highest observed value of (T_{max} - T_{min}), it may be assumed that in all the 4 cases DT_{max} value is probably higher than 1.09 K. In the non-stirring case (curve 1), the thermocouple receives the heat flow coming from surroundings at different degree of recalescence, its temperature increases relatively strongly (1.09 K) but nucleation has stopped much before. In high stirring conditions (curve 4), recalescence is very sharp, as predicted by the model, and the pseudo-plateau starts very early, expressing a steady-state balance between local latent heat release and heat losses toward ambience. In the intermediate conditions (curves 3 and 4), the change in stirring intensity may have changed the thermal flow transporting heat toward ambience, so that the difference (Tmax - Tmin) is lower, and the slope of the pseudo-plateau is less negative. Unfortunately, the theoretical model cannot simulate this variation.because it treats the melt as isothermal.

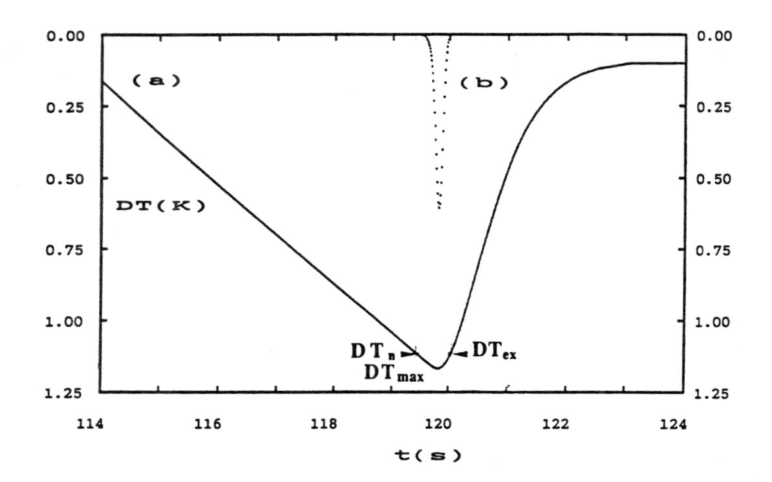

Fig. 9 - Curve a) temperature-time curve calculated for alloy 5182 (DT_{MH} = 17,5 K). Initial cooling rate $P_L = 0.16\ Ks^{-1}$.
Curve b) nucleation rate dN/dt. Nucleation starts at DT_{nucl}, and stops at DT_{ex}. From Desnain [1990].

Concerning the effect of alloy composition, when the model is applied to industrial compositions of aluminium alloys [Desnain 1988], the calculated number of grains gives the same rating as grain size measurements. Typically a "bad alloy" (technical aluminium noted 1050) has a low DT_{MH} value (1,82 K), whereas a "good alloy" (alloy 5182, used for beveradge cans) has a relatively high DT_{MH} value (17,5 K). The model gives to both of them the same value of DT_n and DT_{ex}, but alloy 5182 has a slightly larger DT_{max} value, and a much larger number of grains.The reason is that the growth rate is controlled by the diffusion layer. It is 3 to 5 times lower for the good alloy, so recalescence is delayed, supercooling can be a little higher and additional classes of crystals can be formed.

Conclusion

1) The grain structure of a solidification product results from several mechanisms operating in a range of time and a range of space. So special care is needed to get some structure worth to be submitted to theoretical discussion.
2) In usual conditions of solidification, hydrodynamical phenomena always operate, and they print their mark on the grain structure. But the theoretical discussion has to be specific for the considered structural element or defect, in relation to specific thermal, geometrical and flow conditions.
3) Stirring is an efficient mean to control hydrodynamical phenomena occuring during solidification. If it is reasonably used, it can give a very uniform temperature-time history in all parts of the product, and a very uniform grain structure, to which theoretical discussion can be applied conclusively.
4) In the theoretical discussion, the supercooling effects must be taken into account, because the grain structure depends on the nucleation and growth history. Numerical models for hydrodynamical and thermal phenomena should be implemented by kinetic modules in order to simulate grain structure formation.
5) The convective effects on the grain structure in ingot geometry involve different scales of hydrodynamical phenomena. Large scale vortices transport the superheat or the latent heat release to the solid wall. Homogenization phenomena are performed by small scale eddies. More generally the specific discussion mentionned above must combine the relevant scale physical phenomena to the relevant scale of hydrodynamical phenomena.

REFERENCES

Biloni, H., "Relationship between segregation substructures and casting structures", dans "The Solidification of Metals", ISI publication 110 (1968), 74-82.

Chalmers B., "The structure of ingots" J. Aust. Inst. Metals 8 (1963), 255-263.

Desbiolles, J.L., Thevoz, Ph., Rappaz, M. "Modelling of equiaxed dendritic microstructure formation in castings", in 'Modeling of Casting and Welding Processes' . Ed A.F. Giamei and G.J. Abbaschian, T.M.S. Publ. Warrendale Pensylvania, 1989, pp. 625-634.

Desnain, Ph., Durand, F., Fautrelle ,Y., Bloch, D., Meyer, J.L., Riquet, J.P. "Effects of the electromagnetic stirring on the grain size of industrial aluminium alloys : experiments and theoretical predictions", in "Light Metals" (1988), L.G. Boxall ed., The Met. Soc. AIME (1988), pp. 487-494.

Desnain, Ph., Fautrelle, Y., Meyer, J.L., Riquet, J.P., Durand, F. "Prediction of equiaxed grain density in multicomponent alloys, stirred electromagnetically". Acta Met., 38, n° 8, (1990), 1513-1523.

Lacaze, J., Castro, M., Aichoun, N., Lesoult, G. "Influence de la vitesse de refroidissement sur la microstructure et la cinétique de solidification de fonte GS. Experience et simulation numérique de solidification dirigée". Mem Et. Sci. Rev. Metallurgie, 1988, 85-97.

Maxwell, I., Hellawell, A., "A simple model for grain refinement during solidification". Acta Met., 23, (1975), 229-237.

Meyer, J.L., Durand, F., "Analysis of the transient effect of convection during solidification, with or without electromagnetic stirring". dans "Modeling of Casting and Welding Processes-2". J.A. Dantzig, J.T. Berry ed. The Met. Soc. AIME (1984), 179-191.

N'Guyen Thanh L. Durand F., "Effet du brassage sur la structure de solidification d'un alliage Al-Cu", Mem Et Sci Rev Met allurgie (1990) 749-759.

Ohno, A., "Compositional depression of undercooling and formation of segregation between columnar and equiaxed crystal zones", in "The Solidification of Metals" ISI Publ.110 (1968), 349-355.

Witzke, S., Riquet, J.P., Durand, F., "Visualisation de la convection lors de la cristallisation basaltique et équiaxe d'une solution transparente". Mem. Sci. Rev. Met.allurgie (1979) 701-714.

BUOYANCY-DRIVEN MELT POOL CONVECTION DURING LASER SURFACE TREATMENT

P. EHRHARD and CH. HÖLLE
Kernforschungszentrum Karlsruhe GmbH
Institut fuer Angewandte Thermo- und Fluiddynamik
Postfach 3640, D-7500 Karlsruhe, GERMANY

ABSTRACT. A laser spot scans at constant velocity across a workpiece, which is positioned on a constant-temperature copper table. Below a solid coating of hard particles a melt pool is formed, which during resolidification incorporates the particle layer at the surface forming a composite layer of improved wear resistance.

A model for this process is derived. Experimental results, similarity conclusions and finite element simulations, three dimensional and steady within a moving reference frame, are conducted.

1. Definition of the Problem

We focus on a specific process, which is employed to incorporate hard particles (e.g. *TiN*, *TiC*) at the surface of a workpiece and, thus, to improve wear resistance. To achieve this, in a first step particles are deposed on the substrate in a thickness of about 30 μm. During a second step a rectangular laser spot travels with constant velocity U and power density q across the workpiece. Due to heat input a melt pool of substrate material is formed below the solid coating. The workpiece is positioned on a copper table at constant temperature. A more detailed description of the process and the advantageous processing results is given by Schüssler & Zum Gahr (1990) and Schüssler et al. (1992).

2. Mathematical formulation and solution method

To treat this problem theoretically one has to solve the heat transport equation, the momentum equation and the continuity equation within the liquid domain (melt pool) - coupled through the free liquid/solid interface to the conduction problem in the solid portion of the workpiece. It is convenient to use a laser-attached coordinate system, together with a non-dimensionalized version of the conservation equations. The problem is indeed three-dimensional, since generally the extension of the laser spot in cross track direction covers only a part of the workpiece.

The appropriate boundary conditions are as follows: In scanning direction an adiabatic boundary condition is posed in a far enough distance. The most important heat sink is represented by the bottom plane, where a Nusselt-type law is employed for the contact heat transfer. The vertical boundaries are taken to be adiabatic. The top

S. H. Davis et al. (eds.), Interactive Dynamics of Convection and Solidification, 217–220.

plane features a rectangularly-shaped laser spot with constant heat flux, surrounded by an adiabatic region. The solid/liquid interface is at melting temperature ϑ_m, i.e. we ignore any alloy effects and assume the interface to be at equilibrium. Any consumption/production of latent heat at the melting/solidifying part of the solid/liquid interface is likewise neglected. Kinematically, at both the solid/liquid interface and the top (covered) boundary of the liquid region a no-slip boundary condition is posed.

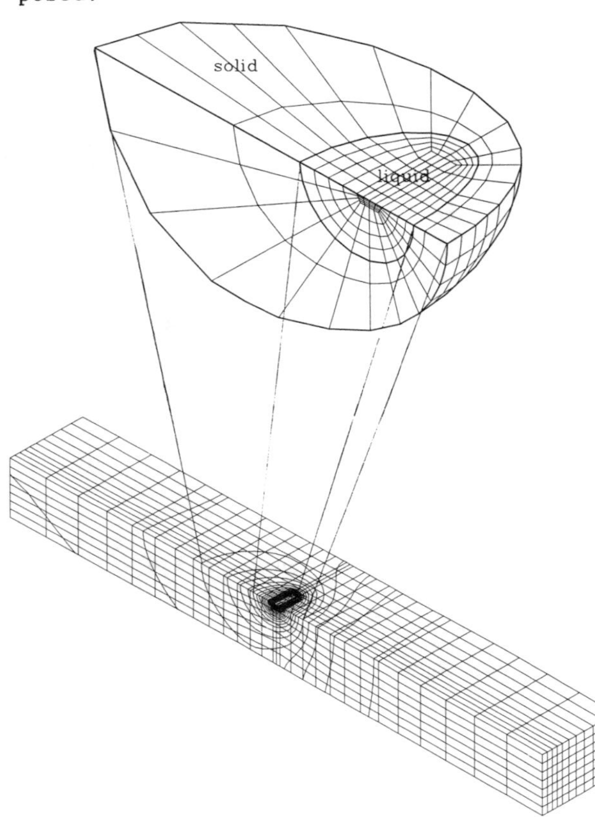

Figure 1: Mesh for numerical solution.

Figure 1 illustrates the mesh on which the above problem is solved. In a first step we solve the pure conduction problem in one half of the workpiece (lower mesh). This is done numerically by means of a standard finite element code (FIDAP) for general boundary conditions in the bottom plane. Alternatively, following Rosenthal (1946), for an adiabatic bottom plane (*Nu=0*) an analytical solution can be constructed. Once a temperature field has been obtained we switch to the combined conduction/flow problem by using a second mesh. We pick an outer isothermal surface and use this as constant temperature boundary for this mesh. The isothermal surface at melting temperature ϑ_m is used as first guess for the position of the solid/liquid interface. The combined conduction/flow problem then again is solved using finite elements. Within a few iterations the solid/liquid interface is adjusted to collapse with the isothermal surface at melting temperature.

3. Results

3.1. Conduction problem

Figure 2 illustrates the results obtained for the depth of the molten pool t/h as function of the Peclet number $Pe=Uh/\kappa$ for a fixed power density. Symbols h, κ denote thickness of the workpiece and conductivity of the solid. We recognize two limiting situations: (i) If Pe is increased beyond $Pe \simeq 10$ the depth of the molten pool approaches zero. This

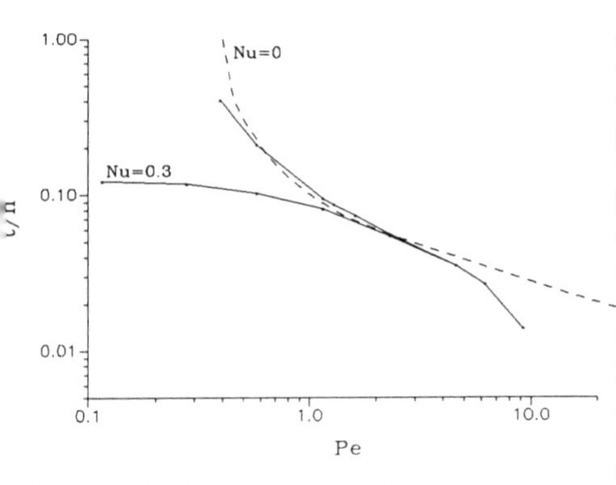

Figure 2: Pool depth from pure conduction analysis.

corresponds to an effective heat removal due to advection which restricts the maximum temperature below ϑ_m. (ii) If *Pe* is decreased towards *Pe*≃*0.5* and *Nu=0* (adiabatic bottom plane) the removal of heat is not possible by advection and consistently the melt pool penetrates through the workpiece (*t/h>1*). The behaviour for small *Pe* is correctly reflected by both the numerical results (solid line) and the analytical results (dashed line). The analytical solution fails, however, to capture the behaviour for large *Pe* as *t/h* approaches zero. This is an artifact of the point heat sources within the analytical solution. There is a second numerical curve for *Nu=0.3* given in figure 2, corresponding to a non-adiabatic bottom plane. This curve strongly departs from the curve with *Nu=0* when we approach small *Pe* numbers - the melt pool depth approaches a constant value. Physically, these findings correspond to a situation where heat is mainly removed through the bottom plane of the workpiece.

3.2. Combined conduction/flow problem

In figure 3 two cross sections are given for the flow inside the liquid domain. We recognize a three-dimensional flow field rising in the central region and falling at the (cold) liquid/solid interface. Thus, a non-symmetric, torus-type flow is observed, whereas velocity amplitudes are considerably below the scanning speed *U*. We find that the melt pool geometry is only weakly affected by the flow - in all cases the changes are less than *2%*. Physically, the weak flow hardly affects the heat transport and therefore no significant change of the temperature field occurs. This result is specific for this particular process, which, due to coverage of the liquid by a solid layer, prevents driving by thermocapillary effects.

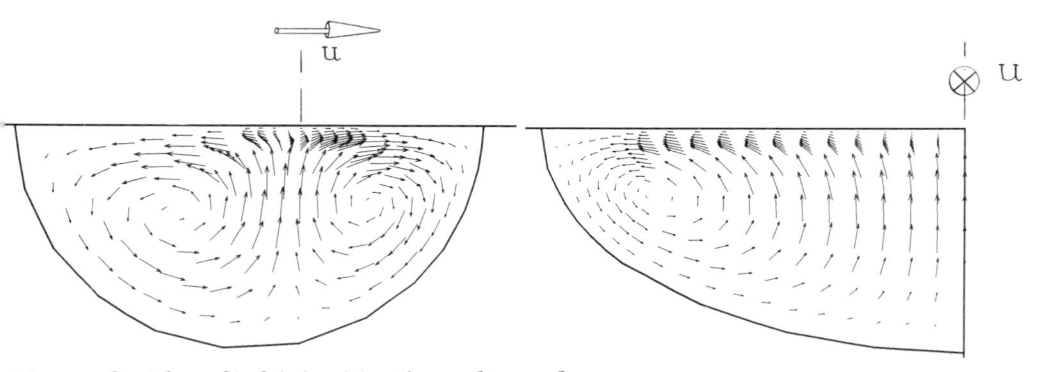

Figure 3: Flow field inside the melt pool.

From the above findings one can, indeed, infer that a reasonable prediction, e.g. of the melt pool depth, is possible purely on basis of a heat conduction analysis. There remains, however, one aspect which obviously is linked to the flow strength: During "normal" processing the micrographs demonstrate the particle layer being homogeneously embedded at the surface of the resolidified substrate. In contrast, e.g. by choosing higher laser power, the particle layer is swirled into and covered by the resolidified substrate. This is obviously due to normal and/or shear stresses caused by the flow below the layer. By evaluating all experiments with respect to an effective Grashof number we find consistently a breakdown of the coating layer if $Gr>600$, independent of any further parameter group (cf. Schüssler et al. 1992). The numerical results on the pressure field below the coating layer support the following explanation: Pressure variations of up to $15\ N/m^2$ allow for a penetration of the liquid substrate through the porous particle layer. Thus, a liquid substrate level is established prior to solidification. These findings are likewise confirmed qualitatively by the micrographs.

4. Summary

We have demonstrated the derivation of a model on the basis of finite elements which allows us to compute in a three-dimensional and quasi-steady fashion the combined conduction/flow problem as present during laser surface treatment. We hereby restrict ourselves to a process which is employed in incorporating hard particles at the surface of a workpiece. Thus, the melt pool remains covered by a solid coating layer during processing - and thermocapillary effects are supressed.

For this situation we find the flow to be very weak and to hardly affect the temperature field (and melt pool geometry). This result is different from results obtained e.g. for powder feeding processes (see e.g. Mazumder et al. 1990) where strong thermocapillary effects are responsible for an intense flow inside the pool. Under several conditions we find a breakdown of the coating layer, caused by normal stresses inside the flow. A characterization of the flow "strength" by means of a physically-correct Grashof number proves to be a reasonable tool to predict the parameter range for which such failure of the coating occurs. Thus, a prediction of favourable processing parameters (laser power q, scanning speed U) as required for the actual processing is possible.

5. References

Rosenthal, D. (1946) 'The theory of moving sources of heat and its application to metal treatments', *Trans. ASME* **68**, pp. 849-866.

Mazumder, J., Chen, M.M., Chan, C.L., Zehr, R., Voelkel, D. (1990) 'Bath convection in laser melting', *Proc. ECLAT*, Erlangen, pp. 37-53.

Schüssler, A. and Zum Gahr, K.-H. (1990) 'Incorporation of *TiC/TiN* hard particles into steel surfaces using laser radiation', *Proc. ECLAT*, Erlangen, pp. 581-592.

Schüssler, A., Steen, P.H., Ehrhard, P. (1992) 'Laser surface treatment dominated by buoyancy flows', *J. Appl. Phys.* **71**, pp. 1972-1975.

LASER SURFACE ALLOYING OF Al99.99 AND AlSi10Mg WITH Cr AND Ni

E. W. Kreutz, N. Pirch, M. Rozsnoki

Lehrstuhl für Lasertechnik
Rheinisch-Westfälische Technische Hochschule Aachen
Steinbachstraße 15
5100 Aachen
Fed. Rep. Germany

ABSTRACT.The surfaces of Al99.99 and AlSi10Mg are alloyed with laser radiation using Cr and Ni,which are blown into the melt pool within the interaction zone by powder feeding technique. The alloyed zone was investigated by optical microscopy, electron microscopy, hardness measurements and wear testing. The alloyed zone is crack and pore free either in the case of single or overlapping tracks. The microhardness and wear resistance are significantly improved by the formation of intermetallic compounds.

1. INTRODUCTION

Laser surface alloying (LSA) with rapid self-quenching and solidification /1/ enables the formation of a wide variety of chemical and structural phases such as different concentration profiles of the alloying element throughout the melted region, extended solid solutions, metastable crystalline phases and metallic glasses. LSA may become a key technology for reducing consumption of strategic materials by surface alloying instead of bulk alloying /1/. The width and depth of the alloyed track typically in the range of 1 to 3 mm are strongly dependent on the processing variables.

The advantages such as single-step processing and real-time control of alloying element supply make the codeposition of alloying elements by particle injection to the most popular processing method /2/. The particles become preheated during passing the laser beam. The absorption and scattering of optical energy by the particles result in a transmitted power density distribution significant different from the initial one. The temperature rise of the particles, trapped at the melt surface with their subsequent melting, instantaneously occurs on the time scale of the interaction time /3/. The processing variables such as power density distribution, dwell time and powder feed rate have to be matched in order to rise up the temperature of the near-surface region for melting and mixing of the particles avoiding evaporation of the different phases.

This paper reports on LSA of Al99.99 and AlSi10Mg with Cr and Ni using the powder injection technique. The single and overlapped tracks are investigated by optical microscopy, electron microscopy, hardness measurements and wear testing.

S. H. Davis et al. (eds.), Interactive Dynamics of Convection and Solidification, 221–224.

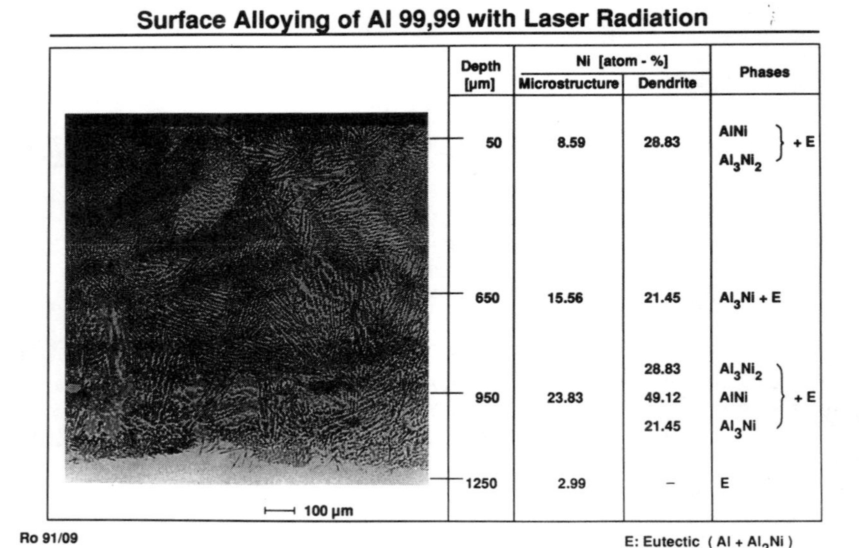

Depth [µm]	Ni [atom - %] Microstructure	Dendrite	Phases
50	8.59	28.83	AlNi, Al_3Ni_2 } + E
650	15.56	21.45	Al_3Ni + E
950	23.83	28.83 49.12 21.45	Al_3Ni_2 AlNi Al_3Ni } + E
1250	2.99	–	E

Fig. 1 Microstructure, average concentration profile and phases of Ni alloyed Al99.99 throughout the alloyed zone perpendicular to the surface.

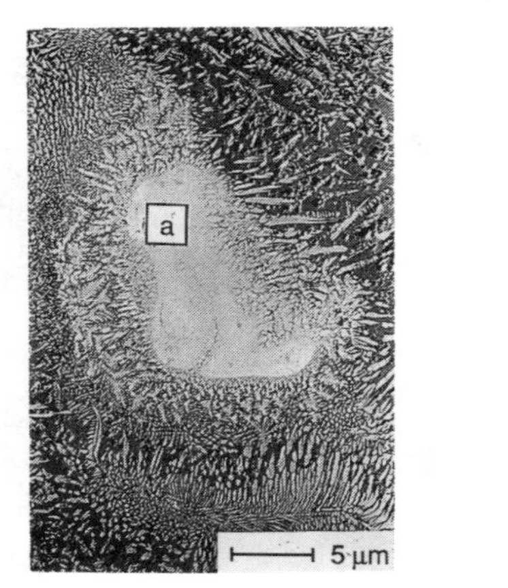

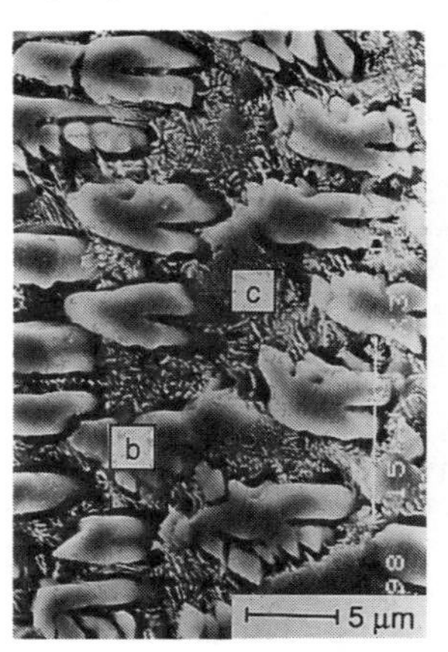

Fig. 2 Microstructure of the Ni alloyed Al99.99 zone showing AlNi agglomerates (a), intermetallic dendrites (b) and interdendritic eutecticum (c)

2. RESULTS AND DISCUSSION

The diffuse and convective energy transport in the molten pool results in an mixing of the alloying elements yielding an average Ni concentration profile throughout the laser alloyed zone exhibiting a maximum near the solid liquid boundary (Fig. 1). LSA originates in various phases and solidification structures, which show the strong interaction of the convective flow in the molten pool /4/ and the rapid solidification depending on the overall temperature distribution generated in combination with the processing variables used. In spite of the low average Ni concentration the phase AlNi is precipitated as agglomerate (Fig. 2) with subsequently dendritic growth of Al_3Ni_2 due to the peritectic reaction

$$L + AlNi \qquad Al_3Ni_2 \tag{1}$$

with the formation of AlNi due to the supressed mixing of dissolved particles. The dendritic structure exhibits a refining from the boundary

substrate Al99.99

substrate AlSi10Mg

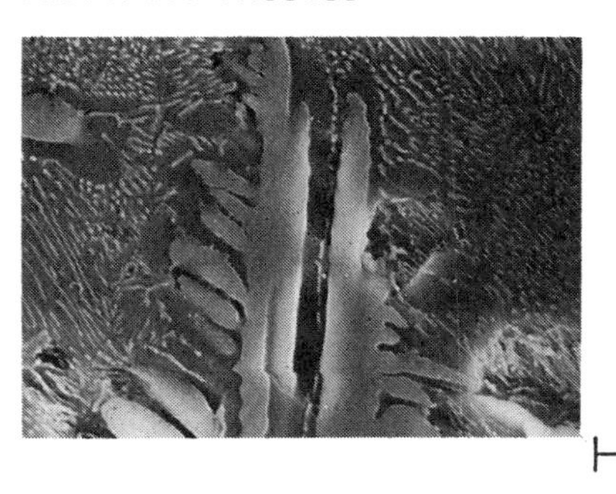

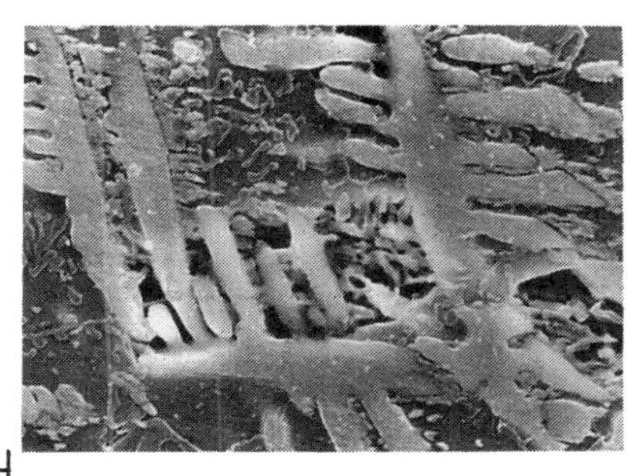

Al_3Ni, Al_3N_2 dendrites,
matrix: Al-Al_3Ni eutecticum

Al_3Ni, Al_3Ni_2 dendrites,
matrix: Al(84.12%),

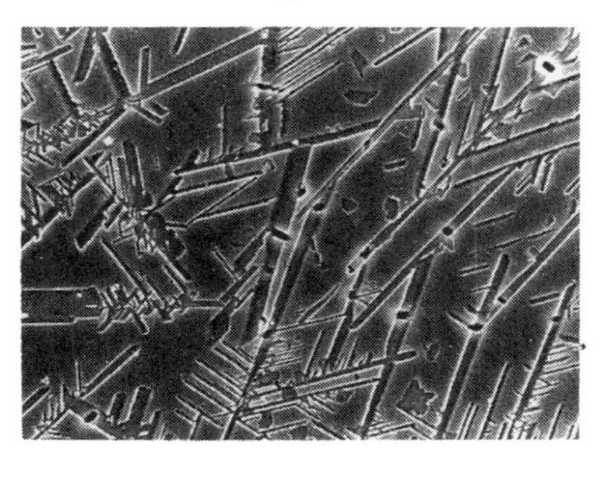

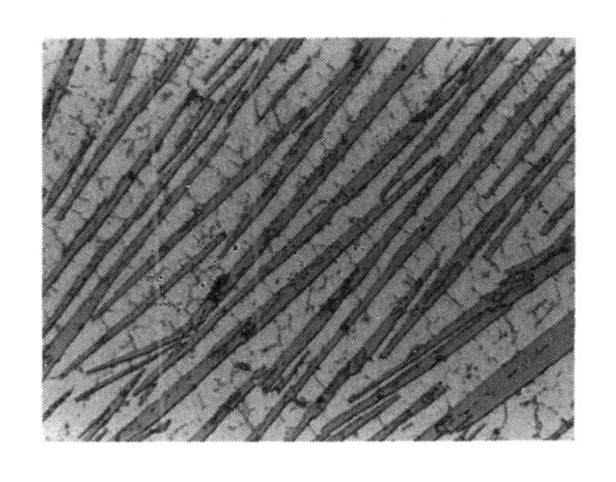

Cr_2Al_{11}, $CrAl_7$ dendrites,
matrix: Al

Cr_2Al_{11}, $CrAl_7$ needles,
matrix: Al-Si

Fig. 3 Microstructure of Ni and Cr alloyed Al99.99 and AlSi10Mg.

to the middle of the alloed track and a preferential orientation of the dendrites normal to the solid/liquid interface indicative for a textural growth of crystallites and dendrites dominated by the solid liquid boundary /5/. The interdendritic eutectic structure (Fig. 2) is in the range of 250 nm. For AlSi10Mg the determination of the microstructure within the interdendritic spacings was not rendered possible by REM and

TEM analysis. The alloying with Cr leads to the formation of the intermetallics Cr_2Al_{11} and $CrAl_7$, in the form of fine dendrites for Al99.99 and coarse needles for AlSi10Mg (Fig. 3). The microhardness is increased from 55-65 HV (0.1) to 350 HV (0.1) for Ni(20%) and to 250 HV(0.1) for

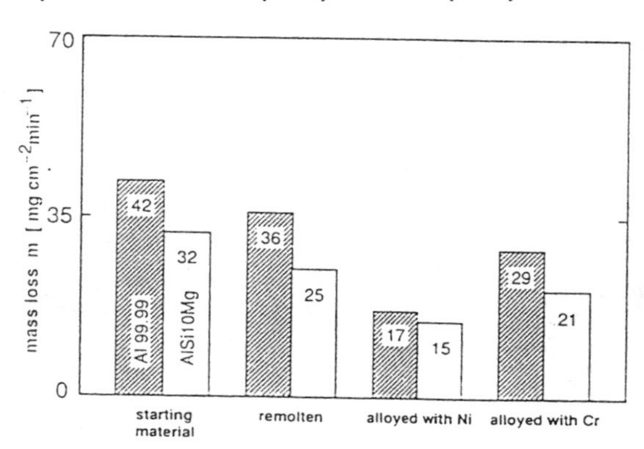

Fig. 4 Wear resistance of the untreated, remolten and alloyed Al99.99 and AlSi10Mg surfaces.

Cr(15%). The alloyed surfaces show a significant improvement of the wear resistance (Fig. 4) compared to the untreated ones /5/.

3. CONCLUSION

Laser alloying with Cr and Ni is an appropriate means for surface modifications of aluminium alloys. The alloyed zone is crack and pore free either in the case of single or overlapping tracks. The formation of intermetallics leads to a significant improvement of the microhardness and wear resistance.

4. REFERENCES

/1/ Draper, C.W. and Poate, J.M. (1985)'Laser Surface Alloying', Int. Metal Reviews 30, 85-108.

/2/ Ariely, S., Shen, J., Bamberger, M., Dausinger, F. and Hügel, H. (1991)'Laser Surface alloying of steel with TiC', Surface and Coatings Technology 45, 403-408.

/3/ Chande, J. and Mazumder, J. (1984) 'A Two-Dimensional Model for Mass Transport in Laser Surface Alloying', in Proc. of 5th Conf. on Application of Lasers and Electrooptics, 140-150.

/4/ Kreutz, E.W. and Pirch, N. (1990) 'Melt Dynamics in Surface Processing with Laser Radiation: Calculations and Applications', SPIE Proceedings Series 1276, 343-360.

/5/ Kreutz, E. W., Rozsnoki, M., and Pirch, N. (1992) 'Surface Remelting and Alloying of Al-Based Alloys with CO_2 Laser Radiation', Proceedings 2nd Intern. Conf. Laser Advanced Materials Processing, Nagaoka, Japan, in press.

TIN RIBBONS OBTAINED BY A MELT SPINNING PROCESS

P.JUILLARD
J.ETAY
Madylam
ENSHMG BP95
38402 St Martin d'Hères Cedex
France

ABSTRACT. In the near-net-shape casting processes, the strip of metal is solidified in a shape as close as possible to that of the final product. We are concerned with the "melt-spinning" process in wich a sheet of liquid metal impinges on the rim of a rotating and cooled wheel where it solidifies. Some tin ribbons were elabored in this way at Madylam.

The macroscopic aspect of these ribbons is good as soon as the impact is stable. But for high rotation speeds, some porosities may be observed on the surfaces of the ribbons. The temperature and chemical state of the cooling substratum also influence the final aspect of the ribbon.

Some micrographs were made on transverse and longitudinal sections :

- on cross-sections, the grains are upright. They cross the ribbon.
- on longitudinal sections : the grains start to grow normally to the substratum until 1/12 of the ribbon thickness. Then, they are tilted to 57° angle.

1. Introduction

Three main types of near net shape processes exist : twin-belts, twin-rolls and melt overflow processes. The process we have been studying suppress the expensive high aspect ratio nozzle, using electromagnetic shaping (Garnier[1]). In electromagnetic shaping, several initially cylindrical jets are elongated along a chosen direction and then made to fuse in a single sheet. Then, this sheet impinges on the rim of a rotating, water-cooled wheel where it solidifies. The impact could also be stabilized by electromagnetic means. The elaborated ribbon is then salvaged and wound.

This paper presents the experimental set-up with its associated variable parameters and measurement techniques. The limitations of the process are mentioned. A description of the aspect of the elaborated ribbons follows.

2. Experimental set-up

In our facility the tin is molten in a tank by electrical resistances. This operation takes two to three hours. When molten, a piston pushes the

S. H. Davis et al. (eds.), Interactive Dynamics of Convection and Solidification, 225–227.

liquid tin into a small channel. The piston down-stroke is such that the level of the tin in the channel is constant. A converging nozzle, whose shape can be chosen at will, is put under the channel. The tin flows through this nozzle then impinges on the rim of a rotating, water cooled, copper wheel where it solidifies. The exterior diameter of the wheel is 40 cm. The copper of the wheel is 4.65mm thick. The cooling water flows at 1.3 $l.s^{-1}$. The speed of the wheel can vary from 200 to 1600 rpm.

Four measurement techniques are used : visualisation, temperature measurements, roughness measurements and micrographs

3 - Conditions of experiments

The hydraulic study made with water shows that it is possible to define the conditions, i.e. an impact angle and a rotation speed, where the impact is stable. Using the hydraulic similitude laws between water and tin, we establish the conditions for a tin impact to be stable (Julliard -Etay [3]).

The solidification modifies the impact condition in the following ways : solidification *does not stabilize* an unstable impact, the *length of contact* L_s between the tin ribbon and the wheel must be long enough to solidify the ribbon entirely and when the *initial temperature* of the tin is too *high* or when the wheel rotates too fast, the ribbon can stick to the wheel.

4. Aspect of the ribbons

4.1 - RELEVANT PARAMETERS

Various phenomena influence the appearance of the elaborated ribbons :

-because of its viscosity the *surrounding air* is swept by the wheel. Air bubbles can be trapped between the substratum and the elaborated ribbon

-the *temperature and the chemical state* of the wheel influence the crystallization. When the wheel is cold, the grains of the ribbon are numerous and small, when the wheel is hot they are bigger. When the wheel is not chemically clean, marks having the same shapes as the wheel defects appear on both faces of the ribbon. Both temperature and chemical effects are related to the wetting of the copper by the tin. The copper substratum is wetted better when it is hot and clean. Therefore a clean, hot substratum leads to better ribbons.

The limitations related to these two points can be avoided if the copper is highly polished.

4.2 - OBSERVATIONS

4.2.1 - *Macrographic aspects.* Briefly stated, the aspect of the ribbons is better when the *hydrodynamic stability* of the impact is perfect, the *tangential speed* of the wheel is less than 10 $m.s^{-1}$ (to reduce centrifuge effets), the temperature of the copper is sufficiently high, but not too high and the copper is polished, and roughness is less than 1.3μm .

4.2.2 - *Micrographics aspects.* Micrographs are made to control the ribbon quality, i.e to check for the presence or the absence of inclusions and porosities.

- cross section : On all the photographs, the grains are straight. They cross the ribbon : their growth goes from the wheel to the exterior. The grain is clearer near the sides of the ribbon. The explanation of these phenomena is that tin is a faceted metal; its crystal grows along preferential directions. In our case, the grains are tilted in opposite direction to the thermal gradient contrary to what was expected.

- longitudinal section : The grains start to grow normally to the wall until 1/12 of the thickness of the ribbon then they tilt to a 57° angle. This has been observed by many other authors [4], [5] : the metallurgical quality of a near-net-shape casted ribbon is different in a thin layer ($\cong$ 30 μm) near the wheel (chilled zone) than in the remainder of the ribbon. The grains develop, as expected, in the opposite direction of the flow in the reference frame of the substratum.

4.2.3 - *Microscopic observations.* Some observations were performed using an electronic microscope. On the wheel face of the ribbon, the grains are elongated in the direction of the velocity of the wheel. The ribbon structure looks like a stack of fish scales.

5 - Conclusions

Because tin is a pure faceted metal, it is difficult to transfer our observations to alloys. Nevertheless, we can say that in a process where a metal solidifies on a moving substratum, the quality of this substratum is of great importance to the quality of the final product. Part of the hydrodynamic stability and the heat transfer coefficient depend on the polish of the substratum.

References :

[1] GARNIER F. (1989) "Formage électromagnétique et coulée directe des métaux"- thèse INPG - France
[2] JUILLARD P. (1990) "Coulée directe de métaux en bandes minces : aspects hydrauliques et thermiques" - thèse INPG - France
[3] JUILLARD P. and ETAY J. (1991) "Impact d'une lame liquide sur la génératrice d'une roue en rotation : expériences et analyse" - C.R. Acad. Sci. Paris, t.312, Série II, pp 169-175.
[4] MURTY Y.V. and ADLER R.P.I. (1982) "High-speed casting of metallic foils by the double-roller quenching technique" - Journal of Materials Science, vol 17, pp 1945-1954
[5] JECH R.W. et al. (1984) "Rapid solidification via melt-spinning : equipment and techniques" - Journal of Metals - pp 41-45

SOLIDIFICATION BY PLANAR-FLOW SPIN-CASTING

P. H. STEEN
Cornell University
School of Chemical Engineering & Center for Applied Mathematics
Ithaca, NY 14853
USA

ABSTRACT. Asymptotic analysis supported by experiment uncovers operating conditions for which the planar-flow process exhibits: (i) a planar solidification front over much of the thickness of the ribbon product, (ii) solidification that is only weakly coupled to the fluid mechanics, and (iii) convection at the solidification front whose influence is a small perturbation. These results are summarized.

1. Introduction

This contribution highlights features of solidification by planar-flow spin-casting that are dominant for operating conditions of low solidification rate relative to wheelspeed and quenching just below the melting temperature. Note that this operating regime *does not* favor amorphous microstructures but *can* achieve solidification rates that reach the limit of absolute stability for some alloys (10 cm/s).

A sketch of the process is shown in the Figure. Hot molten metal is forced under applied pressure Δp from the feed nozzle into the gap between the nozzle face and substrate wheel where it is quenched and spun off as a thin ribbon. Typically, a gap size G is 1mm, linear wheelspeed 10 m/s, with hot nozzle T_h = 700° C (for aluminum) and chill wheel T_c = 30° C.

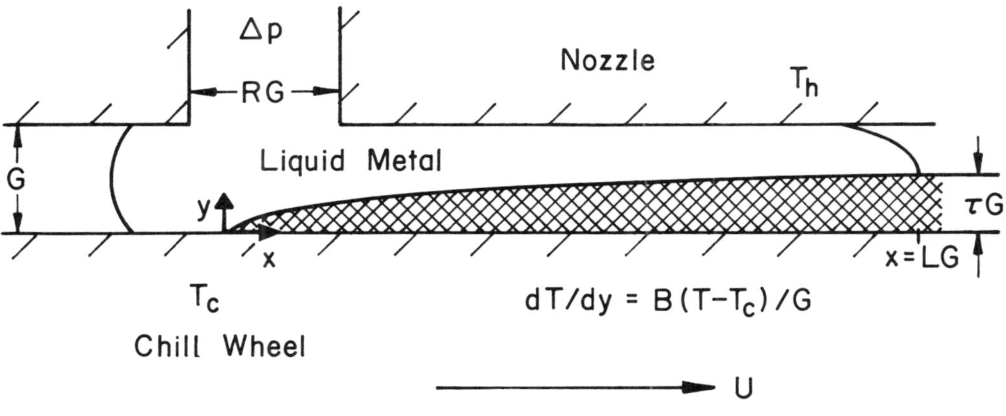

Figure. Process Schematic

S. H. Davis et al. (eds.), Interactive Dynamics of Convection and Solidification, 229–231.

R	3	slot breadth
τ	10^{-1}	ribbon thickness
$\Delta P \equiv \Delta p\ G/2\sigma$	2.5	applied pressure
$W \equiv \rho_l U^2 G/2\sigma$	10^2	Weber number
$C \equiv \mu U/2\sigma$	10^{-2}	capillary number
$\varepsilon \equiv V^*/U$	10^{-2}	solidification parameter
$P_s, P_l \equiv \rho_s Cp_s UG/k_s, \rho_l Cp_l UG/k_l$	10, 10^2	Peclet number (solid, liquid)
$(T_h - T_m)/(T_h - T_c)$	0.14	superheat fraction
$(T_m - T_c)/(T_h - T_c)$	0.86	quench fraction
$B, B_l \equiv H_w G/k_s, H_w G/k_l$	1, 3	Biot number (solid, liquid)
$\rho_{ls} \equiv \rho_l/\rho_s$	0.85	contraction ratio
$k_{ls} \equiv k_l/k_s$	0.45	conductivity ratio
$\Gamma/(G\rho_s\Delta\eta)$	10^{-6}	interfacial energy parameter
$V^* \equiv k_s (T_h - T_c)/\rho_s \Delta\eta\ G$	10^{-1} m/s	solidification rate (dependent)

Table. Dimensionless groups with values typical of our experiments with Al[1].

2. Solidification

All the (latent) heat of solidification must flow through the already solidified material to the chill-wheel. This sets the scale V^* for the solidification rate. If the solidification is sufficiently slow relative to the wheelspeed U (still 'rapid' by metallurgical standards), the molten zone will be long and thin with a small-slope solidification front. This motivates the parameter $\varepsilon \equiv V^*/U$ which controls the asymptotics in much the same way as disparate length scales in a boundary layer analysis. Indeed, solidification kinematics require a molten zone with $L^{-1} \sim \varepsilon$ in the limit of slow solidification ($\varepsilon \to 0$).

Horizontal and vertical scales in ratio L are introduced into the governing equations for the liquid and solid within the gap. Material properties for each phase are assumed constant. At the freeze interface, attachment kinetics and interfacial energies are unimportant for conditions of interest[2]. The advection coupling in the energy equation has strength $L^{-1}P_l$ and $L^{-1}P_s$ for the liquid and solid, respectively. For slow enough solidification, this coupling is negligible and the heat-transfer is independent of the fluid mechanics.

Once decoupled, the heat-transfer problem reduces to a problem of conduction across the liquid and solid layers with a solidification front of unknown shape separating them. This is still a nonlinear problem. If resistance to latent heat removal is dominated by ribbon/wheel contact, the temperature within the solidified ribbon approaches the melting temperature T_m and the thickness τ vanishes ($\tau \to 0$) in a way that maintains finite heat-flux ($\varepsilon/\tau \to 0$ as $\varepsilon \to 0$). For this case, called contact-limited, the front is linear in downstream coordinate. In contrast, for the conduction-limited case, the front grows with the square root of distance. The contact-limited case is appropriate to our apparatus. Details of this section are in [2].

3. Fluid Flow

Inertia and pressure forces balance in the fluid flow (inertia >> surface tension >> viscous force, cf. Table). The appropriate limit is large Weber ($W^{-1} \to 0$) and small Capillary numbers ($CW^{-1/2}L \to 0$) with $\tau^2 W$ order one. The total pressure drop from the feed nozzle upstream to the meniscus downstream is obtained as it depends on solidification length L and nozzle breadth R. The pressure drop arises from an inviscid core flow of constant vorticity with a correction due to domain perturbation. In summary, the flow is like a lid-driven cavity with a leak. The surface tension contains the liquid (acts like sidewalls), the lid is the wheel, and the leak is the ribbon. The flow analysis is sketched in [3,4] and details are forthcoming.

4. Discussion & Summary

The above asymptotics are faithful to experimental conditions with one exception. Heat advection in the liquid ($L^{-1}P_l$) is order one and hence is not negligible *a priori*. Comparison with experiment is informative, nevertheless. The predicted square-root dependence of thickness on pressure drop is confirmed to excellent accuracy against measurement of average ribbon thickness (over cm length scale). The predicted correction due to finite thickness of the ribbon is small and cannot be resolved from these data [4]. Local variations of measured thickness are observed as surface textures (mm scale) [1].

In summary, a regime of slow solidification has been identified where the solidification uncouples from the fluid flow. For contact-limited solidification, a thin ribbon is predicted with linear solidification front. The solidification perturbs the flow domain, within which the liquid metal is found to circulate as an inviscid core driven by the wheel. To first order, the heat-transfer is conduction across the gap and the liquid motion is a plane Couette flow. Measurements of ribbon thickness on a macroscale support the predictions, suggesting that the regime is accessible to experiment.

5. Acknowledgement

This work was supported by NSF Grants MSM 8711824 and CTS 9024461.

6. References

1. Carpenter, J.K. and Steen, P.H. (1992) 'Planar-flow spin-casting of molten metals: process behavior', *J. Mat. Sci. 27,* 215-225.
2. Carpenter, J.K. and Steen, P.H. (1992) 'Heat transfer and solidification in planar-flow melt-spinning', submitted for publication.
3. Carpenter, J.K. (1990) PhD thesis, Cornell University, Ithaca, NY.
4. Carpenter, J.K. Agger, E.C. and Steen, P.H. (1991) 'Fluid mechanics and heat transfer of planar-flow melt-spinning', in M. Rappaz, M.R. Ozgu and K.W. Mahin (eds.), Modeling of Casting, Welding, and Advanced Solidification Processes V, TMS Publishers, Warrendale, PA, pp.621-627.

A NUMERICAL AND EXPERIMENTAL STUDY OF WAVY ICE STRUCTURE IN A PARALLEL PLATE CHANNEL

B. WEIGAND, H. BEER
Institut für Technische Thermodynamik
Technische Hochschule Darmstadt
Petersenstraße 30, 6100 Darmstadt, Germany

ABSTRACT. The paper presents a numerical model for predicting steady-state ice formation inside a cooled, parallel plate channel. The study takes into account the strong interactions occuring between the turbulent flow, the shape of the ice and the heat transfer at the ice-water interface, which lead to the formation of wavy ice layers. The presented model is found to be able to predict realistic variations of the ice layer thickness for a wide range of Reynolds numbers and cooling parameters. The numerical results were verified by comparing with own measurements and good agreement was found.

1. Introduction

Ice formation phenomena of flowing water in a pipe or a channel whose wall is kept at a uniform temperature below the freezing temperature of the water, is a basic engineering problem. It introduces many practical problems, such as pressure drop, diminution of flow rates and sometimes, breakage of the pipe as a result of flow blockage by ice. The phenomenon of freezing of flowing water involves interactions between the turbulent flow, the shape of the ice layer and the heat transfer at the ice-water interface. Under certain conditions these interactions result in an instability of the ice layer. This instability is caused by the strong laminarization of the turbulent flow due to converging ice layers in the entrance region of the cooled test section and results in a wavy ice structure. Wavy ice layers with one wave, occuring in a parallel plate channel, were investigated experimentally by Seki [1] and by Weigand and Beer [2]. No numerical treatment of the conservation equations concerning this type of ice layers was performed in the past because of the complexity of the problem. Therefore, the subject of this paper is the presentation of a numerical calculation of the steady-state ice layers with one wave. The numerical study is supported by a detailed experimental investigation.

2. Analysis

Consideration is given to a turbulent flow entering a cooled parallel plate channel (Fig. 1) with a fully developed velocity profile and with the uniform temperature T_0. In the chill region, the wall temperature is maintained at a constant value T_w, which is below the freezing temperature T_F of the fluid and a frozen layer is generated at the cooled walls. Assuming an incompressible, Newtonian fluid with constant fluid properties, the steady-state boundary layer

S. H. Davis et al. (eds.), Interactive Dynamics of Convection and Solidification, 233–235.

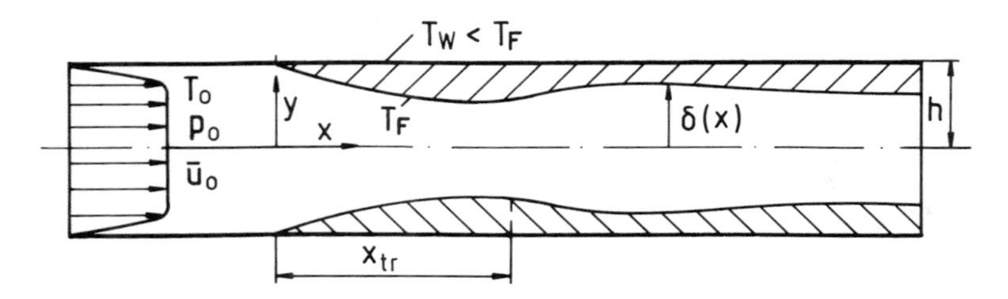

Figure 1. Physical model and coordinate system.

equations for the fluid can be written as [3]

$$\frac{\partial u}{\partial x} + \frac{\partial v}{\partial y} = 0$$

$$u\frac{\partial u}{\partial x} + v\frac{\partial u}{\partial y} = -\frac{\partial p}{\partial x} + \nu_L\frac{\partial}{\partial y}\left[\left(1 + \varepsilon_m^*\right)\frac{\partial u}{\partial y}\right] \quad ; \quad 0 = -\frac{\partial p}{\partial y} \tag{1}$$

$$u\frac{\partial T}{\partial x} + v\frac{\partial T}{\partial y} = a_L\frac{\partial}{\partial y}\left[\left(1 + \frac{Pr}{Pr_t}\varepsilon_m^*\right)\frac{\partial T}{\partial y}\right]$$

where $\varepsilon_m^* = \varepsilon_m/\nu_L$ is the eddy viscosity and Pr_t is the turbulent Prandtl number according to Cebeci [3]. In the solid region, one dimensional heat conduction is assumed. Therefore, the temperature distribution in the solid crust can be shown to be

$$(T_s - T_F) \,/\, (T_w - T_F) = (y - \delta) \,/\, (h - \delta) \;; \quad \delta < y < h \tag{2}$$

Eqs. (1) are coupled with eq. (2) by the interface energy equation

$$k_s \,\partial T_s \,/\, \partial y = k_L \,\partial T \,/\, \partial y \;;\; y=\delta \tag{3}$$

Deissler [4] found out that the turbulent Reynolds shear stress can be taken approximately as constant along a streamline in a highly accelerated flow, which yields a "laminarization"

$$\overline{u'v'}(\psi) = \left(\overline{u'v'}(\psi)\right)_{x=0} \quad , \quad \psi = \text{const.} \tag{4}$$

where ψ is the streamfunction defined in the usual way [3]. Equation (4) was used for calculating ε_m^* in the region $0 \le x \le x_{tr}$. The initial distribution of the turbulent shear stress for $x = 0$ can easily be obtained with the help of a mixing length theory [3].

For $x > x_{tr}$ the acceleration due to converging ice layers ceases and the flow recedes to its originally turbulent state. In this region, where the heat transfer at the solid-liquid interface is strongly enhanced, the eddy viscosity was calculated with a modified mixing length model according to Moffat and Kays [5]. The transition point x_{tr} was correlated from measurements. For more detailed informations concerning the turbulence modelling, the reader is refered to [6]. The calculation of the frozen layer involves an iteration procedure. Initially a distribution of the ice layer is assumed. With this variation of $\delta(x)$ the conservation equations (1) are solved with the help of an implicit finite-difference method, which is known in literature as the Keller box-method [3]. After solving the eqs. (1), a new distribution of $\delta(x)$ is calculated by inserting the yet known temperature gradient at the solid-liquid interface into eq. (4). This iteration process is repeated until $\Delta\delta = |\delta^{(i)} - \delta^{(i-1)}| < 0.01$ h at every axial position for two successive iterations.

3. Results and Discussion

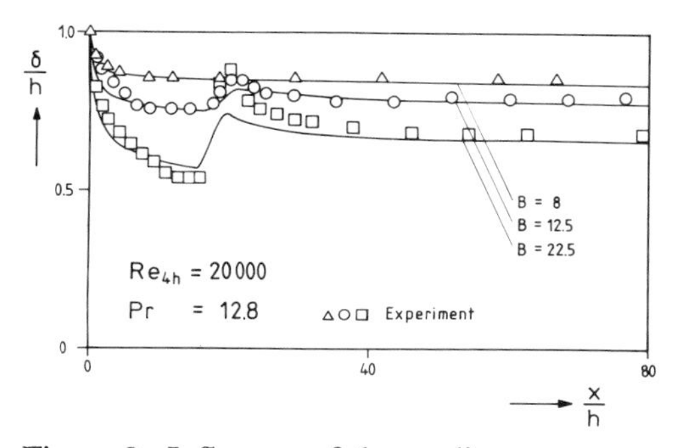

Figure 2. Influence of the cooling parameter B on the shape of the ice layers.

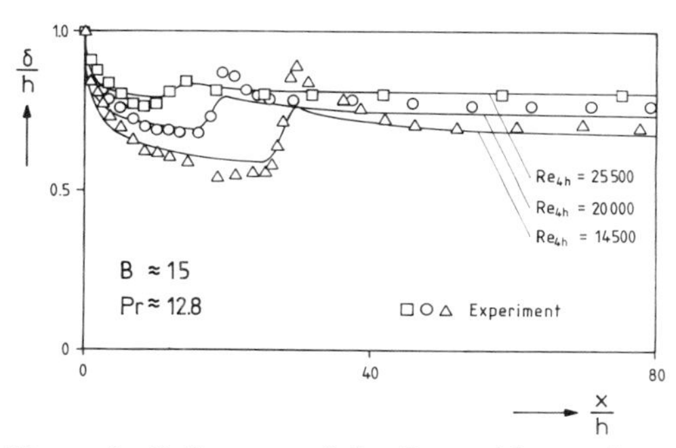

Figure 3. Influence of the Reynolds number on the shape of the ice layers.

Fig. 2 and Fig. 3 show comparisons between calculated and measured ice layers, plotted as a function of the axial coordinate. Fig 2 elucidates that an increasing wall cooling parameter $B = k_s/k_L\ (T_F - T_w)/(T_0 - T_F)$ results in a thicker ice layer for a given value of the Reynolds number $Re_{4h} = \bar{u}_0\ 4h/\nu_L$ (where $\bar{u}_0$ denotes the axial mean velocity at the entrance of the test section). From Fig. 2 it can be seen that the distance x_{tr} between the entrance of the chill region and the point for which the ice layer reaches its maximum thickness, seems to be only a function of the Reynolds number. Fig. 3 visualizes the effect of a variation in Reynolds number on the ice layer thickness for a fixed value of B. It must be stated that an increasing Reynolds number results in a decreasing ice layer thickness. This is due to the intensified heat transfer from the flowing liquid to the solidified crust with growing values of Re_{4h}. As can be observed from Figs. 2 - 3, the agreement between theory and experiment is quite good.

References

1. Seki, N., Fukusako, S. and Younan, G.W. (1984) 'Ice formation phenomena for water flow between two cooled parallel plates', J. Heat Transfer 106, 498 - 505.
2. Weigand, B. and Beer, H. (1991) 'The morphology of ice structure in a parallel plate channel', Proc. 3th Int. Symp. on Cold Regions Heat Transfer, Fairbanks, 167 - 176.
3. Cebeci, T. and Bradshaw, P. (1984) Physical and computational aspects of convective heat transfer, Springer, New York.
4. Deissler, R.G. (1974) 'Evolution of a moderately short turbulent boundary layer in a severe pressure gradient', J. Fluid Mech. 64, 763 - 774.
5. Moffat, R.J. and Kays, W.M. (1984) 'A review of turbulent boundary layer heat transfer research at Stanford, 1958 - 1983,' Adv. in Heat Transfer 16, 242 - 366.
6. Weigand, B. (1992) Erstarrungsvorgänge einer strömenden Flüssigkeit in einem ebenen, geraden Kanal, Doctoral Thesis, TH Darmstadt

DYNAMICAL SOLIDIFICATION OF THE LIQUID FLOWING IN A COOLING CYLINDRICAL CHANNEL

Z. LIPNICKI
Higher College of Engineering
Podgórna 50
65-246 Zielona Góra
Poland

ABSTRAC. The paper examines theoretically the solidification of a viscous and heat-conducting liquid flowing in a cooling cylindrical channel. The considerations are focused on the initial part of the stream liquid which is intensively cooled. Changeable amount of liquid takes part in the motion. It is enclosed both from the frontal area of the stream which is moving with the liquid and from the solidification interface. The flow of the liquid is treated as the set of the Hagen-Poiseuille´s flows. From the conservation equations the differential equation is obtained. The solution of the obtained equation with initial conditions is the function which describes the velocity of the flow of the liquid. The paper discusses the influence of the cooling conditions as well as the conditions of flow on the solidification process.

1. INTRODUCTION

The primary objective at the present work was to provide an analytical solution of the dynamical flow for the solidifying liquid in a cooling cylindrical channel. With the aim to obtain the solution the structural model of the flow was used. It was the set of the Hagen-Poiseuille´s flows.

In the earlier studies the problem of the flow of the liquid with a constant velocity in the cooling channel was considered [1, 2]. In work [3] the problem of the dynamical flow solidifying liquid in the cooling cylindrical channel was experimentaly examined.

In work [2] the analytical solution for the thickness of the freezing layer in a liquid was received. The solution is as follow:

$$\varepsilon = 1 - \sqrt{1 - \frac{4}{\sqrt{\pi}} D \sqrt{z} + F(T_p)} \qquad (1)$$

where:
$z = Z/R$, $\varepsilon = \delta(Z)/R$, Z-distance from the frontal area of the stream, R - radius of channel, $\delta(Z)$ - thickness of freezing layer; $D = B\ Ste/\sqrt{Pr\ Re}$; $B = b_c/b_l$ - aspect ratio; $b = c \cdot \varrho \cdot \lambda$ - coefficient of heat accumulation; c - specific heat; ϱ - density; λ - thermal conductivity; $Ste = c_l(T_m - T_c)/L$ - Stefan number; T_m - melting point; L - latent heat of fusion; $Pr = \eta \cdot c/\lambda$ - Prandtl number; η - coefficient of viscosity; $Re = R V_0 \varrho/\eta$ - Reynolds number; V_0 - velocity; F(Tp) - function of overheating; Tp - temperature of overheating; subscripts; c - channel, l - liquid.

As an illustration of the effect of flow in the cooling cylindrical channel on solidification is shown on fig. 1.

S. H. Davis et al. (eds.), Interactive Dynamics of Convection and Solidification, 237–239.

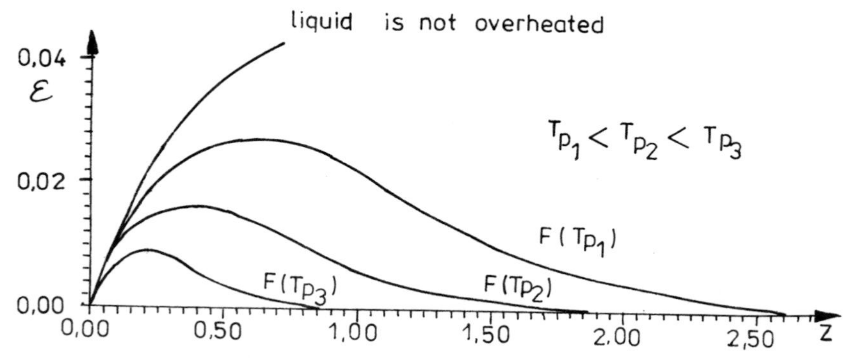

Figure 1. The shape of the interface in the solidified liquid [2]

2. PRESENTATION OF THE PROBLEM AND RESULTS

Changeable amount of the liquid takes part in the motion. It is enclosed both from the frontal area of stream which is moving with the liquid and from the solidification interface. The front of the solidification liquid translocates from the boundary surface of the channel to its axis of symmetry as on the fig. 2. I consider the stream head of length l and temperature T_m. The flow of the liquid is treated as the set of Hagen-Poiseuille´s flows in the space:

$$0 \leqslant r' \leqslant r(t) . \tag{2}$$

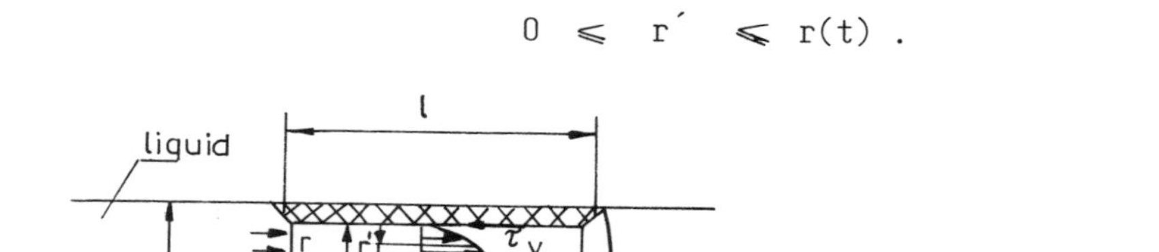

Figure 2. Illustration of the problem

These both values time-t and radius-r are dimensionless.

I assume the equation of the solidification interface as:

$$r(t) = 1 - D \sqrt{t} . \tag{3}$$

The flow of the liquid is determined by two changeable forces. The first is the resistanse force and second the exciting force.

The motion of the liquid is described by the following dimensionless equation

$$Re \frac{d}{dt} (r^2 w) = \frac{Eu\ Re}{l} r^2 - 8 w , \tag{4}$$

where: $Eu = p/ \varrho V_0^2$ is the Euler number, p is the pressure of the liquid, $w = V/V_0$ is the dimensionless velocity.

The solution of the above equation with the initial condition: $w = 1$ for $t = 0$ is the function which describes the velocity of flow of the solidifying liquid in the channel.

$$w = \frac{Eu}{l} r^{k-2} \exp(k/r) \int_0^t r^{2-k} \exp(-k/r)\, dt$$

$$+ r^{k-2} \exp(k/r - k), \qquad \text{where } k = -16/(D^2 Re) . \tag{5}$$

Both the velocity and distance of the flow are presented of the figure 3.

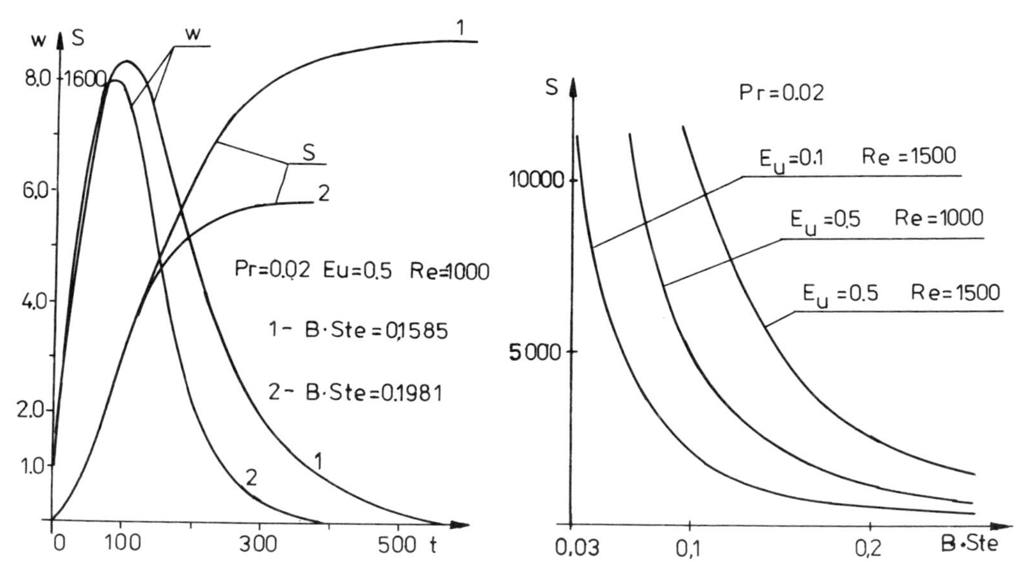

Figure 3. The dependence of the velocity and the distance of flow on time and Be Ste number

The new theoretical structural model of the flow is used in my considerations. It allowed to obtain the four-parameter function of the velocity: w = w (t; B Ste,Re,Eu,l). On the ground of this function the distance of the flow for the liquid in the cooling channel also is determined. The results of this part of the studies are presented on the figure 3. The influence of the solidification on the flow is evident. The obtained findings agree with the experiment [3].

At the begining of the solidification the velocity of liquid increases very fast and the larger the B Ste number is, the larger is the increase. The reason of it is diminution of the flowing liquid. Then the velocity reaches the maximum and next decreases to zero. The maximum of the velocity is smaller for the bigger B Ste number. The distance of flow of the liquid in the cooling channel decreases hyperbolically due to the increase of the B Ste number.

3. REFERENCE

[1] Vijnik A.(1960) ´Tieorija zatwierdivanija otlivki ´Maszgiz, Moskwa.

[2] Bydałek A., Lipnicki Z.(1990)´On the solidification of flowing metal during the mould fulfiling´,Ossolineum: Krzepnięcie metali i stopów t. 15, Wrocław, pp. 77-84.

[3] Mutwil J., Bydałek A.(1986)´On dynamic solidification of metals and alloys´,Simulation Designing and Control of Foundry Processes, FOCOMP´86, Kraków, pp. 339-349.

CONVECTION AND MACROSEGREGATION IN MAGMA CHAMBERS

S.R. TAIT and C. JAUPART
Institut de Physique du Globe
4 place Jussieu
75252 Paris Cedex 05
France

ABSTRACT. Magmas accumulate in large reservoirs where convection occurs, driven by density differences due to fractional crystallization and temperature variations. The observations consist of igneous rocks representing the final crystallized product and volcanic lava series. Chemical layering and stratification occur on all scales, from zoning patterns in individual crystals, to layering in igneous rocks on dimensions ranging from a few tens of centimetres to several tens of metres. There are also lateral heterogeneities along the strike of the layering, the most dramatic example being large cylindrical structures cross-cutting the layering. Important features of magmas are: (1) complicated phase diagrams sensitive to small temperature variations, (2) thick crystallizing boundary layers allowing differential motions between residual liquid and solid, (3) large temperature variations across cooling boundary layers implying variations of physical properties such as viscosity. We show that macrosegregation in magma chambers follows simple principles. Segregation becomes more pronounced with increasing reservoir size and decreasing melt viscosity. Different reservoirs with similar sizes and bulk compositions show almost identical segregation characteristics. Basal zones of intrusions exhibit specific changes of mineralogy at similar distances from the lower boundary. Such changes are therefore governed by local conditions and are consistent with the onset of compositional convection in the cooling boundary layer. They record small temperature and composition changes, and hence can only be reproduced by accurate theoretical models.

1. Introduction

The production, from a relatively homogeneous mantle source, of magmas with a large range of compositions is one of the major concerns of geology. This concern is not only of pure academical interest as many of the earth's natural resources are derived in some way from magmas. Most geological processes go through stages where melting and crystallization occur. Magmas ascend from their sources to form reservoirs of various sizes and shapes, where they lose heat to the colder surroundings. Some of these reservoirs reach enormous proportions, the largest known being the Bushveld complex in South Africa, which is 9 km thick and covers an area of more than 10^5 km^2. It is in these reservoirs that magmas undergo fractional crystallization and generate residual melts of various composition which eventually become the fabric of continents. On a smaller scale, the behaviour of volcanic systems reflects the evolution of the magmatic systems which feed them. Finally, large amounts of rare elements can be found in igneous intrusions which have been concentrated in specific stratigraphic horizons. A famous example is the Merensky Reef horizon in the Bushveld Complex of South Africa, which contains most of the world's platinum resources. Thus, many aspects of Earth Sciences require a knowledge of how magmas crystallize and how long is the crystallization process.

How magmas differentiate is still open to question for two main reasons. (1) Magmas are multicomponent melts whose phase diagrams are extraordinarily varied, and thermodynamic constraints are only available for a limited range of compositions. (2) The available observations are at the same time detailed and yet seldom appropriate for a physical model. In large intrusions, systematic descriptions of chemical variations require

S. H. Davis et al. (eds.), Interactive Dynamics of Convection and Solidification, 241–260.

a lot of time and are only available in a few cases (Hess, 1960; Jackson, 1961; Wager and Brown, 1968). The observations show that the rock composition varies on many different spatial scales, in both the vertical and horizontal directions. By definition, only the final fully-crystallized stage can be documented, without direct evidence of the liquid which once existed. This introduces a difficulty as magma chambers may not be closed systems and may be replenished with new liquids. Thus, a possible explanation for any peculiar feature can be the influx of a different magma with the appropriate composition.

Early studies of magmas and igneous rocks usually emphasized the thermodynamic aspects. It has become increasingly apparent that, to make sense of the observations, thermodynamics alone was not sufficient and that the consideration of disequilibrium phenomena, i.e. melt motion with respect to the crystals, was necessary. The first attempt was to invoke the settling of crystals. Starting with a series of studies in the mid-seventies, convective phenomena have been the focus of attention. Significant progress has been made in explaining the main features of magmatic convection, but the models are yet to achieve a complexity appropriate to reproduce the natural systems. The aim of this paper is to summarize a series of simple observations from a physical point of view which indicate that convective phenomena occur in magma chambers. Another aim is to evaluate the accuracy which is required of physical models to explain petrological observations. We also develop simple scaling arguments to assess some of the features of igneous complexes. The paper starts with a brief review of the main characteristics of magmas as crystallizing melts. The following chapter describes a series of observations on magma chambers emphasizing their common features. In the last chapter, a series of recent results on compositional convection will be brought together.

2. Magmas and magma chambers

2.1. THERMODYNAMICS OF SILICATE MELTS AND THEIR PHASE DIAGRAMS

Magmas are generated with different compositions depending on the geological setting. For the sake of simplicity, we shall restrict the discussion mostly to basaltic systems, which are the best documented and account for the majority of melts generated in the Earth.

In a convective upwelling in the Earth's solid mantle, partial melting occurs by adiabatic decompression, such that the rising material follows a temperature and pressure path which crosses its solidus curve. In this case, the lowest temperature melt is generated, close to a cotectic composition. This melt separates out of its matrix because of its buoyancy and accumulates in a reservoir at shallower depth. There, the phase diagram is slightly different because pressure is lower, but the melt remains close to a cotectic composition because the effect is not large. Figure 1 shows a ternary diagram in which three main crystal components are used as end-members, and illustrates the shift in the various mineral stability curves as pressure decreases. The composition of basaltic liquids is close to a cotectic line, and hence small differences of temperature and composition lead to the precipitation of different minerals, and hence to petrological diversity.

2.2. MAGMA CHAMBERS: STRUCTURE AND IMPLICATIONS

For reference, let us consider the following situation. A volume of liquid is emplaced rapidly, i.e. in a short time which does not allow significant crystallization, to form a reservoir with large aspect ratio. This geometry is appropriate for many igneous complexes and especially for those which will be discussed in chapter 3. This geometrical configuration implies that, as far as the main crystallization sequence developing on the floor is concerned, the side walls can be neglected. The other assumptions are that the temperature and composition of the magma are uniform.

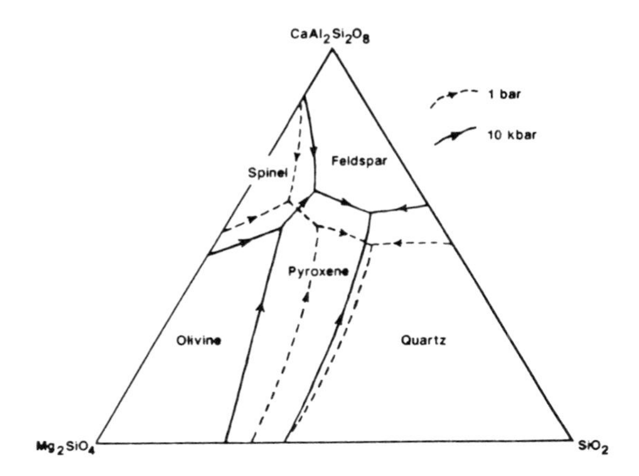

Figure 1. Shows a ternary phase diagram (after Presnall, 1966) which may be used as a simplified system for basaltic magmas. The arrows on cotectic lines show the directions of decreasing temperature. E is the eutectic, A and B are reaction points.

After emplacement, the magma cools through both upper and lower boundaries and thermal boundary layers develop there. Initially, there is a finite amount of time during which heat is being transported by conduction only, with both boundary layers growing at similar rates. The overall temperature difference between the chamber interior and the boundaries takes values of more than 500°C, and the melting range has a typical value of 200°C. Assuming that the initial temperature of the magma is above the liquidus everywhere, both boundary layers can be separated into three parts as follows: a fully crystallized part below the solidus, a partially crystallized zone where porosity varies, referred to as a "mush", and a fully liquid part above the liquidus (Fig.2). The superheat is defined as the difference between the initial magma temperature and the liquidus.

The large dimensions of magma chambers imply that the effect of pressure changes on the phase diagram is not negligible. The liquidus temperature at the roof is smaller than at the floor by an amount which is given by the Clapeyron relation and, to a first approximation, proportional to the chamber thickness. A typical value for the Clapeyron slope is 15°C/kbar, and hence the liquidus temperature difference reaches a value of about 45°C in a 9 km-thick reservoir. This implies that the structure of the two boundary layers at the top and bottom of the chamber are not identical, with different amounts of superheat.

2.3. PHYSICAL PROPERTIES

At low strain rates, silicate melts have Newtonian rheologies, with viscosity depending on temperature according to an Arrhenius law :

$$\mu = \mu_o \exp(E/RT) \tag{1}$$

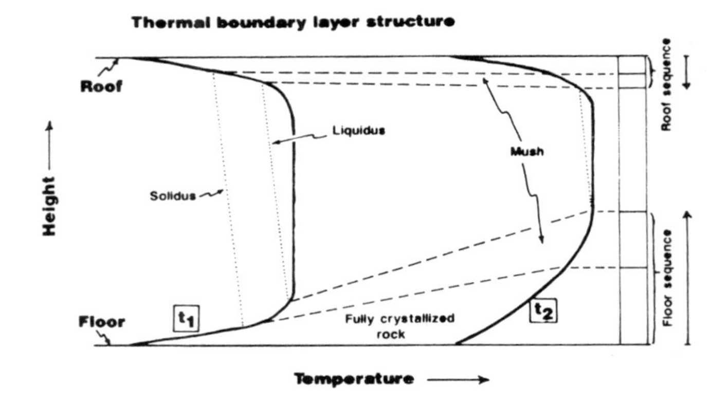

Figure 2. Shows qualitatively the thermal structure of the boundary layers at two consecutive times t_1 and t_2. Thermal convection ensures a higher heat flux through the upper boundary than at the base. This effect causes the rock sequence formed at the floor to be thicker than that at the roof. The increases of solidus and liquidus temperatures with pressure are shown.

The activation energy takes a typical value of 200 kJ/mole, implying that viscosity varies by one order of magnitude for every hundred degrees.

Assuming local thermodynamic equilibrium, the profile of residual melt composition in the mush can be calculated using the phase diagram. The melt composition varies dramatically, which implies large changes of physical properties. In the basalt example, viscosity varies from that of basalt, about 10 Pas, to that of rhyolite, about 10^7 Pas, due to the combined effect of temperature and composition. Further, if we momentarily treat the mush as a single phase with bulk properties, the presence of crystals leads to a further increase of viscosity.

Such large variations of viscosity have implications for thermal convection generated by cooling at the roof of the reservoir. The large temperature difference which may in principle drive powerful convective instabilities is counterbalanced by large viscosity variations. Only a small part of the thermal boundary layer is involved in convective motions: the coldest and most viscous regions remain static and transport heat by conduction only. Recent laboratory experiments at high viscosity contrasts coupled with dimensional analysis have led to scaling relationships for both the effective temperature difference driving convection and the convective heat flux (Davaille, 1991; Davaille and Jaupart, submitted).

2.4. KINETICS OF CRYSTALLIZATION

It is usually assumed that geological processes occur in equilibrium conditions because of the long time-scales involved. However, the kinetics of crystal nucleation and growth are sluggish and the diffusion of chemical species in silicate melts is slow on time-scales which are relevant to magmatic convection (Brandeis and Jaupart, 1986; Kerr et al., 1990). Thus, kinetic considerations are important for a complete understanding of igneous differentiation. Such considerations are outside the scope of the present paper, and we mention only a few important features.

In the growing thermal boundary layer, kinetic effects show up in two important manners. One is that melt remains undercooled and devoid of crystals for some finite amount of time. Thus, at the roof of an intrusion, such a melt can become convectively unstable before the onset of crystallization. Eventually, crystallization will begin and the thermal boundary layer will develop a new structure. However, part of this layer will remain undercooled, which implies again that convection involves undercooled melt. The other effect is linked to crystal growth and diffusion in the solid phase. The small values of diffusion coefficients in solids implies that crystals are usually zoned and that thermodynamic equilibrium is only achieved at the crystal/liquid interface, and not in a finite volume encompassing the whole crystal and surrounding melt. This is important when using mineral chemistry to recover thermodynamic conditions and also when constructing a theoretical model of differentiation aimed at reproducing the observed mineral sequence. An additional difficulty is that different chemical components have different coefficients of diffusion. Thus, some crystals may equilibrate and lose their zonation, and others not: in basaltic systems, olivine belongs to the former category, and plagioclase to the latter.

3. The geological record of magmatic differentiation

Igneous differentiation occurs principally by fractional crystallization involving the separation of the crystals and the chemically evolved residual melt. The mechanisms for separation include crystal settling and compositional convection. In this chapter, we shall mainly discuss convection. Although crystallization and convection are coupled phenomena, it is reasonable, in a gross sense, to think of igneous differentiation as being a competition between the rate at which crystallization proceeds and the rates at which convective processes occur. The time taken for a body of magma to solidify is determined mainly by its size. Of those physical parameters which determine the regime of convection, viscosity is that which exhibits by far the largest range of values. For example, values of the density contrast and temperature difference may vary by a factor of two or three, whereas the range of different magma compositions is such that viscosity can vary by as much as six orders of magnitude. In this overview, we will organize a simple set of observations to show that the degree of segregation in intrusions is a function of melt viscosity and intrusion size. Further, we shall show that a differentiation sequence is a reproducible process. In other words, we shall show that the phenomenon of magmatic differentiation is well-behaved and hence amenable to physical modelling.

3.1. VOLCANIC ROCKS

Volcanic eruptions bring molten rock to the Earth's surface where it is cooled rapidly and quenched to glass. Series of superimposed lava flows and volcanic deposits are easily recognized as an important tool in documenting melt evolution in deep reservoirs as a function of time. They provided the first lines of evidence that magmas were undergoing fractional crystallization with segregation of the melt and crystal phases. We make two simple observations.

Fig.3a shows data from a sequence of eruptions from Santorini volcano, Greece, which repeatedly sampled the same reservoir. A complete set of dates is not available, however the beginning and end of the sequence are well-dated. The magma composition was steadily evolving over a period of a few thousand years, which can be attributed to fractional crystallization.

Large eruptions which eject a substantial fraction of the chamber volume can be used to reconstruct its instantaneous state. Such eruptions commonly show that the liquid composition changes as a function of time implying that the chamber was chemically zoned. As an example, the concentration of K_2O in the liquid ejected by Laacher See volcano gradually decreased as the eruption proceeded and sampled deeper levels in the magma reservoir (Fig.3b). The principle crystallizing phase in this case is alkali feldspar which produces residual melt depleted in K_20. A plausible interpretation of these data is that the buoyant residual melt generated by fractional crystallization had risen to the top of the chamber and, by a "filling-box" mechanism, caused compositional zoning to develop (Tait et al., 1989).

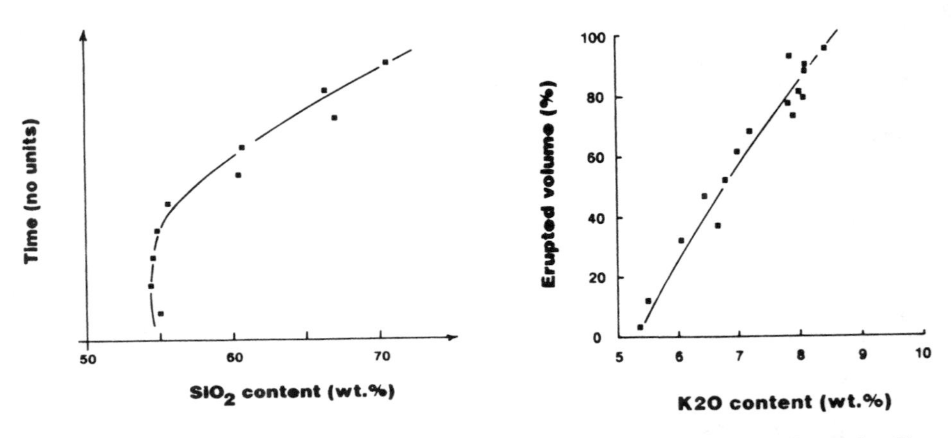

Figure 3. (a) The concentration of SiO_2 in a long sequence of eruptions from the Skaros shield of Santorini volcano, Greece (Huismans and Barton, 1989). The data show that the magma in the chamber which was sampled by the eruptions was evolving chemically with time. (b) The concentration of K_2O in volcanic glasses as a function of cumulative erupted volume ejected in the 11,000 B.P. eruption of Laacher See volcano in Germany (Bogaard and Schmincke, 1985). The data show that the liquid in the chamber had become compositionally zoned prior to the eruption. The total volume ejected was 5 km^3 of liquid + crystals.

3.2. The Nature of Petrological Data

In solidified intrusions which cooled slowly, chemical differentiation of a magmatic liquid has to be reconstructed indirectly. Silicate minerals show quite elaborate solid solutions and change in composition as the melt evolves. Plagioclase feldspar is convenient for comparing the degree of chemical differentiation of different mafic intrusions because it is present virtually throughout the entire volume of such bodies and because it shows complete solid solution between two end members, Anorthite ($CaSi_2Al_2O_8$) and Albite ($NaSi_3AlO_8$). The composition of a plagioclase is simply expressed as the percentage of one of these two end-members.

Trace elements present in minute amounts which can be incorporated in the crystal network of the main minerals offer independent evidence. They have various behaviours during fractional crystallization as a function of their atomic properties. Incompatible trace elements have very low partition coefficients in all the principal minerals present and are incorporated in specific minerals at the end of the crystallization sequence when their concentrations are high enough. If there is no separation of melt and crystals, the concentration of an incompatible element is the same as in the initial magma. Variations of their concentrations therefore provide evidence for melt segregation.

3.3. Gross Features of Intrusions: The Viscosity and Size Factors

A first observation is that intrusions derived from very viscous melts, for example granites, exhibit much less internal differentiation than their lower viscosity counterparts, for example, basaltic ones. Indeed, most of the work carried out on igneous macrosegregation has concentrated on basaltic systems, for they are the ones which show the most developed and conspicuous features.

At constant composition, the size of an intrusion determines the degree of macrosegregation. Figure 4 shows a diagram comparing the plagioclase compositions from several intrusions whose bulk composition is basaltic and which span almost three orders of magnitude in their vertical thickness. All of these have a horizontal extent much larger than their vertical thickness. Although the precise composition of the plagioclase which crystallizes from a given magma can be affected by a number for quite subtle factors, the overall effect is that, as the liquid differentiates towards more evolved compositions, the anorthite content of the plagioclase decreases. This feature can be seen clearly in the profiles from the Bushveld and Skaergaard intrusions, approximately 9 km. and 3 km. thick respectively. This indicates that, as the rocks built up from the intrusion floor, the liquid reservoir from which they were forming was progressively differentiating. The data from the 350 m. thick Palisades Sill show weaker differentiation of the melt. Plagioclase composition hardly varies at all in the thinner (40 m.) Dippin Sill. The important feature is that composition changes gradually over the whole crystallized sequence. On a smaller scale, there are more subtle variations which will be discussed later.

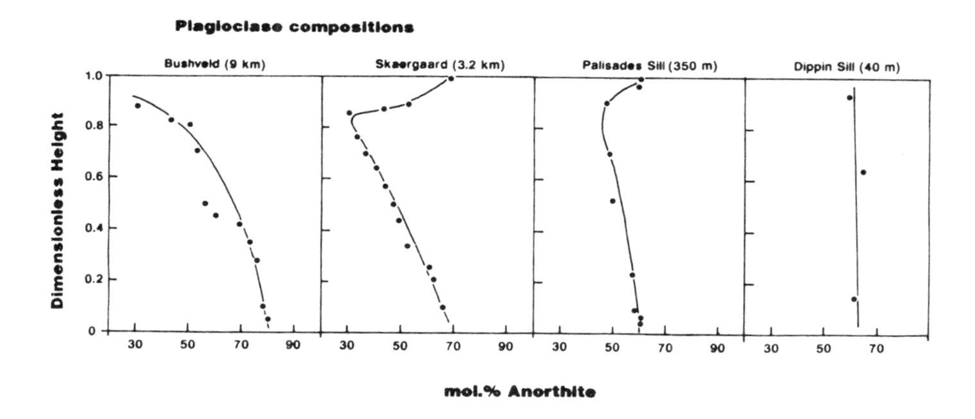

Figure 4: Shows the stratigraphic variations of plagioclase composition for intrusions of widely different sizes. Differentiation becomes more pronounced the larger the intrusion.

Another interesting feature can be seen by comparing the stratigraphic profiles of the concentrations of incompatible trace elements in the rocks. No such data could be found for the Dippin Sill, but note that all of the other profiles show that segregation of residual melt from the crystallizing regions has taken place (Fig.5). All three show a sharp peak in incompatible element concentrations at the stratigraphic level where the floor and roof sequences converge. Below this level, these concentrations remain roughly constant for large thicknesses of rock, indicating that the residual liquid was being removed as the porosity of the mush was being reduced. The roof sequences in the Skaergaard and Palisades intrusions show a steady build-up of incompatible element concentration, showing that the residual liquid was not being removed as crystallization was proceeding in more and more evolved melt.

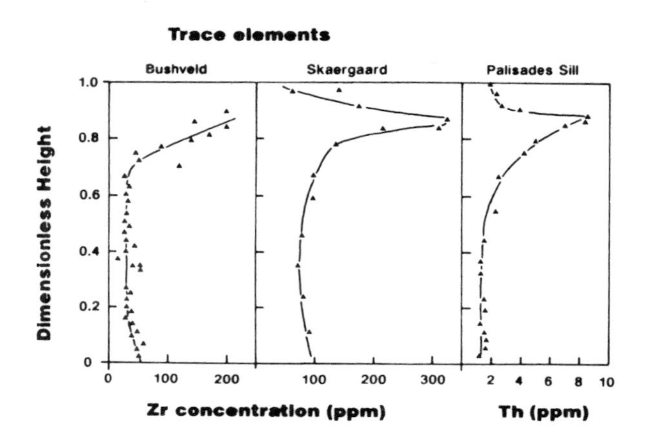

Figure 5: Stratigraphic variations of incompatible trace elements concentrations for the same intrusions as in Fig.4.

Both sets of observations can be accounted for by compositional convection, such that the gradient of residual melt density in the mush is unstable. In basaltic systems, this unstable situation occurs in the lower boundary layer at the floor of the intrusion. Other mechanisms which have been invoked, such as crystal settling or mush compaction for example, cannot account for the whole profiles.

3.4. REPRODUCIBILITY OF THE OVERALL CRYSTALLIZATION SEQUENCE

The overall regularity of the chemical evolution of layered intrusions can be further emphasized by considering some aspects of their mineralogical variations. Figure 6 shows generalized sections of two of the largest known intrusions with similar starting composition which exhibit extreme differentiation, the Bushveld and the Stillwater. Many detailed stratigraphic subdivisions have been made, but we shall restrict the discussion to a few simple features. Adjacent to the lower contact, in both cases is a basal zone containing rocks whose bulk composition is close to that of the original magma. Moving up sequence, the rocks become increasingly enriched in high temperature components, and eventually they are composed almost entirely of olivine and pyroxene, with plagioclase present only in trace amounts. Thick sequences of olivine and pyroxene rocks then occur. Such rocks do not have the composition of any natural magma and imply that a concentration process has been active. Starting from a basaltic magma, closure of the global chemical budget of the intrusion then requires that plagioclase minerals should form, and indeed this mineral becomes a major component of the rocks higher up in the stratigraphic sequence. This change of mineral composition occurs at remarkably similar levels, and hence for similar proportions of crystallization, in the two cases.

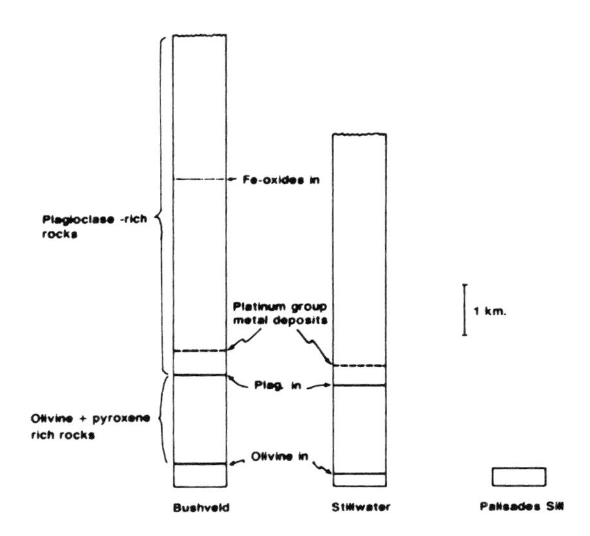

Figure 6. Comparative stratigraphy of the Bushveld and Stillwater intrusions emphasizing a few large scale features of the mineralogic sequences, from Jackson (1961) and Vermaak (1976).

Another feature of the two successions is the presence of major sulphide horizons which are strongly enriched in precious metals such as platinum and palladium. We draw attention to the remarkably similar stratigraphic heights of these metalliferous horizons. In this figure, we also show the Palisades sill in order to compare its thickness to some of the major stratigraphic features of the Bushveld and Stillwater complexes. Note that the entire sill is about the same thickness as the basal zones of the two large complexes. This may be another hint that the extreme mineralogical differentiation exemplified by the formation of olivine and pyroxene rocks is associated with the size of intrusion. The suggestion is that the Palisades sill crystallized too rapidly for the convective processes which cause extreme differentiation to have time to operate.

4. The onset of compositional convection: experimental work

4.1. Introduction

In order to account for the large-scale features of igneous differentiation, the proper framework is that of averaged two-phase models, such as those of Hills et al.(1983) and Worster (1986, 1992). The governing equations have been derived using averaging techniques and have been closed by constitutive relations describing the interactions between the two phases. The difficulty is to validate these equations by experimental verification. We have designed a set of laboratory experiments to address this problem (Tait and Jaupart, this volume).

4.2. Experimental Set-up

Our experiments were designed to study the fractional crystallization of a binary melt under simple laboratory conditions. To simulate the situation at the floor of a magma reservoir, we cooled a solution from below which caused the growth of a mush layer at the base of the experimental tank. The solution was an aqueous solution of ammonium chloride with a supereutectic composition, such that fractional crystallization produced light residual fluid and hence an unstable density gradient in the fluid. Part of the phase diagram is shown in Fig.7a. For super-eutectic compositions the liquidus has a constant slope Γ (=4.9°C/wt.%), such that:

$$T_L(C) - T_S = \Gamma.(C-C_S) \qquad (2)$$

where $T_L(C)$ is the liquidus temperature of composition C and T_S is the melting point of pure salt (C_S=1).

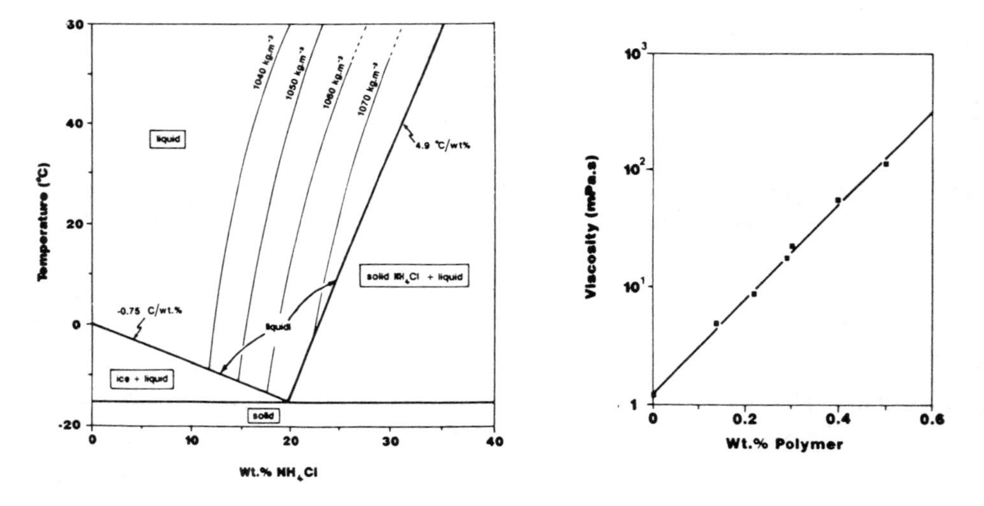

Figure 7. (a) Showing part of the NH4Cl-H2O phase diagram. (b) Showing the viscosity of 28 wt% NH_4Cl-H_2O solutions as a function of cellulose polymer concentration.

The interaction between crystallization and compositional convection can be elegantly studied by using the recently devised technique of adding trace amounts of a polymerizing agent to the solutions in order to increase their viscosities (Fig.7b; Tait and Jaupart, 1989, 1992). This permits us to vary the strength of convection independently of other physical and chemical variables. In a typical experiment, a solution at a uniform temperature (T_o) and uniform composition (C_o) was prepared and the lower boundary was brought rapidly to a constant temperature (T_B) below the liquidus of the starting composition but above the eutectic. The final equilibrium state of the system was known: uniform temperature (T_B) and liquid composition ($C_L(T_B) = C_B$). The total amount of crystals is independent of initial temperature and solution viscosity. All experiments were carried out with initial solution concentrations of 28 wt.% NH_4Cl of which the liquidus temperature is 23.7°C. The base plate temperature was 5±1°C. For these conditions, the mass of crystals formed was about 4% of the initial mass of liquid.

Let us set the experiments in their dynamical context. We assume that both liquid and solid have the same density, thermal conductivity and heat capacity. These assumptions are not restrictive in our experiments because of the small amount of crystallization. Physical properties are: L the latent heat per unit mass, k the thermal conductivity, D the chemical diffusion coefficient, C_p the heat capacity. The equation of state for the liquid is:

$$\rho = \rho_o \{ 1 - \alpha (T - T_S) + \beta (C - C_S) \} \tag{3a}$$

where α and β are constant coefficients. The assumption of local equilibrium relates temperature and composition in the mush (equation 1), so that liquid density can be written as:

$$\rho_o - \rho = \gamma (T - T_S) \tag{3b}$$

Π is permeability and μ is viscosity. We scale coordinates with the total height of liquid H, time and velocity with the conduction scales H^2/κ and κ/H respectively, composition by $\Delta C=C_o-C_L(T_B)$, temperature by $\Delta T=T_L(C_o)-T_B=\Gamma.\Delta C$ and permeability by some reference value Π_o. This leads to the following dimensionless numbers:

Porous medium Rayleigh number $$Ra_p = \frac{g\,\gamma\,\Delta T\,H\,\Pi_o}{\kappa\,\mu} \tag{4a}$$

Thermal Rayleigh number $$Ra_T = \frac{g\,\alpha\,\Delta T\,H^3}{\kappa\,\mu} \tag{4b}$$

Compositional Rayleigh number $$Ra_C = \frac{g\,\beta\,\Delta C\,H^3}{\kappa\,\mu} \tag{4c}$$

Stefan number $$\sigma = \frac{L}{C_p\,\Delta T} \tag{4d}$$

Prandtl number $$Pr = \frac{\mu}{\rho_o\,\kappa} \tag{4e}$$

Lewis number $$Le = \frac{\kappa}{D} \tag{4f}$$

In the experiments as well as in magma chambers, both the Prandtl and Lewis numbers can be considered as being infinite. Thus, inertial effects are small, and chemical diffusion can be neglected except in a thin boundary layer at the top of the mush (Worster, 1986; Tait and Jaupart, 1989). The Stefan number takes a value of about 5, however latent heat is a small component of the heat balance, because the amount of solids formed is small, and can be neglected. The Rayleigh numbers Ra_T and Ra_C are both large with minimum values of 6×10^4 and 8×10^4 respectively. In contrast the porous medium Rayleigh number

Ra_p was varied between values of approximately 10 and 10^4, i.e. from values somewhat below the critical threshold for the onset of convection in fluids of constant physical properties (Lapwood, 1948; Nield, 1968) to highly supercritical values. The other non-dimensional quantities are the function describing permeability as a function of porosity and the amount of superheat. The permeability function is hard to determine, however, our experimental technique allows us to vary the dynamic permeability Π/μ without affecting this function from one experiment to the next.

4.3. CONDUCTIVE REGIME AND THE ONSET OF COMPOSITIONAL CONVECTION

After cooling was begun, a horizontal layer of crystals formed at the base of the tank. This layer initially thickened in a conductive regime. Two different convective instabilities can occur given the profile of density which is shown in Fig.8. These are the instability of the thin chemical boundary layer at the top of the advancing crystallization front, and the instability of the interstitial liquid within the mush. We indeed observed both of these instabilities and have been able to show that both can be understood in terms of local critical Rayleigh number criteria (Tait and Jaupart, 1989, 1992).

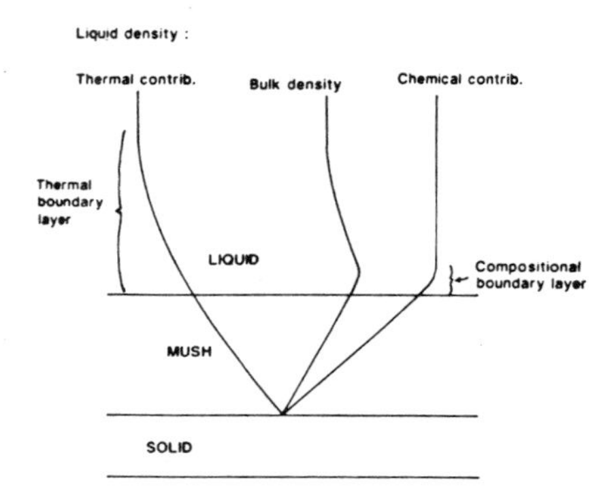

Figure 8: Density profile in the liquid resulting from the vertical variations of temperature and composition in the NH_4Cl-H_2O solutions cooled from below.

For the chemical boundary layer just above the top of the mush, the appropriate local Rayleigh number is:

$$Ra_1 = \frac{g\,\Delta\rho\,\delta^3}{D\,\mu} \tag{5}$$

where δ is the thickness of the boundary layer, $\Delta\rho$ is the difference in fluid density across the boundary layer. For the interstitial fluid in the mush, the appropriate local Rayleigh number is:

$$Ra_2 = \frac{g\,\Delta\rho_m\,h\,\Pi}{\kappa\,\mu} \tag{6}$$

where h is the instantaneous mush thickness and $\Delta\rho_m$ the density difference in the fluid across the mush.

In the conductive regime, as long as the total thickness of the fluid layer present is effectively infinite compared to the thickness of the mush, growth of the mush can be described by a similarity solution of the governing equations (Roberts and Loper, 1983; Huppert and Worster, 1985; Worster, 1986). The thickness (h) at time (t) is given by:

$$h = 2 \lambda \sqrt{D\, t} \tag{7}$$

where λ is the value of the similarity variable at the mush-liquid interface. As long as the value of λ is greater than about 3, a condition verified in our experiments, it can be shown that the thickness of the chemical boundary layer at the top of the mush is close to that predicted in the case of constant velocity of crystallization:

$$\delta \approx \frac{D}{\frac{dh}{dt}} \tag{8}$$

where dh/dt is the rate of mush growth. δ can be calculated as a function of h:

$$\delta = \frac{1}{2\,\lambda^2}\, h \tag{9}$$

To make comparison with our experimental data straightforward, we express the mush thickness at the onset of instability as a function of viscosity for each of the two instabilities. For the chemical boundary layer, instability occurs when the mush thickness is h_1:

$$h_1 = \left(\frac{R_{c1}\, D}{g\, \Delta\rho}\right)^{1/3} 2\,\lambda^2\, \mu^{1/3} \tag{10}$$

Similarly, for the interstitial fluid, instability occurs at a thickness h_2 such that:

$$h_2 = \left(\frac{R_{c2}\, \kappa}{g\, \Delta\rho_m\, \Pi}\right) \mu \tag{11}$$

In these expressions, R_{c1} and R_{c2} are the critical values of the two local Rayleigh numbers. The different forms of the definitions of these two numbers give different functional dependences of the critical mush thickness on viscosity.

There are two possibilities; ($h_1 < h_2$), i.e. that the first instability involves only the boundary layer at the top of the mush, with instability of the interstitial fluid occurring later; ($h_2 < h_1$), i.e. that the interstitial fluid becomes unstable first, in which case the boundary layer at the top of the mush will be destroyed by convective exchange of fluid between the mush and the overlying reservoir.

In a set of experiments with aqueous NH_4Cl solutions polymerized to relatively high viscosities ($\geq$100 mPas), Tait and Jaupart (1989) showed that the scaling suggested by (10) is verified for the first instability and that the boundary layer at the top of the mush became unstable first. In a second set of experiments conducted at lower viscosities and continued until longer times, we studied the instability of the interstitial fluid (Tait and Jaupart, 1992). This could be distinguished visually because it led to low amplitude corrugations of the upper surface of the mush and to fluctuations of mush porosity. We also observed that the first instability to develop was that of the chemical boundary layer at the top of the mush. At some greater value of mush thickness, we subsequently observed the onset of porosity fluctuations corresponding to instability of interstitial fluid. The mush thickness at which this occurred followed the linear scaling with viscosity implied by equation (11). From these experiments, the critical values R_{c1} and R_{c2} were determined to be 3 and 25 respectively for low values of superheat.

Figure 9 shows the fully developed stage of compositional convection. A critical effect of compositional convection is to modify the porosity structure of the crystallizing mush in order to satisfy the constraint of local thermodynamic equilibrium between crystals and migrating interstitial liquid (Copley et al., 1970; Roberts and Loper, 1983). This leads to horizontal variations of porosity which may ultimately become focussed into "chimney" structures. Chimneys are devoid of crystals and channel the upward flow of residual melt from the mush. Both the chimney diameter and spacing depend on the liquid viscosity (from Tait and Jaupart, 1992).

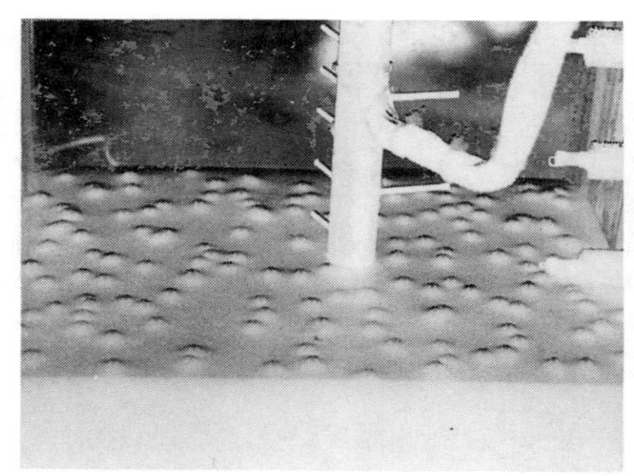

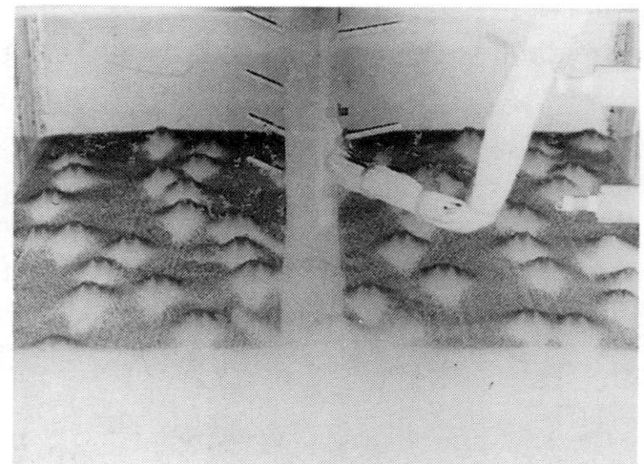

Figure 9. Photographs showing the size and spacing of chimney structures formed in two laboratory experiments with NH_4Cl-H_2O solutions. Chimneys have larger diameters and more widely spaced at the higher viscosity. (a) Viscosity at 24°C: 1.5 mPas, (b) Viscosity at 24°C: 6 mPas.

5. The basal zones of basaltic intrusions

5.1. THE ONSET OF DIFFERENTIATION IN AN IGNEOUS INTRUSION

The onset of convective differentiation occurs at some finite time, when unstable density gradients have built up over a critical thickness. Thus, one expects to see a transition in the rock sequence from early formed units for which there was no motion of residual melt to later units which formed when compositional convection was active. We thus look in more detail at the internal stratigraphy of basal zones taking the Stillwater intrusion as an example.

Adjacent to the basal contact with country rock, the rocks are fine-grained. Moving up the stratigraphy, they become coarser-grained, indicating that the cooling rate was decreasing, and pass through a sequence of plagioclase/pyroxene bearing rocks to those containing mainly pyroxene (Fig.10). The top of the basal zone is taken to be where olivine becomes the main constituent of the rock (Zientek et al., 1985). This stratigraphy can be explained using the phase diagram shown earlier and in the light of the experimental results which we discuss in subsequent sections. Fig.11a shows an enlarged portion of the phase diagram and Fig.11b shows the structure of the conductive boundary layer which would be predicted to grow prior to the onset of compositional convection - see comparison with phase diagram. Basal zones should record the structure of this boundary layer. Right adjacent to the contact, the liquid can be fully crystallized in situ by rapid conductive heat loss. If no, or negligible mass transport phenomena occur, the same final rock composition should be found at all levels. As we move up the stratigraphic sequence, however, the rocks become progressively enriched in minerals containing the high melting point components of the melt (Fig.10). We interpret this progression as recording the action of compositional convection in the interstitial fluid. The advection of fresh melt from the uncrystallized interior of the reservoir down the temperature gradient provides the flux of high melting point components necessary to form a final rock highly enriched in olivine and/or pyroxene relative to the original melt.

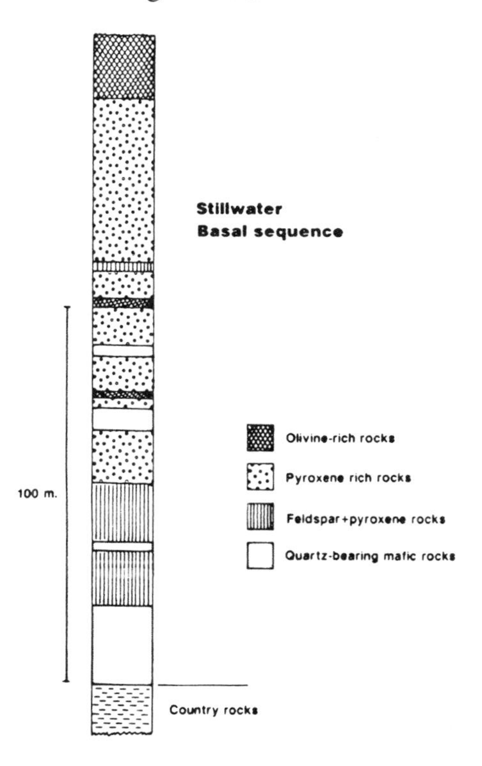

Figure 10: A simplified profile of the mineralogic variations in the Basal Zone of the Stillwater Complex, modified from Page (1979).

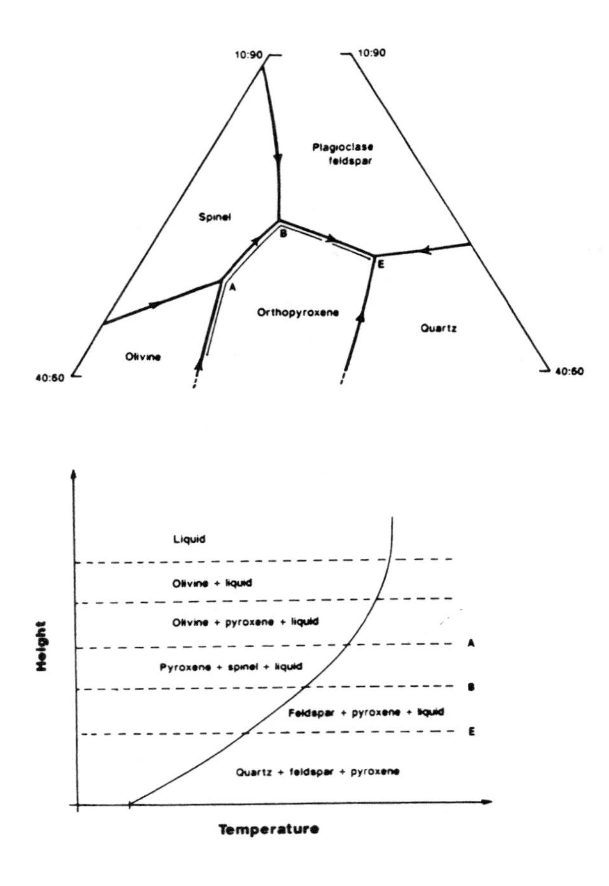

Figure 11. (a) Enlarged portion of the phase diagram of Fig.1 showing only 10 kbar boundaries. (b) The predicted structure of the conductive boundary layer prior to the onset of compositional convection, showing coexisting mineral phases for an initial liquid composition lying in the olivine field.

5.2. MAGMATIC CHIMNEY STRUCTURES

By continuity, the downward flow of melt from the overlying reservoir requires an upward flow of the residual melt. Thus, the liquid part of the reservoir evolves chemically at a small rate because of its large size, which leads to the gradual variation of composition documented above (Fig.10). A related question is to find field evidence for this upward flow. An important observation is that the igneous rocks appear to be homogeneous over large distances. Thus, if they are due to downward flow from the reservoir interior, the implication is that the return flow must have been localized in narrow zones which are rarely found on the outcrop. The chimney structures of convection in a crystallizing mush have the required properties. Both the Bushveld and Stillwater intrusions show many vertical or sub-vertical structures which cross-cut the horizontally layered rocks. The most striking examples of such structures are the cylindrical platiniferous pipes which cut through the ultramafic rocks of the Bushveld complex (Fig.12) (Schiffries, 1982; Stumpfl and Rucklidge, 1982). Viljoen and Scoon (1985) describe a large range of cross-cutting structures such as iron-rich pegmatites derived from very evolved melts which pervade the whole intrusion. Their observations indicate that the rocks in these structures crystallized from more differentiated melts than did the host rocks which they cut, and furthermore, that the higher up in the overall stratigraphy they are found, the more differentiated are their compositions.

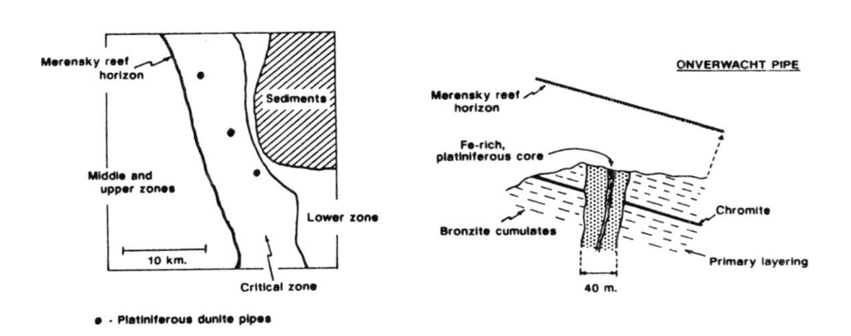

Figure 12. Showing the stratigraphic relations of the platiniferous dunite pipes of the Bushveld Complex (from Schiffries, 1982 and Stumpfl and Rucklidge, 1982).

5.3. QUANTITATIVE RESULTS

In order to apply the models of mushy layer convection to magmas, a large source of uncertainty is the permeability coefficient Π appropriate for the complex crystal morphologies involved in magmas. However, it is possible to test the hypothesis in two ways. One is that there should be a specific sequence of events recorded in the crystallized rocks near the floor: it is likely that the compositional boundary layer above the growing mush will become unstable first, followed by the instability of the interstitial melt within the mush. The reasons are that magmas have high viscosities, and that the permeability of silicate mushes is much lower than those of our experiments (see equation 11). The other requirement is that these instabilities occur when local conditions are reached in the growing mush independently of the whole intrusion, and should thus be recorded at a distance from the lower boundary which does not depend on the whole intrusion size.

The boundary layer at the top of the mush is extremely thin at marginal stability, and the compositional and temperature differences across it are, as a result, small. Consequently, compositional convection generated from this boundary layer has a small impact on the thermal and chemical evolution of the system (Tait and Jaupart, 1989). The same is not true for compositional convection generated from the interior of the mush (Tait and Jaupart, 1992).

Taking typical calculations from Brandeis and Jaupart (1987), the crystallisation front h(t) moves according to $h=2\sqrt{\kappa t}$. For basalts, we take μ in the range $10\text{-}10^2$ Pa.s, $\kappa=10^{-7}$ $m^2.s^{-1}$, and D to be $10^{-12}\text{-}10^{-13}$ $m^2.s^{-1}$ (Henderson et al., 1985). Taking $\Delta\rho$ to be 10 $kg.m^{-3}$ (Sparks & Huppert, 1984) and Ra_c to be 3, equation (10) predicts h_c to be in the range of approximately ten metres to 100 m. A common observation in basaltic sills is that, at a certain distance from the lower contact, some mineralogical layering is observed - commonly a layer with increased olivine content. The important points are that this increase (1) is moderate: in the Palisades sill, the olivine concentration increases from an average of 2% to about 22% (Walker, 1940), and (2) starts at a distance from the lower contact which is independent of the intrusion size: 9 meters in the 43 m-thick Dippin sill (Gibb & Henderson, 1978), 12 meters in the 350 m-thick Palisades sill (Walker, 1940) and 11 meters in the 600 m-thick Lambertville sill (Jacobeen, 1949). The latter strongly suggests that the increase of olivine content is due to a local mechanism operating at the bottom of the reservoir independently of the rest of the intrusion. The instability criterion (10) fulfills this condition and, furthermore, is quantitatively consistent with the observations.

Large igneous complexes show more extreme differentiation, to the point that rocks made almost entirely of olivine crystals are formed. Such rocks are observed at distances between 100 and 300 meters from the floor contact in several different intrusions: Stillwater (Jackson, 1961), Bushveld (Vermaak, 1976) and Muskox (Irvine, 1970). These distances are remarkably similar to each other and point to a mechanism which starts operating when appropriate local conditions are reached. We suggest that this is the onset of interstitial melt convection, which leads to a much larger compositional flux and hence to a greater concentration of high temperature components. To apply our criterion (11) for the onset of this form of convection, we use the Kozeny-Carman equation with a representative porosity of 0.5 and a crystal size of 1 mm implying a permeability of 10^{-8} m^2. We take $\Delta\rho_c$=100 kg.m^{-3}, and $\mu=10^3$ Pa.s. These values suggest that porous medium convection would begin after a mush approximately 200 m. thick had formed. On the basis of a simple proportion of the overall temperature drop in the lower thermal boundary layer, this implies that the fully crystallised rock below the mush had a thickness of 400 m. Clearly, these estimates are subject to large errors, but are of the correct order of magnitude.

6. Conclusions

Compositional variations in igneous intrusions and erupted lavas have long been recognized as due to complex physical processes which involve the separation of melt and crystals (Bowen, 1928). These pioneering petrologists had to struggle with data they could not interpret because the dynamics of crystallization were not understood in their day. Petrological observations and chemical measurements are now available in many different intrusions, which offers the opportunity to study the mechanisms of liquid/solid segregation, including compositional convection, as a function of the size and shape of the reservoir, and of the starting liquid viscosity. These observations show that the sequence of igneous rocks of different intrusions are strikingly similar, and indicate that many features are governed by local conditions in the thermal boundary layer growing at the floor. Other features such as cyclic layering remain unexplained. The physical models of fully-developed compositional convection required to explain them must be accurate for two main reasons. The first is that petrological observations can constrain variables such as temperature to within a few tens of degrees, which is small compared to the overall variations involved. Second, open system behaviour is always possible in these natural systems, with important implications for the behaviour of the deeper source of magma. In order to tackle this fascinating problem, we must be in a position to say which features of a complex crystallization sequence result from closed system behaviour, and identify the signature of reinjection processes.

References

Bogaard, P.v.d., and Schmincke, H-U. (1985). Laacher See Tephra: a widespread isochronous late Quaternary tephra layer in central and northern Europe. Geol. Soc. Am. Bull. 96, 1554-1571.

Bowen, N.L. (1928). The Evolution of Igneous Rocks. Princeton University Press, Princeton, New Jersey, pp.

Brandeis, G., and Jaupart, C. (1986). On the interaction between convection and crystallization in cooling magma chambers. Earth Planet. Sci. Lett. 77, 345-361.

Copley S.M., Giamei A.F., Johnson S.M. and Hornbecker M.F. (1970). The origin of freckles in uni-directionally solidified castings. Metall. Trans. 1, 2193-2204

Davaille, A. (1991). La convection thermique dans un fluide à viscosité variable. Applications à la Terre. Unpub. Ph.D. Thesis, Université Paris 6, 262 pp.

Gibb, F.G.F., and Henderson, C.M.B. (1978). The petrology of the Dippin sill, Isle of Arran. Scott. J. Geol. 14, 1-27.

Henderson, P., Nolan, J., Cunningham, G.C. and Lowry, R.K. (1985). Structural controls and mechanisms of diffusion in natural silicate melts. Contrib. Mineral. Petrol. 89, 263-272.

Hess, H.H. (1960), Stillwater igneous complex. Geol. Soc. Am. Mem. 80, 230 pp.

Huijsmans, J.P.P., and Barton, M. (1989). Polybaric geochemical evolution of two shield volcanoes from Santorini, Aegean Sea, Greece: Evidence for zoned magma chambers from cyclic compositional variations. J. Petrol. 30, 583-625.

Huppert H.E., Worster M.G. (1985). Dynamic solidification of a binary melt. Nature 314, 703-707

Irvine, T.N. (1970). Crystallization sequences in the Muskox intrusion and other layered intrusions, 1. Olivine-pyroxene-plagioclase relations, Geol. Soc. South Africa Sp. Pub. 1, 441-476.

Jackson D.E. (1961). Primary textures and mineral associations in the ultramafic zone of the Stillwater Complex, Montana. U.S. Geol. Surv. Prof. Pap. 358, 106.pp

Jacobeen, F.H., Jr. (1949). Differentiation of Lambertville diabase, Unpub. B.S. Thesis, Princeton University, Princeton, New Jersey, 25 pp.

Kerr, R.C., Woods, A.W., Worster, M.G. and Huppert, H.E. (1990). Disequilibrium and macrosegregration during solidification of a binary melt. Nature 340, 357-362.

Lapwood E.R. (1948). Convection of a fluid in a porous medium. Proc. Camb. Phil. Soc. 44, 508-521

McBirney, A.R. (1989). The Skaergaard layered series: I. Structure and average compositions. J. Petrol. 30, 363-397.

Nield D.A., (1968). Onset of thermohaline convection in a porous medium. Water Res. Res.4, 553-560

Page, N.J. (1979). Stillwater Complex, Montana - Structure, mineralogy, and petrology of the Basal Zone with emphasis on the occurrence of sulphides. U.S. Geol. Surv. Prof. Paper 1038, 69 pp.

Presnall, D.C. (1966). The join forsterite-diopside-iron oxide and its bearing on the crystallization of basaltic and ultramafic magmas. Am. J. Sci. 264, 753-809.

Roberts P.H., and Loper D.E. (1983). Towards a theory of the structure and evolution of a dendrite layer. In "Stellar and planetary magnetism" (ed. A. Soward), Gordon and Breach, London, pp.329-349.

Schiffries C.M. (1982). The petrogenesis of a platiniferous dunite pipe in the Bushveld Complex: Infiltration metasomatism by a chloride solution. Economic Geology 77, 1439-1453

Shirley, D.N. (1987). Differentiation and compaction in the Palisades sill, New Jersey. J. Petrol. 28, 835-865.

Sparks, R.S.J., Huppert H.E. (1984). Density changes during the fractional crystallisation of basaltic magmas: implications for the evolution of layered intrusions. Contribs. Mineral. Petrol. 85, 300-309

Stumpfl, E.F., and Rucklidge J.C. (1982). The platiniferous dunite pipes of the Eastern Bushveld. Economic Geology 77, 1419-1431

Tait S., and Jaupart, C. (1989). Compositional convection in viscous melts. Nature 338, 571-574

Tait S., and Jaupart, C. (1992). Compositional convection in a reactive crystalline mush and melt differentiation. J. Geophys. Res., in press.

Tait, S., Worner, G., Bogaard, P.v.d., and Schmincke, H.-U., Cumulate nodules as evidence for convective fractionation in a phonolite magma chamber, J. Volc. Geotherm. Res. 37, 21-37.

Vermaak, C.F. (1976). The Merensky Reef - Thoughts on its environment and genesis. Economic Geology, 71, 1270-1298.

Viljoen, M.J. and Scoon, R.N. (1985). The distribution and main geologic features of discordant bodies of iron-rich ultramafic pegmatites in the Bushveld Complex. Economic Geology 80, 1109-1128.

Wager, L.R. and Brown, G.M. (1968). Layered igneous rocks. Oliver & Boyd, Edinburgh, 588 pp.

Walker, F. (1940). Differentiation of the Palisade diabase, New Jersey. Geol. Soc. Am. Bull. 51, 1059-1106.

Worster M.G. (1986). Solidification of an alloy from a cooled boundary. J. Fluid Mech. 167, 481-501

Worster M.G. (1991). Natural convection in a mushy layer. J. Fluid Mech. 224, 335-359.

Worster M.G. (1992). Onset of compositional convection within and above a mushy layer during solidification of a binary alloy. J. Fluid Mech. (in press).

Zientek, M.L., Czamanske, G.K. and Irvine, T.N. (1985). Stratigraphy and nomenclature for the Stillwater Complex. Montana Bureau of Mines and Geology Special Pub. 92, pp. 21-38.

ROTATING CONVECTION

T. MAXWORTHY and S. NARIMOUSA
University of Southern California
Los Angeles, California 90089-1191
USA

ABSTRACT. The original motivation for this study came from a number of observations that revealed small-scale (O(1–10 km)) vortex motion in the neighborhood of intense cooling and freezing events in the Arctic Ocean and its adjacent seas. In this case the freezing of sea-water results in the rejection of the majority of its salt-content which is then available to generate an intense convective flow. At a certain stage of this motion the effect of the Earth's rotation must become important and changes the characteristics of the turbulence. Also the turbulent plumes that are generated interact with the bottom either before or after this condition is reached. As a result vortex structures can be generated by a number of mechanisms and our aim here is to try to understand these both qualitatively and quantitatively. Clearly, there are also important technological applications in which the same considerations must apply, e.g. Czochralski types of process.

Here we attempt, using scaling argument plus an experimental study, to show under what circumstances these various flow regimes can be found. Consider a source of buoyance of finite diameter at the center of a large rotating tank. The source strength is characterized by a buoyancy flux $Qg\Delta\rho/\rho \equiv B_0$, where g is the acceleration of gravity, ρ the ambient density, $\rho + \Delta\rho$ the source fluid density and Q the volume flux/unit source area.

An experiment starts by turning on the source at $t = 0$. Initially a three-dimensional, turbulent region forms under the source for which velocity, length and buoyancy fluctuations scale as $(B_0 z)^{1/3}$, z and $(B_0^2/z)^{1/3}$ respectively, where z is the local depth. The local turbulent Rossby number is of order unity i.e. rotation begins to affect the turbulence, at a depth z_c given by $z_c \sim (B_0/f^3)^{1/2}$ where $f = 2\Omega$ and Ω is the tank rotation rate. At this stage the turbulence below z_c changes character and becomes concentrated into long, axially-oriented vortices with velocity, length and buoyancy scales that are of order $(B_0/f)^{1/2}$, $(B_0/f^3)^{1/2}$ and $(B_0 f)^{1/2}$ respectively. Our measurements have shown that the constants of proportionality for the first two of these quantities are 1 and 15.9, while Fernando et al. (1991) gives a value of approximately 0.5 for the last of these. We have also determined that the constant of proportionality giving the scaling of the transition depth (z_c) from "normal" to rotationally dominated turbulence is 13.2. Thereafter the vortices continually evolve with these same characteristics until they interact with the bottom of the tank. Information about the existence of the bottom propagates vertically upward and the vortices begin to "fill-out". Eventually the whole column under the source takes on the

S. H. Davis et al. (eds.), Interactive Dynamics of Convection and Solidification, 261–263.

form of a truncated cone. Finally the base diameter becomes so large or alternatively the slope of the side of the cone becomes so great that the whole column becomes unstable to a baroclinic instability of a scale (D) given by:

$$D \sim [(B_0 f)^{1/2} H]^{1/2}/f \quad \text{or} \quad D/H \sim [B_0/f_3 H^2]^{1/4} = R_0^{*1/2}$$

where H is the depth of the tank and $R_0^* = (B_0/f)^{1/2}/fH$ is a "natural" Rossby number based on the vortex velocity.

The resulting vortices have a heavy core of density of order $(B_0 f)^{1/2}$ and rotate anticyclonically, while the fluid that has converged above them is rotating cyclonically.

A different sequence of events occurs if $z_c > H$. In this case the turbulence under the source is not affected by rotation but flows from under the source as a gravity current once it impinges on the bottom. Here the outflow can become baroclinically unstable but now with a diameter given by

$$D \sim [(B_0^2/H)^{1/3} H]^{1/2}/f \quad \text{or} \quad \frac{D}{H} \sim [B_0^2/f^3 H^2]^{1/3} \equiv R_0^{*2/3}$$

If we define D to be the diameter of the maximum swirl velocity within these vortices then the resulting measurements are shown in figure 1.

It appears that the actual transition between the two regimes, as reflected in these measurements, is a smooth one which occurs over a range of R_0^* centered on the location at which $z_c = H$. Also shown in the figure are the diameters of the vortices that initially form under the source before baroclinic instability occurs. It appears that the diameter of the two types of vortex becomes approximately equal when $H = z_c$. Application of these results to conditions observed in the Arctic and elsewhere gives vortex diameter in the range from 4–6 km, in agreement with observations.

This work is an extended abstract of results that appeared in Maxworthy and Narimousa (1991) and are presently being prepared for publication in the archival literature.

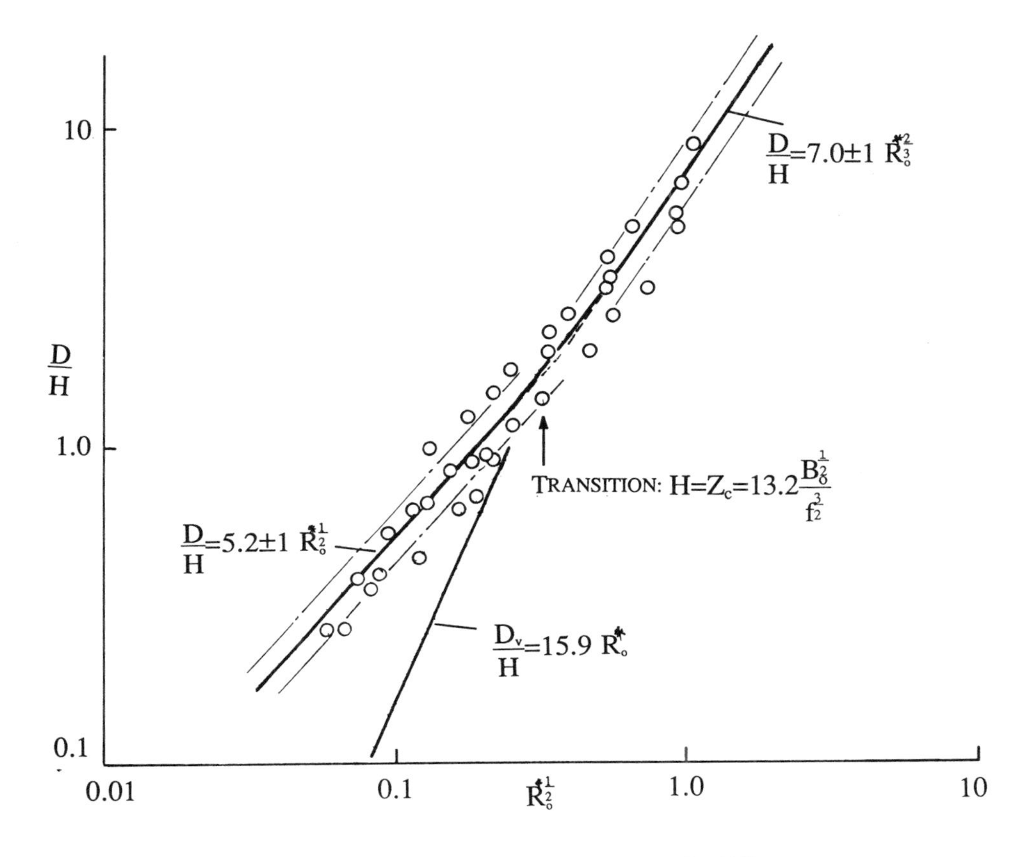

figure 1: Dimensionless vortex diameter (D/H) versus characteristic Rossby Number $(Ro^*)^{1/2}$.

References

Fernando, H. J. S., Chen, R.-R. and Boyer, D. L. 1991, "Effects of Rotation on Convection Turbulence", *J. Fluid Mech.*, 228, 513–547.

Maxworthy, T. and Narimousa, S. 1991 "Vortex Generation by Convection in a Rotating Fluid", *Ocean Modeling*, 92 (unpublished manuscript).

SOLIDIFICATION AND CONVECTION IN THE CORE OF THE EARTH

H.E. HUPPERT, B.A. BUFFETT, J.R. LISTER and A.W. WOODS
Institute of Theoretical Geophysics
Department of Applied Mathematics and Theoretical Physics
University of Cambridge
Silver Street
Cambridge CB3 9EW
UK

The Earth can be divided into three regions: a solid inner core, whose current radius is 1221 km; a liquid outer core, which extends to 3480 km; and a generally solid mantle. Concomitant with heat being lost from all these regions of the earth, the fluid outer core gradually solidifies to extend the radius of the inner core which is made up of almost pure iron. The residual fluid released by the solidification is poor in iron and is thereby less dense than the original melt. This compositional difference, along with thermal differences across the boundary between the core and the mantle, drives vigorous convection in the liquid outer core at a Rayleigh number which is extremely large, and may even exceed 10^{25}. The motion in the outer core is influenced by the rotation of the earth and, because the fluid in the outer core is a good conductor, it is also influenced by electromagnetic effects. Indeed, it is generally believed that the all-important external magnetic field of the earth is maintained by the interaction of the convection within the rotating outer core and the magnetic fields which permeate it. As described, this is a very complicated, highly nonlinear coupled system, concerning which there have been a vast number of intensive studies over many decades (see, for example, Gubbins 1991 and references therein).

The new approach to the solidification problem, which is described in this contribution, is based on global heat conservation and quantifies the dominant features of the thermal evolution of the core without requiring a detailed knowledge of the convective processes involved. Our model, which is described in more detail by Buffett, Huppert, Lister & Woods (1992), is constructed using the idea that the convection in the outer core is sufficiently strong to keep the fluid there well mixed. Because of the slow growth of the inner core, its surface will be in thermodynamic equilibrium with the (spatially uniform) temperature, say T, of the much more voluminous outer core. This temperature, which decreases steadily with time, will be identical to the (prescribed) solidification temperature of the inner core, which is a function of the pressure at the inner core boundary and hence of the radius, say r_i, of the inner core.

The heat extracted from the cooling outer core is transported through the overlying mantle to the surface of the Earth by thermal convection. The relatively sluggish and more massive mantle controls the cooling of the core by regulating the heat flux, say $f_m(t)$, across the core-mantle boundary. While the magnitude and time dependence of $f_m(t)$ will

S. H. Davis et al. (eds.), Interactive Dynamics of Convection and Solidification, 265–267.

depend on the details of mantle convection, in order to focus on the dominant processes involved, we simply prescribe $f_m(t)$ here and investigate the consequent thermal evolution of the core. In spherical geometry, conservation of heat may be expressed as

$$\frac{4\pi}{3}(r_0^3 - r_i^3)c\frac{dT}{dt} - 4\pi r_i^2(L+B)\frac{dr_i}{dt} = 4\pi r_i^2 f_i(t) - 4\pi r_o^2 f_m(t), \tag{1}$$

where c and L are the assumed constant specific heat and latent heat per unit volume, r_0 is the constant radius of the fluid outer core and B, which is due to the change in gravitational energy that results from the rearrangement of mass within the core, can be approximated as (Loper, 1984)

$$B = \Delta\rho g(r_o)r_o\left[\frac{3}{10} - \frac{1}{2}\left(\frac{r_i}{r_o}\right)^2\right], \tag{2}$$

where $\Delta\rho$ is the compositional density jump across the inner core boundary and g is the local acceleration due to gravity.

The conductive heat flux f_i from the inner core, which is given by Fourier's law, is determined from the solution of the thermal diffusion equation in the inner core. Two limiting cases immediately present themselves. These correspond physically to the limits of a perfectly insulating and a perfectly conducting inner core. The difference between these two extremes can be shown to be of order $(r_i/r_o)^3$, which is currently negligible. We hence restrict attention to the simpler case of a perfectly insulating inner core, for which no heat is conducted from the inner core, and $f_i \equiv 0$. A first integral of (1) can then be written as

$$c\int_o^{r_i(t)} \frac{dT}{dr}(r_o^3 - r^3)dr - (L+B)r_i^3 = -3r_0^2\int_0^t f_m(t)dt, \tag{3}$$

where $t = 0$ is the time at which the solid inner core first begins to form.

Over the range of pressures in the core, the solidification temperature is a function of pressure which, by using the hydrostatic equation and Newton's law of gravitation, can be shown to be well approximated by a quadratic function of the radius from the centre of the Earth. Thus we can write

$$T(r) = T(0) - Ar^2, \tag{4}$$

where A is a (fairly well) known constant. Inserting (4) into (3) and carrying out the resulting integrations, we find that

$$\eta^2 + S\eta^3 + B\eta^3(1-\eta^2) = D\int_0^t f_m(t)dt, \tag{5}$$

where

$$\eta = r_i/r_o, \quad D = 3/(Acr_o^3) \tag{6a,b}$$

and the Stefan number

$$S = \frac{1}{3}DLr_o. \tag{7}$$

Using currently accepted values of physical parameters of the Earth, we find that $S \sim 0.4$ and $B \sim 0.6$. Thus since $\eta \sim \frac{1}{3}$ the second and third terms in (5) are significantly smaller than the first. This indicates that at the moment the growth of the inner core is almost entirely controlled by a balance between the cooling of the outer core and the heat flux into the base of the mantle; the latent heat released from the solidification of the inner core, the heat transfer from the inner core and the gravitational energy release play secondary roles. Thus to lowest order, the radius of the inner core can be well represented by

$$r_i(t) = r_o \left\{ D \int_0^t f_m(t) dt \right\}^{\frac{1}{2}}. \tag{8}$$

Figure 1 presents the nondimensional radius of the inner core as a function of time, as determined both from the full relationship (5) and from the analogous relationship obtained by assuming that the inner core is a perfect conductor (in which case its temperature would be spatially uniform). At the present time, marked with a $*$, there is very little difference between the two curves. Not until the final stages of the solidification, in approximately another 10^{10} years, will there be a significant difference.

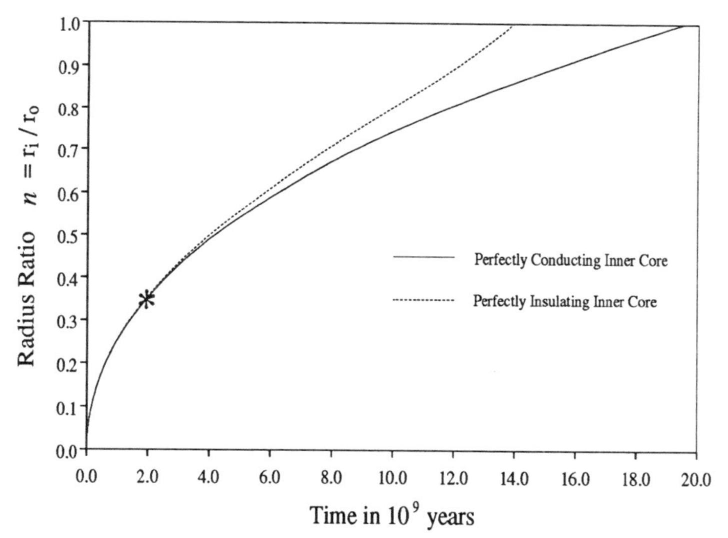

Figure 1: The ratio of the radius of the solid inner core to the outer core, η, as a function of the time since the initiation of the solid inner core. The current values are denoted by a $*$.

References

Buffett, B.A., Huppert, H.E., Lister, J.R. & Woods, A.W. 1992 A simple analytical model for solidification of the Earth's core. *Nature* (in press).

Gubbins, D. 1991 Harold Jeffreys Lecture 1990: Convection in the Earth's core and mantle. *Quarterly Journal of the Royal Astonomical Society*, **32**, 69–84.

Loper, D.E. 1984 Structure of the core and lower mantle. *Advances in Geophysics,* **26**, 1–34.

CONVECTION DRIVEN PHASE CHANGE AND MASS TRANSFER

ANDREW W. WOODS
Institute of Theoretical Geophysics
Department of Applied Mathematics and Theoretical Physics
Cambridge, England

ABSTRACT. There are a number of geophysical processes in which the convective transport of initially unsaturated fluid either through an ambient of different composition or temperature, or through which the pressure changes with position, causes the fluid to become supersaturated and precipitate solid. Such solidification can lead to a redistribution of the mass by either gravitational settling of the precipitate and subsequent dissolution or position dependent melting, convective transport of the melt, and subsequent freezing. We consider two examples herein.

Release of buoyant fluid from a source can result in the formation of a buoyant plume which entrains and mixes ambient with the source fluid; if the saturation curve is non-linear, then the mixture may be supersaturated, even if the ambient and source fluids are unsaturated (figure 1 and Phillips, 1991). This can lead to precipitation in the fluid mixture, as occurs with sulphide and iron oxide particulate in black-smoker plumes at mid-ocean ridges (Feely *et al.*, 1987) and crystallisation as buoyant magma intrudes the base of a magma chamber and mixes with the dense ambient magma as it rises through the chamber (Huppert *et al.*, 1986).

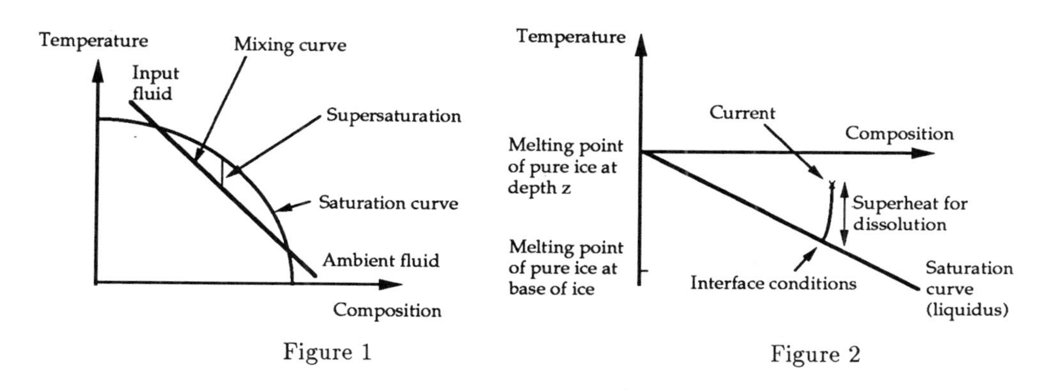

Figure 1 Figure 2

As the plume continues to be diluted through entrainment, the mixture may eventually become unsaturated as a result of any earlier precipitation. In a hydrothermal plume, precipitate may form rapidly as the hot and buoyant, upward rising plume mixes and cools. However, subsequent dissolution of the precipitate in the colder plume may be rate limited by dissolution kinetics. Thus the fluid mixture may remain unsaturated and loaded with precipitate for some time, τ_d; for example, chalcopyrite and pyrite particles, which are

S. H. Davis et al. (eds.), Interactive Dynamics of Convection and Solidification, 269–271.

formed in smoker plumes and are of size of order 100 μm, take $10^6 - 10^7$s to dissolve (Feely *et al.* 1987) in sea water. Therefore, as the plume material spreads out at the neutral buoyancy height, precipitate may be carried laterally from the source, and settle, under gravity, into the ambient some distance from the source. If the settling time exceeds the dissolution time then mass is transferred to otherwise unmixed ambient.

Dynamic Ice Pumping.

In our second example we consider phase changes associated with ice-sheets extending over ambient shelf water. A buoyant current on the lower side of the ice sheet may be initiated by either melt water issuing from the grounding line or enhanced melting near the grounding line where tidal mixing is most intense; such a current passes through a pressure gradient as it rises along the ice, and may thereby become saturated, even if initially unsaturated. In such a situation, an ice-pump may develop in which ice is ablated at depth but may form nearer to the surface (Lewis and Perkins, 1986). Typically, such currents travel 100-1000 km laterally while rising through a vertical distance of 100-1000 m from the grounding line to the surface (Jenkins, 1991). They are characterised as being anomalously fresh, and contain a few percent melt water.

We have developed a simple model of this pumping process, in which we analyse the heat and mass transfer at the ice-current interface, and then incorporate the associated fluxes of heat and mass into a dynamic model of the the motion of the sub-ice shelf current. We assume that the melt water issuing from the groundline is initially in equilibrium with the ice. As it runs along the ice face, entraining ambient water and melting or freezing the ice, its buoyancy and mass fluxes evolve. We assume that the buoyancy force is resisted by friction from the ice sheet and also through transfer of momentum to the entrained ambient (Ellison and Turner, 1959, Jenkins, 1991). The large convective heat and mass transfer at the plume-ice interface, coupled with the large Stefan number $L/C_p(T_a - T_o + \Gamma(S_a - S_o))$ of phase change and relatively small diffusive heat flux in the ice, requires the salinity at the interface to be approximately equal to that of the plume; otherwise, the interface cannot remain in equilibrium (figure 2). Therefore, the ice sheet effectively *dissolves* into liquid whenever the plume fluid is superheated and the ice sheet grows whenever the plume fluid is supercooled. This observation is important since the rate of phase change depends upon this superheat (Woods, 1992a). This present situation of forced dissolving is quite different from that in which the solid only melts at one fixed temperature; in that case, a superheated plume which is colder than the solid melting point does not cause any phase change (Woods, 1992b). It is therefore possible that the current is of lower temperature than the ice face, yet it still causes ablation of the ice, as a result of its superheat (figure 2).

The buoyancy flux of the current rapidly increases through melting fresh ice in the lower part of the ice sheet. Hence, once it is in motion, the current loses memory of the initial flux of melt water at or near the grounding line, although this flux is required to initiate the current (figure 3). As the fluid ascends, it mixes with the ambient and becomes warmer, more saline and hence superheated; the superheat of the current is then able to dissolve the lower part of the ice sheet. The melt water added to the current causes the buoyancy flux of the current to increase; this dominates any decrease in the buoyancy due to cooling of the current as it melts the ice (figure 3).

However, higher up the ice face, the pressure decreases, and hence the saturation temperature increases. As a result, the current may become supercooled, and, rather than ablating the ice face, new ice is formed (figure 4). The buoyancy of the current then decreases since the residual fluid is saline (figure 3), and so the current begins to slow down.

If the deep ambient water is very cold or there is a strong stable ambient stratification then the buoyancy of the current may decrease to zero, thereby leading to the arrest of its motion and termination of the ice pump. Otherwise the current surfaces at the leading edge of the ice, causing phase change all along the ice sheet (figure 4).

In summary, we have discussed the effects of non-linear saturation curves and pressure variations with depth, to show how convective transport of fluid, possibly driven by phase change at one point in space, can lead to supersaturation and solidification elsewhere and hence cause mass transfer.

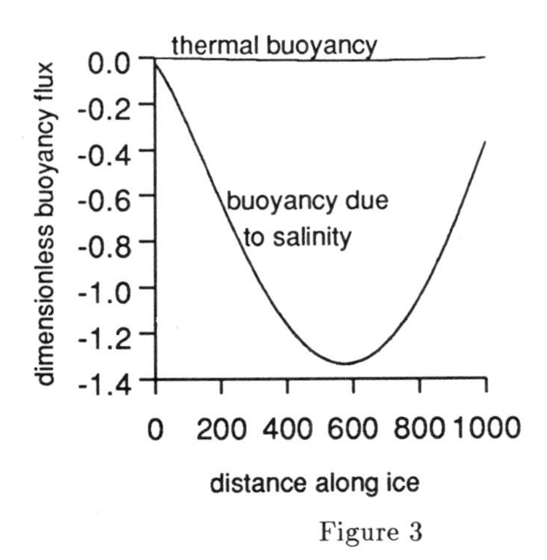

Figure 3

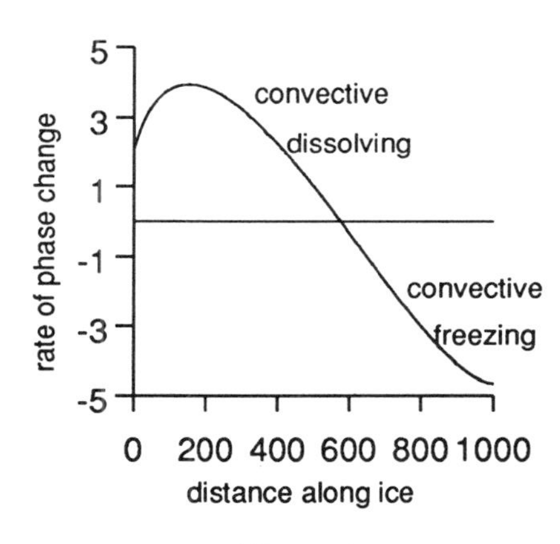

Figure 4

References.

Feely, R.A., and others, 1987, J. Geophys. Res., 92, 11347-11363.
Ellison, and Turner, 1959, J.F.M., 6, 423-428.
Huppert, H.E., Sparks, R.S.J., Whitehead, J.A. and Hallworth, M., 1986, J. Geophys. Res., 6113-6122.
Jenkins, A., 1991, J. Geophys. Res., 96, 20671-20677.
Lewis, E.L. and Perkins, R.G., 1986, J. Geophys. Res., 11756-11762.
Phillips, O.M., 1991, Flow and reactions in permeable rocks, CUP
Sparks, R.S.J., Carey, S.N., Sigurdsson, H., 1991, Particle settling from gravity currents, Sedimentology, 38, 839-856.
Woods, A.W., 1992a, Melting and Dissolving, J. Fluid Mech., 239.
Woods, A.W., 1992b, Convective dissolution and the dynamic ice-pump, in prep.

CLOSURE

The Grand Roundtable was organized by H.E. Huppert and U. Müller, and included the invited panelists B. Billia, R.A. Brown, S.R. Coriell, J.A. Dantzig, R.F. Sekerka and M.G. Velarde. The aim was to discuss the status of the field(s) and to peer into the future. The discussion was lively, ranging from specific models to the integrated behavior of full systems and to strategies for system optimization. Similarities in issues and models emerged in geology and in nuclear engineering. This closure summarizes the discussion at the Grand Roundtable and, as well, borrows heavily from the proceedings. The attached cartoon by T. Maxworthy gives a minimalist rendition of the workshop.

The question of scale re-emerged as crucial in identifying and quantifying behavior. On the microscale, studies of cells and dendrites involve pattern and scale selection. At the mesoscale, mushy zones are treated as composite "media" that obey certain constitutive laws. The mesoscale links the microstructure to the macroscale systems that are often treated phenomenologically; physically-based models must reflect their microscale origins. The macroscale studies involved full systems and large-scale dynamics that include complex, even turbulent convective flows.

On the microscale, it is clear that the field is developing rapidly, with impressive progress having been made in the last decade in bringing to bear powerful analytical and numerical tools on model systems. New phenomena have been uncovered in the linear and nonlinear regimes.

On the one hand, there was a sentiment expressed that some theoretical investigations have outpaced the experiments. Verification of the predictions of asymptotic, nonlinear analyses requires more experiments of high quality and resolution. The conducting of experiments with commercially-interesting materials (metals and semi-conductors) is limited by their high melting points and their opacity. The substitution of transparent organics in systems that involve flow exchanges materials of low Prandtl number Pr for those with large Pr. The character of flows and their transitions may differ substantially in different Pr ranges. Further, the scales that need to be resolved may be 100 mm or less. Collaborations may be required that bring into the field new techniques of measurement, say x-ray or acoustic instruments, that allow flow "visualization" on these small scales.

On the other hand, the theoretical investigations have been principally applied to pared-down models. Clearly, other physical effects must be confronted. These include anisotropy of surface and bulk properties, point defects, dislocations, facets, and impurity clusters.

One type of microscale study centers on PROTOTYPE FLOWS that involve simple flow geometries that correspond to well-studied hydrodynamic instabilities. If one of the rigid boundaries is replaced by a crystal interface, the effect of phase change can be dramatic in significantly altering the critical conditions for instability and/or modifying the selected pattern of the instability.

A second type of microscale study involves DIRECTIONAL SOLIDIFICATION of a binary liquid. A major aim of microscale studies has been to find the means of delaying or

S. H. Davis et al. (eds.), Interactive Dynamics of Convection and Solidification, 273–275.

eliminating the morphological instability of the planar front so that one could grow crystals without radial segregation. Much understanding of the systems has been gained but the means remains elusive. One may have to impose new control parameters, such as magnetic and/or electric fields, to obtain such an ability.

Another major aim is to carry the studies into the regime of strongly nonlinear cells and their possible transition to dendrites. The cellular/dendritic transition supplies a link to the prediction of microstructure. Presently, cellular microstructure near the low-speed branch of the morphological neutral curve cannot be predicted very well since the neutral curve is very flat there, and the bifurcation is usually sub-critical. Complex dynamics appears very near the neutral curve here. The neutral curve is not very flat on the high-speed (absolute-stability) branch and the bifurcation is supercritical there so that progress in this parameter regime has been made. It is near the upper branch that the assumption of local thermodynamic equilibrium may break down.

A third type of microscale study involves UNCONSTRAINED GROWTH of a pure material into an undercooled melt. One studies single, isolated dendrites. When an Ivantsov dendrite, which lacks surface energy, grows downward into a buoyancy field, the tip increases in speed if the undercooling ΔT and Pr are are not too small, consistent with experiments with succinonitrile but if ΔT and Pr are small enough, the opposite occurs. New experiments using succinonitrile to grow dendrites in a forced flow of magnitude U show that $\rho^2 V$ decreases with U, where the tip speed is V and the tip radius is ρ. This is the opposite behavior to that reported by others, the differences here being that a pure liquid, rather than a mixture, is used, and the speeds vary over a much larger range. Such behavior contradicts that predicted by "microscopic solvability theory" of dendrites with surface energy. An alternate theory of dendrite dynamics and stability for interfaces with surface energy involves turning-point theory and seems to give the correct sense of changes with U.

On the mesoscale, important interactions occur between crystal growth and the convective flow of interdendritic melt in mushy regions. Major progress in understanding has been made in this area by the systematic development of mathematical models guided and tested by concurrent experimental studies. Novel experiments using x-ray tomography or growing crystals in Hele Shaw cells allow greater visualization of mushy layers. Careful measurement of the critical conditions for the onset of convection have been made by altering the viscosity of the melt. Although convection can be driven solely by the buoyancy of the interstitial fluid in the mushy layer, account must also be taken of the convective state of the wholly-liquid region external to the mush. While significant advances have been made in the theoretical models of mushy regions that evolve through states that are close to thermodynamic equilibrium, many important effects of disequilibrium remain to be adequately described. The ripening and ultimate break-off of dendritic side branches may cause the production of equiaxed crystal fragments, which are transported by flow out of mushy regions. Such fragments may occur in such profusion that they form a slurry.

Interactions were described between forced flow and the freezing of liquid in pipes. While explanations for the shape of the solid–liquid interface were given, these were partially based on empirical findings. Planar-flow spin casting was examined asymptotically and shown to exhibit solidification that is metallurgically rapid but slow when compared to fluid-mechanical processes. By this latter criterion, the solidification decouples from the fluid flow. The substrate finish can be the controlling factor in the production of uniform tin ribbons by melt spinning. Laser surface alloying of aluminum or Al-Si-Mg surfaces by Cr and Ni using a powder-feeder technique can lead to alloyed zones that are crack and pore free. Laser surface treatment creates molten pools that generally have intense thermocapillary-driven flows that couple with the resolidification. However, when particles cover the molten zone, thermocapillarity is effectively eliminated and the remaining buoyancy-driven flow is weak enough to leave the thermal field unaffected.

As espoused by Brown, one may have to "get on with one's life" and confront full-scale crystal-growth devices, continuous-casting systems, and ultra-large scale geological and engineering complexes. This requires the quantitative description of phenomena that extend over large distances and involve all of the smaller-scale phenomena already discussed: the interaction of flow with columnar arrays, the evolution of two-phase structure and microscale modeling. The characteristics that cause one alloy to produce mostly equiaxed grains while cause another to form predominantly columnar structures are not yet well understood. Often, a large-scale melt will change configuration depending on the local freezing/melting that occurs; crust formation and the melting-through of containers by hot liquids play a large role in geology and nuclear-safety engineering.

On the macroscale, one studies segregation via averaged models and integrates a multitude of effects. The validation of such models is needed before reliable prediction with these models can be expected. Some aspects of slurries have been incorporated into numerical models of casting processes. The study of fluid-mechanical effects in two dimensions often requires difficult and costly numerical analysis, so it is prudent to use dimensional analysis first to chart the possible physical balances in different regions of parameter space.

There are a number of geophysical phenomena in which convection and solidification interact. It was evident that geophysical events occur on a megascale, and that careful interpretation and extrapolation of laboratory-scale phenomena is required. For example, could dunite pipes, some tens of meters in diameter, in the Bushveld intrusion be caused by the same processes that produce freckles of about 1 mm diameter in laboratory experiments and industrial castings? Not only large lengths but also long times characterize geophysical processes such as the slow mineral alteration of crustal rocks by the flow of groundwater. On terrestrial length scales, for example, when studying the Earth's core, it is important to comprehend the big picture before adding refinements of detail to theoretical models. In a rather different context, the freezing of the world's oceans in polar regions is an important process affecting global climate, for example. The evolution of polar ice caps can be significantly affected by rotation and by pressure-induced melting.

Finally, the meeting gave rise one clear admonition. Unless there is a close coupling between theory and experiment, both endeavors can suffer. On the microscale new experiments are needed to uncover phenomena and guide the theoretician. On the macroscale complex computer codes must be based on physical principles and validated in detail before predictions can be regarded as reliable.

The papers in this volume represent the latest work of researchers in a broad range of fields. These were seen to be overlapping in the types of phenomena seen and in the approaches used. This workshop was aimed at being a catalyst in developing a unity in the field as a whole.

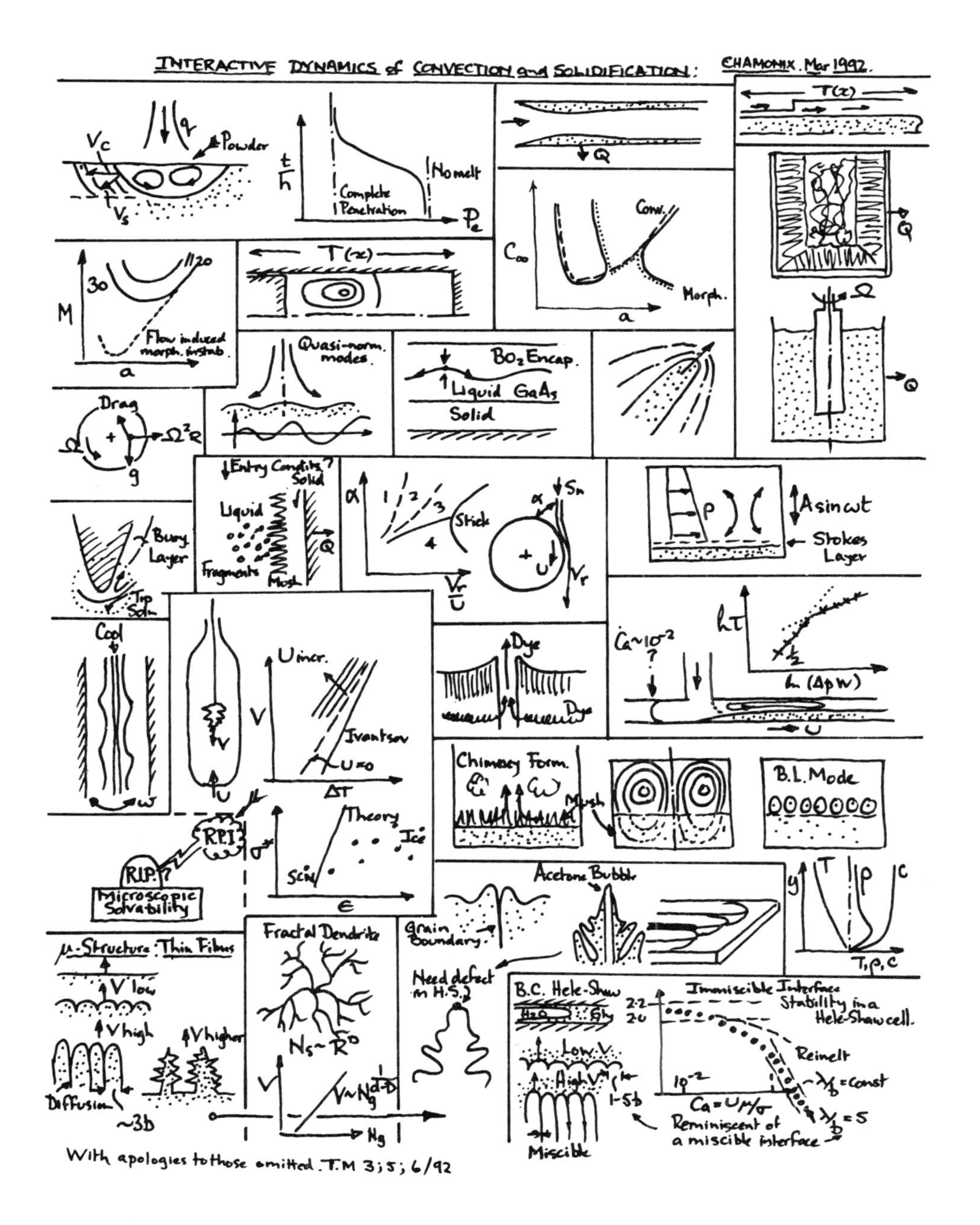
INTERACTIVE DYNAMICS of CONVECTION and SOLIDIFICATION: CHAMONIX, Mar 1992.
With apologies to those omitted. T.M 3; 5; 6/92